LINEAR ALGEBRA

Ideas and Applications

ABOUT THE COVER

Richard Penney's devotion both to communicating with students and to conveying the beauty and utility of linear algebra are evident on every page—and the cover—of this remarkable text.

One of the text's major themes, the parallel development of Linear Algebra ideas and applications, is reflected on its cover: Clouds suggest ideas; the building represents the genuine applicability that these ideas have in the real world; and reflections of clouds on the building's surface illustrate the text's parallel development of these themes.

ABOUT THE AUTHOR

Richard C. Penney (Purdue University) earned a BS degree from Tulane University in 1968 and a PhD from the Massachusetts Institute of Technology in 1971. He is a noted researcher in the field of representation theory and its applications to complex analysis, and has published over thirty research articles. Professor Penney's research work is supported by the NSF, and he has been an invited speaker at numerous international conferences, including many in the USA, Poland, Australia, France, Japan, and Germany (including Oberwolfach).

Professor Penney, voted by students at Purdue University as the Most Outstanding Teacher in the School of Science for the academic year 1995–1996, couples his understanding of students' needs and abilities in the Linear Algebra course with his deep devotion to imparting the beauty and utility of the subject. In addition to his regular teaching, he has taught summer workshops in AP calculus for in-service high school teachers; designed undergraduate courses at Purdue in Differential Equations, Honors Calculus, and Introduction to Analysis for prospective high school teachers; and is also the founder of Purdue's undergraduate career seminar, as well as Career Night, where invited speakers from various industries discuss career paths with students. Professor Penney served as a consultant on the NSF sponsored research project, "The Development of Proof Understanding, Production and Appreciation (PUPA) with Undergraduate Mathematics Majors" (Professor Guershon Harel is the Principal Investigator), which studies how college students learn Linear Algebra.

LINEAR ALGEBRA

IDEAS AND APPLICATIONS

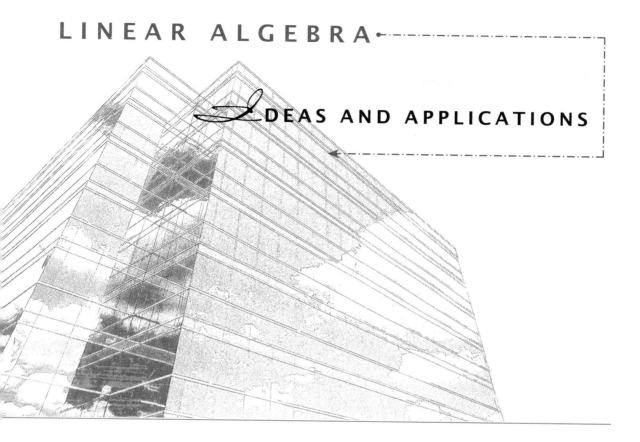

RICHARD C. PENNEY

Purdue University

John Wiley & Sons

New York • Chichester • Weinheim • Brisbane • Singapore • Toronto

MATHEMATICS EDITOR	Barbara Holland
MARKETING MANAGER	Leslie Hines
PRODUCTION MANAGER	Lucille Buonocore
SENIOR PRODUCTION EDITOR	Tracey Kuehn
DESIGN DIRECTION	Karin Gerdes Kincheloe
COVER DESIGNER	Karin Gerdes Kincheloe
TEXT DESIGNER	Nancy Field
SENIOR PHOTO EDITOR	Hilary Newman
PHOTO EDITOR	Kim Khatchatourian
ILLUSTRATION COORDINATOR	Jaime Perea
ILLUSTRATION STUDIO	Tech-Graphics
COVER PHOTO	Tony Stone Images/Chris Thomaidis

The book was set in 10/12 Times Roman by Eigentype Compositors, and was printed and bound by Quebecor Printing, Martinsburg. The cover was printed by Lehigh Press.

Recognizing the importance of preserving what has been written, it is a policy of John Wiley & Sons, Inc., to have books of enduring value published in the United States printed on acid-free paper, and we exert our best efforts to that end.

The paper in this book was manufactured by a mill whose forest management programs include sustained yield harvesting of its timberlands. Sustained yield harvesting principles ensure that the number of trees cut each year does not exceed the amount of new growth.

Library of Congress Cataloging-in-Publication Data

Penney, Richard C.
Linear algebra : ideas and applications / Richard C. Penney.
p.   cm.
Includes index.
ISBN 0-471-18179-X (alk. paper)
1. Algebras, Linear.   I. Title.
QA184.P46   1997
512'.5–dc21
97-17169
CIP

Printed in the United States of America

10 9 8 7 6 5 4 3 2 1

PREFACE

I wrote this book because I have a deep conviction that mathematics is about ideas, not just formulas and algorithms, and not just theorems and proofs. The text covers the material usually found in a one-semester linear algebra class. It is written, however, from the point of view that knowing *why* is just as important as knowing *how*.

To ensure that the readers see not only why a given fact is true, but also why it is important, I have included a number of the beautiful applications of linear algebra.

Most of my students seem to like this emphasis. For many, mathematics has always been a body of facts to be blindly accepted and used. The notion that they personally can decide mathematical truth or falsehood comes as a revelation. Promoting this level of understanding is the goal of this text.

R. P.

FEATURES OF THE TEXT

Parallel Structure

Most linear algebra texts begin with a long, basically computational, unit devoted to solving systems of equations and to matrix algebra and determinants. Students find this fairly easy and even somewhat familiar. But, after a third or more of the class has gone by peacefully, the boom falls. Suddenly the students are asked to absorb abstract concept after abstract concept, one following on the heels of the other. They see little relationship between these concepts and the first part of the course or, for that matter, anything else they have ever studied. By the time the abstractions can be related to the first part of the course, many students are so lost that they neither see nor appreciate the connection.

This text is different. We have adopted a parallel mode of development in which the abstract concepts are introduced only as they are needed to understand the computations. We never introduce a new concept without first justifying its importance and relating it to something already in the students' sphere of experience. In this way, the students see the abstract part of the text as a natural outgrowth of the computational part.

The advantages of this kind of approach are immense. The parallel development allows us to introduce the abstractions at a slower pace, giving students a whole semester to absorb what was formerly compressed into two-thirds of a semester. Students have time to fully absorb each new concept before taking on another. Since the concepts get used as they are introduced, the students see *why* each concept is necessary. The relation between theory and application is clear and immediate.

Gradual Development of Vector Spaces

One special feature of this text is its treatment of the concept of vector space. Most modern texts tend to introduce this concept fairly late. We put it early because we need it early. Initially, however, we do not develop it in any depth. Rather, we slowly expand the reader's understanding by introducing new ideas and examples as they are needed. Usually, the vector space concepts are introduced in the context of answering a specific question or solving a concrete problem, such as designing a music synthesizer.

   This approach has worked extremely well for us. When we used more traditional texts, we found ourselves spending endless amounts of time trying to explain what a vector space is. Students felt bewildered and confused, not seeing any point to what they were learning. With the gradual approach, on the other hand, the question of what a vector space is hardly arises. When the vector space concepts are introduced in a concrete context, the abstractions come across as not only meaningful, but essential.

Treatment of Proofs

It is essential that students learn to read and produce proofs. Proofs serve both to validate our results and to explain why they are true. For many students, however, linear algebra is their first proof-based course. They come to the subject with neither the ability to read proofs nor an appreciation for their importance.

   Many introductory linear algebra texts adopt a formal "definition-theorem-proof" format. In such a treatment, a student who has not yet developed the ability to read abstract mathematics can perceive both the statements of the theorems and their proofs (not to mention the definitions) as so much "abstract nonsense." Often, such students only master the computational techniques, since this is the only part of the course that has any meaning for them. In essence, we have only taught such students to be nothing more than slow, inaccurate computers.

   Our point of view is different. Most results are completely proved, but often the proofs are presented as discussions in order to encourage the students to read and understand them. Before stating a particular property, we usually explain why we would expect such a property to hold and its significance. If this explanation leaves little doubt that the property is always true, then our explanation is an informal, but perfectly valid, proof.

   If our explanation does seem to leave some doubt, then we will typically give a formal proof. Our proofs also get more formal in the later sections, after the students have had more experience in coping with abstractions. In this way, the students are allowed to grow in mathematical maturity as the course progresses. They (hopefully) come to see that mathematics is based on ideas as well as gaining some understanding of how mathematics is created.

Applications Fully Integrated into the Text

Teaching a linear algebra class without including a number of the beautiful applications is like writing a mystery novel and leaving out the ending. Our point of view toward applications, once again, is nontraditional. Most linear algebra texts introduce a body of material and afterward provide an example or a "real-world application." We feel that this is backwards. We often begin with the application and show how the necessity to solve a real world problem forces us to develop certain mathematical tools. Not only is this kind of development more interesting for the student, it is also more realistic. Engineers or mathematicians will be required to do exactly this kind of thinking when they get into the workplace.

Conceptual Exercises

Most texts at this level have exercises of two types: proofs and computations. We certainly do have a number of proofs and we definitely have lots of computations. The vast majority of the exercises are, however, "conceptual, but not theoretical." That is, each exercise asks an explicit, concrete question, which requires the student to think conceptually in order to provide an answer. Such questions are both more concrete and more manageable than proofs, and thus are much better at demonstrating the concepts. They do not require that the student already have facility with abstractions. Rather, they act as a bridge between the abstract proofs and the explicit computations.

Early Eigenvalues Option

We have designed the text so that eigenvalues and eigenvectors may be covered early, before most of the material on orthogonality. (See the "Sample Syllabus (Early Eigenvalues) on page xi.

True–False Questions

We have included a section of True–False questions in both the Student and Instructor Resource Manuals.

Attention to Pedagogy

Our style is informal and conversational. We usually begin each section with an example from which we extract a general principle. Most theorems are completely proved. The proofs are often presented, however, as discussions so as to encourage the student to read and understand them.

We also emphasize geometry from beginning to end. We want our students to understand linear algebra "computationally, analytically, and geometrically."

Student Tested

This text has been used over a period of years by numerous instructors at both Purdue and at other universities nationwide. We have incorporated comments from instructors, reviewers, and (most importantly) students.

Technology

Most sections of the text include a selection of computer exercises under the heading "On Line." Each exercise is specific to its section and is designed to support and extend the concepts discussed in that section.

These exercises have a special feature: they are designed to be "freestanding." In principle, the instructor should not need to spend any class time at all discussing computing. Everything most students need to know is right there. In the text, the discussion is based on MATLAB. However, translations of the exercises into various other platforms (such as Maple, Mathematica, and TI calculators) are contained in the Student Resource Manual. They are also available free on the World Wide Web at "http://www.wiley.com/college/penney."

Chapter Summaries, Glossary, and Selected Answers/Hints

The glossary contains carefully stated definitions of the major concepts in the text. The answer section contains answers to selected exercises as well as hints for others. Also, at the end of each chapter there is a chapter summary which brings together the major points from the chapter so that the students can get an overview of what they just learned.

Student Supplement

There is a student supplement available which contains more worked out examples, further hints for some of the exercises, and translations of the computer exercises into various platforms (such as Maple, Mathematica, and TI calculators).

Meets LACSG Recommendations

The Linear Algebra Curriculum Study Group (LACSG) recommended that the first class in linear algebra be a "student oriented" class which considers the "client disciplines" and which makes use of technology. The above comments make it clear that this text meets these recommendations. The LACSG also recommended that the first class be "matrix oriented." We emphasize matrices throughout.

Richard Penney
West Lafayette, IN
1997

$\mathcal{A}$CKNOWLEDGMENTS

I thank many people who have had a substantial impact on this work. I would especially like to acknowledge both the students and faculty who acted as reviewers, problem solvers, accuracy checkers, class testers, and software translators. Their comments have been crucial to the final form of this text. Specifically, I'd like to thank the following: Neil Wigley, University of Windsor; Luz M. DeAlba, Drake University; Larry Grove, University of Arizona; Manil Suri, University of Maryland, Baltimore County; Irwin Pressman, Carleton University; Justin J. Price, Purdue University; David Saltman, University of Texas at Austin; John Rossi, Virginia Polytechnic Institute and State University; Duane Porter, University of Wyoming; Paul Binding, University of Calgary; Jane Day, San Jose State University; Richard Orr, Rochester Institute of Technology; Donald Bailey, Trinity University; Bill Hager, University of Florida; Dale Rohm, University of Wisconsin, Stevens Point; Wolodymyr Madych, University of Connecticut; Louis deBranges, Purdue University; David Goldberg, Purdue University; Mark Farris, Midwestern State University; Peter Mercer, St. Mary's College of Maryland; Karin Reinhold, State University of New York at Albany; Bill Emerson, Metropolitan State College of Denver; Robert Lopez, Rose-Hulman Institute of Technology.

I owe a particular debt to Guershon Harel. His influence can be seen on every page of the text. I owe an even greater debt to my wife, Mary Ann, who has supported me faithfully throughout my whole career. Without her support, this work certainly would not have been possible.

I also thank the people in the Mathematical Division of John Wiley & Sons. In particular, my editor, Barbara Holland, has provided me with the freedom to create the text I wanted to write and the support to ensure that it is as good as we can possibly make it. Finally, I thank all of my students, both past and present. I have learned much more from them than they will ever realize.

R. P.

SAMPLE SYLLABUS AND CHAPTER DEPENDENCIES

SAMPLE SYLLABUS (STANDARD COURSE)

The following outline represents what we cover in a standard, one-semester course (fourteen weeks with three fifty-minute lectures per week). One certainly could cover more. However, we feel that there is a general tendency to "overstuff" courses with material, resulting in *less* learning and less understanding.

There are no computer exercises for the determinant chapter. This omission is intentional, for we want to stress that determinants are not a viable computational technique. Classes that incorporate a regular computer lab could use this time to do one of the application sections such as 1.5.1 or 3.4.1.

Chapter 1   (10 classes)
All sections. Section 1.4.1 is assigned as outside reading and Section 1.5.1 is assigned as a computer project.

Chapter 2   (6 classes)
All sections, except 2.2.1.

Chapter 3   (8 classes)
Sections 3.1 to 3.4. Also Section 3.4.1 may be assigned as outside reading.

Chapter 4   (6–8 classes)
Sections 4.1, 4.2, 4.4 and (time permitting) one of 4.3 or 4.5.

Chapter 5   (3 classes)
Sections 5.1 and 5.2.

Chapter 6   (6 classes)
Sections 6.1,6.2 and 6.5 (omit proof of the Spectral Theorem).

Total: (39–41 classes)

SAMPLE SYLLABUS (EARLY EIGENVALUES)

After following the standard syllabus through Chapter 3, cover 5.1 and 5.2. Then do (in order) 6.1, 4.1, 6.2, 4.2, 4.4, and 6.5.

Students may find the "skipping around" slightly confusing. However, the payoff is that (a) the important topics of eigenvectors and eigenvalues are not postponed to the end of the class and (b) the students can better appreciate the connections between chapters 4 and 6.

SECTION DEPENDENCIES

Chapters 1,2, and 3   The sections in Chapters 1–3 are linearly ordered: each depends upon the previous. The only sections here which are not required for the rest of the text are the application sections (those with 3 digit numbers) and Section 3.5.

If one wishes a "formal" discussion of function spaces, then either 2.2.1 or 4.3 should eventually be covered. However, there is enough "informal" discussion of function spaces in the exercise sets for other sections that students should be able to get a good feel for function spaces without covering either 2.2.1 or 4.3.

Chapter 4   All sections presume 4.1. Sections 4.2, 4.4 and 4.5 are independent of each other, while 4.3 requires 4.2. The material on least squares (Section 4.5) is (in our opinion) quite beautiful. One might consider covering the notions of dot product and orthogonality from 4.1 and then doing 4.5. The instructor should note that 4.1 is required for 6.2 while 4.2 and 4.4 are required for 6.4.

Chapter 5   This chapter depends only on Chapters 1–3 and is required only for Chapter 6. The material in 5.3 is optional.

Chapter 6   All sections depend only on 6.1 and 6.2, with the exception that 4.2, 6.3, and 6.4 are required for a complete proof of the spectral theorem in 6.5. We have, however, written this proof so that it is possible to get a good understanding of it without covering either 6.3 or 6.4.

Chapter 6 assumes 5.1 and 5.2. Additionally, 6.2 requires 4.1 and 6.5 assumes 4.4. The proof of the spectral theorem also requires 4.2.

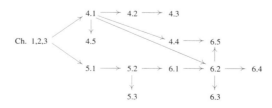

HOW TO SUCCEED IN LINEAR ALGEBRA

A Note to the Student

In beginning any new class, most students wonder, *"What will it take for me to succeed?"* The answer to this question for linear algebra may be very different from what it was for your earlier math classes. Often, in elementary math classes, the instructor tells you what is true and (perhaps) why it is true. You then study how to apply what the instructor tells you in various contexts.

In learning linear algebra (at least from this text) knowing *why* is just as important as knowing *how*. You will discover, for example, that the majority of the exercises cannot be solved simply by imitating the examples in the text. Instead, you must understand what the problem is asking and think about how it relates to what was said in the text.

Thus, to be successful in using this text you must read "for meaning." You should pay particular attention to the definitions. (Each new term is indicated in boldface. Many of the definitions may also be found in the glossary.) If the instructor shows you how to solve a particular exercise, you should not be satisfied until you understand the explanation to the point of saying, "I should have thought of that myself!"

The students who have used this text have generally appreciated the emphasis on "why." They like the feeling of mastery of the subject that understanding gives. We hope that you will share their enjoyment.

R. P.

CONTENTS

CHAPTER 1
SYSTEMS OF LINEAR EQUATIONS

$\mathscr{C}$HAPTER 1

SYSTEMS OF LINEAR EQUATIONS

1.1 VECTORS

"Linear algebra" may be defined as the study of the equations that describe lines, planes, and their higher dimensional generalizations. Such equations are called "linear equations." The methods of linear algebra are a mixture of algebra and geometry. In this section we describe some of the geometry.

To begin, consider the statement "The set of solutions to the equation

$$y = 2x + 1$$

is a line." This is a geometric statement about an algebraic equation. A solution is just a pair of numbers (x, y) that satisfy the equation. Thus, $(1, 3)$ and $(-1, -1)$ are both solutions. When we say that the set of solutions forms a line, we are interpreting pairs of numbers as representing points in the plane. Specifically, we choose two coordinate axes (called x and y) and plot the points as usual. Then the set of solutions, when interpreted as points, forms the line indicated in Figure 1.

There is another important geometric interpretation of pairs of numbers. Recall that in physics, a vector is thought of as an arrow that begins at some initial point

1

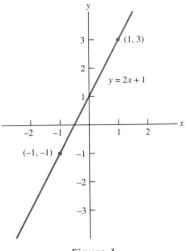

Figure 1

and ends at some terminal point. We shall often visualize a pair of numbers (x, y) as representing the vector from the origin to the point (x, y). Thus, for example, the pair $(2, 1)$ can be thought of as representing the vector indicated in Figure 2.

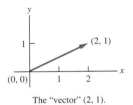

The "vector" $(2, 1)$.

Figure 2

Thinking of pairs of numbers as vectors is important because we can do algebra with vectors. Specifically, if vectors A and B have the same initial point, then their vector sum $A + B$ is defined to be the diagonal of the parallelogram that has A and B as two of its sides (see Figure 3). Figure 4 shows that if we interpret (x_1, y_1) and (x_2, y_2) as representing vectors originating at the origin, then the sum of these vectors is represented by the pair $(x_1 + x_2, y_1 + y_2)$. This suggests that we define addition of pairs of numbers by the formula

$$(x_1, y_1) + (x_2, y_2) = (x_1 + x_2, y_1 + y_2)$$

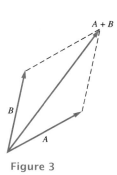

Figure 3

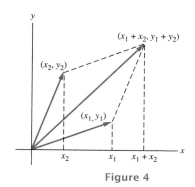

Figure 4

▶ EXAMPLE 1:   Compute the sum of the vectors represented by $(-1, 2)$ and $(2, 3)$ and draw a diagram illustrating your computation.

Solution: The vectors (along with their sum) are plotted in Figure 5. Their sum is computed as follows:

$$(-1, 2) + (2, 3) = (-1 + 2, 2 + 3) = (1, 5).$$

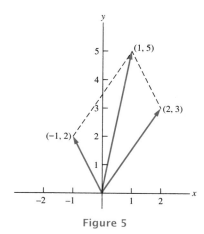

Figure 5

◀

Remark: Students sometimes find the idea that the pair $(1, 2)$ can be thought of as both a point and a vector confusing. They might ask, "I have always been told that $(1, 2)$ is a point. Now you are telling me that it is a vector. Which is it?" The answer is that it is *neither*. A pair of numbers is just "a pair of numbers." At times it helps our understanding to visualize the pair as a point in the plane. At other times, it is easier to visualize it as a vector. Sometimes it is easiest not to visualize it at all.

We will denote the set of all pairs (x, y) *of real numbers by* $\mathbb{R}^2$. Thus, for example, if we say that A is a point in $\mathbb{R}^2$, or a vector in $\mathbb{R}^2$, we mean that A is a pair of numbers. Whether we call it a point or a vector depends only on how we happen to be visualizing it at the moment.

In physics, vectors are often used to describe forces. The direction the vector points indicates the direction in which the force is pulling, and the length of the vector indicates the magnitude of the force. A physicist would say that two different vectors define the same force if they have the same direction and magnitude, regardless of their initial point. We say that such vectors are **equivalent**. Geometrically, two vectors are equivalent if one of them can be "slid" so as to exactly coincide with the other. In Figure 6, vector A is equivalent with vector B. Neither is equivalent with vector C.

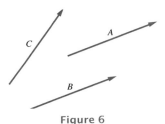

Figure 6

Evidently, a vector is equivalent with one and only one vector having its initial point at the origin. Finding this vector involves **subtraction** of points. We define subtraction of points in $\mathbb{R}^2$ by

$$(x_1, y_1) - (x_2, y_2) = (x_1 - x_2, y_1 - y_2)$$

Figure 7 indicates that if A and B are points in $\mathbb{R}^2$, then the vector from B to A is equivalent with the vector described by $A - B$.

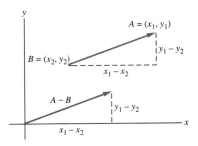

Figure 7

In physics vectors indicate direction and magnitude.

▶ EXAMPLE 2: Let $B = (-2, 5)$, $A = (4, -3)$, $D = (-4, 10)$, and $C = (2, 2)$. Is the vector from B to A equivalent with the vector from D to C?

Solution: The vector from B to A is equivalent with the vector described by

$$A - B = (4, -3) - (-2, 5) = (6, -8)$$

The vector from D to C is equivalent with the vector described by

$$C - D = (2, 2) - (-4, 10) = (6, -8)$$

Thus, the two vectors are equivalent.   ◄

There is a nice geometric picture one can draw for subtraction (Figure 8). Let A and B be points in $\mathbb{R}^2$. From the preceding comments, the vector from B to A is

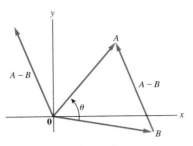

Figure 8

equivalent with the vector described by $A - B$. If we take the point of view that only the direction and magnitude of our vectors matter, we can say that $A - B$ *is* the vector from B to A.

We can also multiply points by numbers. We define

$$c(x, y) = (cx, cy)$$

for all (x, y) in $\mathbb{R}^2$ and all real numbers c. This operation is called **scalar multiplication**. (In linear algebra, "scalar" and "number" mean essentially the same thing. However, what we mean by "number" may change depending upon context. For example, in Section 6.3, our scalars will be *complex* numbers. For the most part, however, all of our scalars will be *real* numbers.)

Multiplication of a point by positive scalars will either stretch or shrink the corresponding vector, and negative scalars will reverse the direction of the vector, as well as change the length. It follows that the set of all multiples of a given point describes a line through the origin.

Lines not containing the origin may be obtained by adding a fixed vector to a line through the origin, as in the diagram on the right in Figure 9. Clearly, if A and B are points, then the set of points of the form $B + tA$, where t ranges over all real numbers, will form a line that is parallel to the vector A and passes through B. This representation of the line is called the **parametric representation**. The vector B is

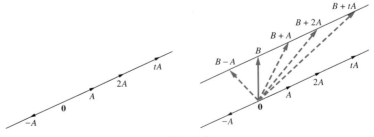

Figure 9

called the **translation vector** and the vector A is called the **spanning vector**. Note that the line is made up of *points*, not vectors. The arrows in Figure 9 certainly do not form part of the line: they merely *point to* points on the line. Otherwise put, the line is the set formed from the tips of the vectors, not the vectors themselves.

▶ EXAMPLE 3: Find a parametric equation for the line $\mathcal{L}$ passing through the points $A = (-1, 2)$ and $B = (2, 3)$.

Solution: From Figure 10, the vector $B - A = (3, 1)$ is parallel to $\mathcal{L}$. We use this vector as the spanning vector. The translation vector can be chosen to be any vector whose tip is on $\mathcal{L}$. We choose $(-1, 2)$. Hence, the desired line should have parametric form $t(3, 1) + (-1, 2)$. As a check on our work, we note that $(-1, 2)$ corresponds to $t = 0$ while $(2, 3)$ corresponds to $t = 1$.

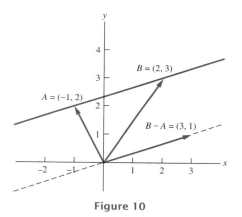

Figure 10

Remark: A given line can have many different parametric descriptions. In Example 3, we could equally well have translated by the point $(2, 3)$, obtaining $t(3, 1) + (2, 3)$ as our answer. The relation between this answer and the previous is simple. Let $s = t + 1$ so that $t = s - 1$. Then

$$t(3, 1) + (2, 3) = (s - 1)(3, 1) + (2, 3)$$
$$= s(3, 1) - (3, 1) + (2, 3) = s(3, 1) + (-1, 2)$$

This represents the same line as that found in Example 3.

▶ EXAMPLE 4: Let $\mathcal{L}$ be the line with parametric representation $(1, 1) + t(2, 1)$. Express this line in $y = mx + b$ format and prove that both expressions define the same set of points in $\mathbb{R}^2$.

Solution: Let

$$(x, y) = (1, 1) + t(2, 1) = (1 + 2t, 1 + t) \tag{1}$$

Thus

$$x = 1 + 2t \text{ and } y = 1 + t \tag{2}$$

We eliminate the parameter t by solving for t in the first equation and substituting into the second. We find that $t = (x - 1)/2$, and hence

$$y = 1 + t = 1 + \frac{x - 1}{2} = \frac{1}{2}x + \frac{1}{2}$$

Thus, every point of the form $(1, 1) + t(2, 1)$ satisfies the equation $y = \frac{1}{2}x + \frac{1}{2}$.

To finish our example, we need to show that, conversely, every point (x, y) which satisfies $y = \frac{1}{2}x + \frac{1}{2}$ may be written in the form of equation (1). However, if we let $t = (x - 1)/2$, then (as the reader may verify), x and y satisfy the equations in (2), which proves that (x, y) may be written in the form of equation (1).   ◄

Remark: In the previous example, we proved the equality of two sets: the set of points of the form of equation (1) and the set of points which satisfy the equation $y = \frac{1}{2}x + \frac{1}{2}$. In general, proving the equality of two sets A and B requires proving two things. We must show that each element of A is an element of B *and* that each element of B is an element of A. This explains the necessity for the second paragraph in our solution.

In geometry, distance and angle are obviously fundamental concepts. From Figure 11, the length of the vector $A = (x, y)$ is

$$|A| = \sqrt{x^2 + y^2}$$

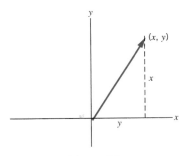

Figure 11

From Figure 8, the distance between two points A and B is the length of the vector $A - B$. If $B = (x_2, y_2)$ and $A = (x_1, y_1)$, then the distance between A and B is

$$|A - B| = |(x_1 - x_2, y_1 - y_2)| = \sqrt{(x_1 - x_2)^2 + (y_1 - y_2)^2} \tag{3}$$

This, of course, is just the usual distance formula in $\mathbb{R}^2$.

▶ **EXAMPLE 5:** Compute the distance between the points $A = (2, 4)$ and $B = (4, 3)$.

Solution: The distance is

$$|A - B| = |(-2, 1)| = \sqrt{(-2)^2 + 1^2} = \sqrt{5}$$

◀

If we square both sides of formula (3), we see that

$$\begin{aligned}
|A - B|^2 &= (x_1 - x_2)^2 + (y_1 - y_2)^2 \\
&= x_1^2 + y_1^2 + x_2^2 + y_2^2 - 2(x_1 x_2 + y_1 y_2) \\
&= |A|^2 + |B|^2 - 2(x_1 x_2 + y_1 y_2)
\end{aligned} \tag{4}$$

The expression in parentheses on the right is called the **dot product** of A and B. Specifically, if A and B are as above, we define

$$A \cdot B = x_1 x_2 + y_1 y_2$$

Thus, for example,

$$(2, 3) \cdot (-3, 5) = (2)(-3) + (3)(5) = -6 + 15 = 9$$

Then, formula (4) may be stated

$$|A - B|^2 = |A|^2 + |B|^2 - 2A \cdot B \tag{5}$$

We refer to this identity as the "law of cosines." To understand why, consider Figure 8 again. The law of cosines from geometry states that

$$|A - B|^2 = |A|^2 + |B|^2 - 2|A| \, |B| \cos \theta$$

Comparing this with formula (5), we see that

$$A \cdot B = |A| \, |B| \cos \theta \tag{6}$$

In many texts (in particular, in many physics texts), this expression is used as the definition of the dot product, in which case formula (5) is exactly the same as the law of cosines.

We say that two non-zero vectors in $\mathbb{R}^2$ are **perpendicular** if the angle between them is $\pi/2$ radians. We also consider the zero vector to be perpendicular to every vector. Since $\cos \pi/2 = 0$, an immediate consequence of formula (6) is perpendicular.

Theorem 1.   *Two vectors A and B in $\mathbb{R}^2$ are perpendicular if and only if $A \cdot B = 0$.*   ■

Thus, for example $(2, 3)$ and $(-3, 2)$ are perpendicular because

$$(2, 3) \cdot (-3, 2) = (2)(-3) + (3)(2) = -6 + 6 = 0.$$

Dealing with only two dimensions is unrealistic; we live in three dimensions. Fortunately, all that we have said carries over to three dimensions with only slight modification. Points in three dimensions are described using three coordinates (x, y, z) (see Figure 12). Just as in $\mathbb{R}^2$, the point (x, y, z) may also be thought of as the endpoint of the vector from the origin to (x, y, z). *The set of all triples (x, y, z) of real numbers is denoted $\mathbb{R}^3$.*

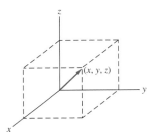

Figure 12

Virtually every formula stated thus far has an obvious analogue in $\mathbb{R}^3$:

$$(x_1, y_1, z_1) + (x_2, y_2, z_2) = (x_1 + x_2, y_1 + y_2, z_1 + z_2)$$
$$(x_1, y_1, z_1) - (x_2, y_2, z_2) = (x_1 - x_2, y_1 - y_2, z_1 - z_2)$$
$$c(x, y, z) = (cx, cy, cz)$$
$$|(x, y, z)| = \sqrt{x^2 + y^2 + z^2}$$
$$(x_1, y_1, z_1) \cdot (x_2, y_2, z_2) = x_1 x_2 + y_1 y_2 + z_1 z_2$$

Formula (6) remains valid in $\mathbb{R}^3$. In particular, two vectors in $\mathbb{R}^3$ are perpendicular if and only if their dot product is 0. Also, the distance between two points A and B in $\mathbb{R}^3$ is $|A - B|$.

▶ EXAMPLE 6:   Let $A = (2, 4, 1)$ and $B = (5, -3, 7)$. Compute:

(1) $A - B$

(2) The distance between the points A and B.

Are the vectors defined by A and B perpendicular?

Solution:

$$A - B = (2, 4, 1) - (5, -3, 7) = (-3, 7, -6)$$

The distance between A and B is

$$|A - B| = |(-3, 7, -6)| = \sqrt{(-3)^2 + 7^2 + (-6)^2} = \sqrt{94}$$

To check for perpendicularity we compute

$$A \cdot B = (2)(5) + (4)(-3) + (1)(7) = 5 \neq 0$$

Thus, A and B are not perpendicular.   ◀

EXERCISES

(1) Determine whether the vector from B to A is equivalent with the vector from D to C, where

(a) $A = (4, 5)$, $B = (3, 1)$, $C = (1, 1)$ and $D = (3, 9)$
(b) $A = (3, -8)$, $B = (2, 1)$, $C = (7, 6)$ and $D = (6, 15)$
(c) $A = (1, 2, 3)$, $B = (4, 5, 6)$, $C = (1, 0, 1)$ and $D = (2, 1, 2)$
(d) $A = (1, 2, 2)$, $B = (1, 3, 1)$, $C = (1, 0, 1)$ and $D = (2, 2, 2)$

(2) For each pair of vectors A and B below, (i) compute $A + B$, (ii) plot A, B and $A + B$ on a single graph, (iii) find the distance between A and B, (iv) compute $|A|$ and $|B|$, (v) compute $A \cdot B$.

(a) $A = (3, 4)$, $B = (-1, 2)$
(b) $A = (3, 4, 0)$, $B = (-1, 2, 0)$
(c) $A = (3, 0, 4)$, $B = (-1, 0, 2)$
(d) $A = (-3, 2)$, $B = (-4, 7)$
(e) $A = (3, 4)$, $B = (6, 8)$
(f) $A = (1, 1, 0)$, $B = (1, 1, 1)$
(g) $A = (1, 1, 0)$, $B = (0, 2, 2)$

(3) Which of the following pairs of vectors are perpendicular to each other? Use the dot product to prove that you are right.

    (a) $A = (7, -3)$, $B = (2, 1)$
    (b) $A = (3, -2, 1)$ $B = (1, 1, -1)$
    (c) $A = (a, b)$, $B = (-b, a)$
    (d) $A = (3, 4, 3)$, $B = (2, -1, \frac{-2}{3})$

(4) Let $A = (1, 1)$. Describe geometrically the set of all vectors B in $\mathbb{R}^2$ such that $B \cdot A \leq |B|$. [*Hint*: Look at formula (6).]

(5) Let $A = (1, 1, 1)$. Describe geometrically the set of all vectors B in $\mathbb{R}^3$ such that $B \cdot A \geq \frac{3}{2}|B|$. [*Hint*: Look at formula (6).]

(6) Let $A = (3, -2)$ and $B = (1, 1)$.

    (a) Plot the vector from B to A.
    (b) Show that the vector in (a) is equivalent with the vector from D to C where $C = (4, 1)$ and $D = (2, 4)$.
    (c) List two other vectors in $\mathbb{R}^2$ (reader's choice) that are equivalent with the vector in (a). (Note that *each* vector will be defined by *two* points.)

(7) Let $A = (1, 2, -5)$ and $B = (1, 2, 1)$. List two other vectors in $\mathbb{R}^3$ (reader's choice) that are equivalent with the vector from B to A. (Note that *each* vector will be defined by *two* points.)

(8) Indicate on a graph the set of points in $\mathbb{R}^2$ of each of the following forms:

    (a) $t(1, 1)$ where t ranges over all real numbers
    (b) $(2, 1) + t(1, 1)$ where t ranges over all real numbers
    (c) $(2, 1) + t(-3, -3)$ where t ranges over all real numbers
    (d) $(1, 0) + t(1, 1)$ where t ranges over all real numbers

(9) Express the line in Exercise 8b in the form $y = mx + b$. Prove your answer. (See Example 3.)

(10) Write the line whose parametric form is $(c, d) + t(e, f)$ in $y = mx + b$ form. [The numbers c, d, e, and f are considered as fixed constants. Assume also that $(e, f) \neq (0, 0)$.] Under what conditions will the line be vertical?

(11) How would you describe geometrically the set of all points in $\mathbb{R}^2$ of the form $s(2, 1) + t(1, 1)$ where s and t range over all real numbers? Demonstrate what you say by plotting a number of points. Be sure you plot some points for which s and t have opposite signs as well as those for which s and t have the same sign.

(12) Describe geometrically the set of all points in $\mathbb{R}^3$ of each of the following forms:

(a) $t(1, 1, 0)$ where t ranges over all real numbers
(b) $(1, 0, 1) + t(1, 1, 0)$ where t ranges over all real numbers
(c) $s(1, 0, 1) + t(1, 1, 0)$ where s and t range over all real numbers
(d) $(1, 0, 0) + s(1, 0, 1) + t(1, 1, 0)$ where s and t range over all real numbers

Draw pictures to illustrate your responses.

(13) Find a parametric description of the line $y = 3x + 2$. Prove that your answer is correct. [*Hint:* Start by finding two points on the line.]

(14) Consider the line described by the equation $y = 7x + 16$. Find two different parametric descriptions for this line and prove that both do indeed describe this line.

(15) Find a parametric description for the line in $\mathbb{R}^3$ that contains the points $(3, -3, 2)$ and $(1, -1, 1)$.

(16) Find a parametric description for the line in Exercise 15 such that the translation vector has its z-coordinate 0 and its spanning vector has two negative entries.

(17) Find the point of intersection for the two lines whose parametric descriptions are, respectively, $(1, -2, 1)t + (2, 4, 5)$ and $(2, 4, 4)t + (4, 4, 8)$. (Yes, they DO intersect!) [*Hint:* It might help to change the name of the paramenter in the second line to, say, s.]

(18) Find a parametric description for the first line in Exercise 17 such that the translation vector has its z-coordinate 0 and its spanning vector has two negative coefficients.

(19) Prove that $(2, 4, 4)t + (4, 4, 8)$ and $(1, 2, 2)t + (2, 0, 4)$ are parametric descriptions of the same line in $\mathbb{R}^3$. For your proof, demonstrate that every point on the first line lies on the second and, conversely, every point on the second lies on the first.

(20) These questions refer to the equation

$$2x + 3y - 6z = 0$$

(a) Find three different points in $\mathbb{R}^3$ (reader's choice) that satisfy the equation. Call your points A, B, and C.
(b) Show that the vectors A, B, and C are all perpendicular to $(2, 3, -6)$.
(c) Explain: "A point (x, y, z) satisfies the equation if and only if (x, y, z) is perpendicular to $(2, 3, -6)$. Therefore the set of points in $\mathbb{R}^3$ that satisfy the equation is a plane." (You may want to draw a picture.)

   Remark: The vector $(2, 3, -6)$ is referred to as a normal vector for the plane described by this equation. In this context, "normal" means "perpendicular."

(21) These questions refer to the equation

$$2x + 3y - 6z = 1$$

(a) Find three different points (reader's choice) that satisfy the equation. Call your points A, B, and C.
(b) Show that the vectors $A - B$ and $C - B$ are both perpendicular to $(2, 3, -6)$.
(c) Explain: "A point (x, y, z) satisfies the equation if and only if the vector $(x, y, z) - B$ is perpendicular to $(2, 3, -6)$. Therefore the set of points that satisfy the equation is a plane." (You may want to draw a picture.)
(d) How do these planes and the planes in Exercise 20 relate to each other?

(22) Let A and B be vectors in $\mathbb{R}^2$. Prove that

$$|A + B|^2 + |A - B|^2 = 2|A|^2 + 2|B|^2$$

Interpret this equality geometrically. Is the same equality true in $\mathbb{R}^3$? If so, prove it.

1.2 THE VECTOR SPACE OF MxN MATRICES

It is difficult to go through life without seeing matrices. For example, the 1997 annual report of Acme Squidget might contain the following table, which shows how much profit (in millions of dollars) each branch made from the sale of each of the company's three varieties of squidgets in 1997.

Profits: 1997

	Red	Blue	Green	Total
Kokomo	11.4	5.7	6.3	23.4
Phili	9.1	6.7	5.5	21.3
Oakland	14.3	6.2	5.0	25.5
Atlanta	10.0	7.1	5.7	22.8
Total	44.8	25.7	22.5	93.0

If we were to enter this data into a computer, we might enter it as a rectangular array without labels. Such an array is called a **matrix**. The Acme profits for 1997 would be described by the matrix:

$$P = \begin{bmatrix} 11.4 & 5.7 & 6.3 & 23.4 \\ 9.1 & 6.7 & 5.5 & 21.3 \\ 14.3 & 6.2 & 5.0 & 25.5 \\ 10.0 & 7.1 & 5.7 & 22.8 \\ 44.8 & 25.7 & 22.5 & 93.0 \end{bmatrix}$$

This matrix is a 5×4 matrix (read "five-by-four") in that it has five rows and four columns. We would also say that its "size" is 5×4.

Each row may itself be thought of as a matrix. The rows are numbered from top to bottom. Thus, the second row of the Acme profit matrix is the 1×4 matrix

$$[9.1, 6.7, 5.5, 21.3]$$

This matrix would be called the "profit vector" for the Phili branch. (In general, any matrix with only one row is called a **row vector**. For the sake of legibility, we usually separate the entries in row vectors by commas, as above.)

Similarly, a matrix with only one column is called a **column vector**. The columns are numbered from left to right. Thus, the third column of the Acme profit matrix is the column vector

$$\begin{bmatrix} 6.3 \\ 5.5 \\ 5.0 \\ 5.7 \\ 22.5 \end{bmatrix}$$

This matrix is the "green squidget profit vector."

In general, if a matrix is denoted by an uppercase letter, such as A, then the entry in the ith row and jth column may be denoted by a_{ij}, using the corresponding lowercase letter. We shall refer to a_{ij} as the "(i, j) entry of A." For example, for the matrix P above, the $(2, 3)$ entry is $p_{23} = 5.5$. Note that the row number comes first. Thus, the most general 2×3 matrix is

$$A = \begin{bmatrix} a_{11} & a_{12} & a_{13} \\ a_{21} & a_{22} & a_{23} \end{bmatrix}$$

We will also occasionally write "$A = [a_{ij}]$", meaning that "A is the matrix whose (i, j) entry is a_{ij}."

At times, we want to take data from two tables, manipulate it in some manner, and display it in a third table. For example, suppose that we want to study the performance of each division of Acme Squidget over the two-year period 1996–97. We go back to the 1996 Annual Report, finding the 1996 profit matrix to be

$$Q = \begin{bmatrix} 11.0 & 5.5 & 6.1 & 22.6 \\ 9.0 & 6.3 & 5.3 & 20.6 \\ 14.1 & 5.9 & 4.9 & 24.9 \\ 9.7 & 7.0 & 5.8 & 22.5 \\ 43.8 & 24.7 & 22.1 & 90.6 \end{bmatrix}$$

If we want the totals for the two-year period, we simply add the entries of this matrix to the corresponding entries from the 1997 profit matrix. Thus, for example, over the two-year period, the Kokomo division made $5.5 + 5.7 = 11.2$ million dollars from selling blue squidgets. Totaling each pair of entries, we find the two-year profit matrix to be

$$T = \begin{bmatrix} 22.4 & 11.2 & 12.4 & 46.0 \\ 18.1 & 13.0 & 10.8 & 41.9 \\ 28.4 & 12.1 & 9.9 & 50.4 \\ 19.7 & 14.1 & 11.5 & 45.3 \\ 88.6 & 50.4 & 44.6 & 183.6 \end{bmatrix}$$

In matrix notation, we indicate that T was obtained by summing corresponding entries of Q and P by writing

$$T = Q + P$$

In general, if A and B are matrices of the same size, their sum is defined by the formula

$$[a_{ij}] + [b_{ij}] = [a_{ij} + b_{ij}]$$

For example:

$$\begin{bmatrix} 2 & 3 & 0 \\ 1 & 0 & -1 \end{bmatrix} + \begin{bmatrix} 1 & 2 & 1 \\ 0 & 5 & 1 \end{bmatrix} = \begin{bmatrix} 3 & 5 & 1 \\ 1 & 5 & 0 \end{bmatrix}$$

Addition of matrices of different sizes is not defined.

There is also a notion of subtraction of matrices, defined by subtracting corresponding components. Thus

$$\begin{bmatrix} 3 & 5 & 1 \\ 1 & 5 & 0 \end{bmatrix} - \begin{bmatrix} 1 & 2 & 1 \\ 0 & 5 & 1 \end{bmatrix} = \begin{bmatrix} 2 & 3 & 0 \\ 1 & 0 & -1 \end{bmatrix}$$

What if, instead of totals for each division and each product, we wanted two-year *averages*? We would simply multiply each entry of T by $\frac{1}{2}$. The notation for this is "$\frac{1}{2}T$." Specifically,

$$\frac{1}{2}T = \begin{bmatrix} 11.20 & 5.60 & 6.20 & 23.00 \\ 9.05 & 6.50 & 5.40 & 20.95 \\ 14.20 & 6.05 & 4.95 & 25.20 \\ 9.85 & 7.05 & 5.75 & 22.65 \\ 44.30 & 25.20 & 22.30 & 91.80 \end{bmatrix}$$

In general, if c is a number and $[a_{ij}]$ is an $m \times n$ matrix, we define

$$c[a_{ij}] = [ca_{ij}] = [a_{ij}]c \tag{1}$$

Hence

$$2\begin{bmatrix} 2 & 3 & 0 \\ 1 & 0 & -1 \end{bmatrix} = \begin{bmatrix} 4 & 6 & 0 \\ 2 & 0 & -2 \end{bmatrix} = \begin{bmatrix} 2 & 3 & 0 \\ 1 & 0 & -1 \end{bmatrix}2$$

In linear algebra, the terms "**scalar**" and "number" mean essentially the same thing. Thus, multiplying a matrix by a number is often called "multiplication by a scalar."

We can use our Acme Squidget profit matrices to demonstrate one of the most important concepts in linear algebra. Consider the last column of the 1997 profit matrix. Since this column represents the total profit for each branch, it is just the sum of the other columns in the profit matrix:

$$\begin{bmatrix} 11.4 \\ 9.1 \\ 14.3 \\ 10.0 \\ 44.8 \end{bmatrix} + \begin{bmatrix} 5.7 \\ 6.7 \\ 6.2 \\ 7.1 \\ 25.7 \end{bmatrix} + \begin{bmatrix} 6.3 \\ 5.5 \\ 5.0 \\ 5.7 \\ 22.5 \end{bmatrix} = \begin{bmatrix} 23.4 \\ 21.3 \\ 25.5 \\ 22.8 \\ 93.0 \end{bmatrix}$$

This last column doesn't tell us anything we didn't already know in that we could have computed the sums ourselves. Thus, while it is useful to have the data explicitly displayed, it is not essential. We say that this data is "dependent on" the data in the other columns. Similarly, the last row of the profit matrix is dependent on the other rows in that it is just their sum.

For another example of dependence, consider the two profit matrices Q and P and their average

$$A = \frac{1}{2}(Q + P) = \frac{1}{2}Q + \frac{1}{2}P \tag{2}$$

The matrix A depends on P and Q—once we know P and Q, we can compute A.

These examples exhibit an especially simple form of dependence. In each case, the matrix we chose to consider as dependent was produced by multiplying the other matrices by scalars and adding. In general, if $A_1, A_2, \ldots, A_k$ is some sequence of $m \times n$ matrices, and B is another $m \times n$ matrix, we say that B is **linearly dependent on** the A_i if there are scalars c_i such that

$$B = c_1 A_1 + c_2 A_2 + \cdots + c_n A_n$$

We will often omit the word "linearly" and simply say that "B is **dependent on** the A_i." We also say that "B is a **linear combination** of the A_i."

One particular $m \times n$ matrix is dependent on every other $m \times n$ matrix. This is the $m \times n$ matrix, which has all of its entries equal to 0. We denote this matrix by **0**. Thus, the 2×3 zero matrix is

$$\mathbf{0} = \begin{bmatrix} 0 & 0 & 0 \\ 0 & 0 & 0 \end{bmatrix}$$

The $m \times n$ zero matrix depends upon every other $m \times n$ matrix because, for any matrix A,

$$0A = \mathbf{0}$$

In general, if we are given some set of $m \times n$ matrices, we say that the set is **linearly dependent** if at least one of the matrices in the set is a linear combination of the other matrices in the set. Thus, our three matrices P, Q, and A are, as a set, dependent. We also consider the set consisting of the zero matrix by itself to be dependent. A set is said to be **linearly independent** if it is not dependent. Thus, independence means that *none* of the matrices in the set are combinations of other matrices from the same set.

▶ **EXAMPLE 1:** Let A_1, A_2, A_3, and A_4 be as shown. Are these matrices linearly dependent?

$$A_1 = \begin{bmatrix} 0 & -1 \\ 1 & 1 \end{bmatrix} \quad A_2 = \begin{bmatrix} 3 & 5 \\ 0 & 1 \end{bmatrix} \quad A_3 = \begin{bmatrix} 1 & 2 \\ 0 & 0 \end{bmatrix} \quad A_4 = \begin{bmatrix} 1 & 1 \\ 0 & 1 \end{bmatrix}$$

Solution: By inspection

$$A_2 = 2A_3 + A_4 = 0A_1 + 2A_3 + A_4$$

Then A_2 depends on A_1, A_3, and A_4, showing that the set is dependent. ◀

Remark: The reader should note that this set is dependent, even though A_1 *is not* a combination of the other A_i. This demonstrates that a set is dependent if *any one* of its elements is a combination of the others. Specifically, dependence does not require that each element be a combination of the others.

▶ **EXAMPLE 2:** Let B_1, B_2, and B_3 be as shown. Are these matrices linearly dependent?

$$B_1 = \begin{bmatrix} 2 \\ 0 \\ 0 \\ 1 \end{bmatrix} \quad B_2 = \begin{bmatrix} -7 \\ 2 \\ 0 \\ 0 \end{bmatrix} \quad B_3 = \begin{bmatrix} 1.3 \\ 0 \\ 2.2 \\ 0 \end{bmatrix}$$

Solution: We begin by asking ourselves whether B_1 is dependent on B_2 and B_3. In other words, are there scalars a and b such that the following relation holds?

$$B_1 = aB_2 + bB_3$$

The answer is clearly no. To see this, we need only note that the last entries of both B_2 and B_3 are 0. Hence, every linear combination of these vectors will have a zero in this position. However, the last entry of B_1 is 1.

Similarly, we see that B_2 is not a linear combination of B_1 and B_3 (from the second entries) and B_3 is not a linear combination of B_1 and B_2 (from the third entries). Thus, the given three matrices are independent. ◄

In general, the set of all matrices that depend on a given set of matrices $A_1, A_2, \ldots, A_k$ is called the **span** of the A_i and is denoted "span $\{A_1, A_2, \ldots, A_k\}$." This is just the set of all linear combinations of the given matrices. Thus, for example, if B_1, B_2, and B_3 are as in Example 2, then

$$B_1 + B_2 + 10B_3 = \begin{bmatrix} 8 \\ 2 \\ 22 \\ 1 \end{bmatrix}$$

is one vector in the span of B_1, B_2, and B_3. The span consists of all possible linear combinations of the B_i.

One of the major advantages of matrix notation is that it allows us to treat a matrix as if it were one single number. For example, it follows from formula (2) on page 17 that

$$2A = Q + P$$

and hence that

$$Q = 2A - P$$

showing that Q is dependent on A and P.

The preceding calculations used a large number of properties of matrix arithmetic which we have not proved. In greater detail, our argument was

$$A = \tfrac{1}{2}(Q + P)$$
$$2A = 2(\tfrac{1}{2}(Q + P)) = \tfrac{2}{2}(Q + P) = Q + P$$
$$2A + (-P) = (Q + P) + (-P)$$
$$= Q + (P - P) = Q + \mathbf{0} = Q$$

We certainly used the associative law $(A + B) + C = A + (B + C)$ as well as the laws $A + (-A) = \mathbf{0}$ and $A + \mathbf{0} = A$, as well as several other laws. In the subsection that follows, we list the most important algebraic properties of matrix addition and scalar multiplication. These properties are called the **vector space properties**. Experience has proven that these properties are all that one needs to effectively deal with any computations such as those done with A, P, and Q. For the sake of this list, we shall let $\mathcal{V}$ denote the set of all $m \times n$ matrices for some fixed n and m. Thus, for example, $\mathcal{V}$ might be the set of all 2×3 matrices.[1]

---

[1] We use "$\mathcal{V}$" in order to avoid the necessity of relisting these properties when we define the general notion of "vector space."

Remark: In set theory, an object that belongs to a certain set is called an **element** of that set. The student must be careful not to confuse the terms "element" and "entry." The matrix below is *one element* of $M(2, 2)$. All elements of $M(2, 2)$ have four *entries*.

$$\begin{bmatrix} 1 & 2 \\ 4 & -5 \end{bmatrix}.$$

The Vector Space Properties.   Let A, B, and C be elements of $\mathcal{V}$. Then:

(a) $A + B$ is a well-defined element of $\mathcal{V}$.

(b) $A + B = B + A$ *(Commutativity)*

(c) $A + (B + C) = (A + B) + C$ *(Associativity)*

(d) There is an element denoted $\mathbf{0}$ in $\mathcal{V}$ such that $A + \mathbf{0} = A$ for all $A \in \mathcal{V}$. This element is referred to as the "zero element."

(e) For each $A \in \mathcal{V}$, there is an element $-A \in \mathcal{V}$ such that $A + (-A) = \mathbf{0}$.

Additionally, for all scalars k and l,

(f) kA is a well-defined element of $\mathcal{V}$.

(g) $k(lA) = (kl)A$

(h) $k(A + B) = kA + kB$

(i) $(k + l)A = kA + lA$

(j) $1A = A$ ■

The proofs that the properties from this list hold for $m \times n$ matrices are left as exercises for the reader. However, let us prove property (c) for $m \times n$ matrices as an example of how such a proof should go.

▶ **EXAMPLE 3:**   Prove property (c) for $m \times n$ matrices.

Solution: Let A, B, and C be $m \times n$ matrices. Then

$$\begin{aligned}
A + (B + C) &= [a_{ij}] + ([b_{ij} + c_{ij}]) \\
&= [a_{ij} + (b_{ij} + c_{ij})] \\
&= [(a_{ij} + b_{ij}) + c_{ij}] \quad \text{(from the associative law for numbers)} \\
&= ([a_{ij} + b_{ij}]) + [c_{ij}] = (A + B) + C \qquad \blacktriangleleft
\end{aligned}$$

When we introduced linear independence, we mentioned that for any matrix A, $0A = \mathbf{0}$. This is very simple to prove:

$$0A = 0[a_{ij}] = [0a_{ij}] = \mathbf{0}$$

This proof explicitly used the fact that we were dealing with matrices. It is possible to give another proof that uses only the vector space properties. We first note from Property (i) that

$$0A + 0A = (0 + 0)A = 0A$$

Next, we cancel $0A$ from both sides using the vector space properties:

$$
\begin{array}{ll}
-0A + (0A + 0A) = -0A + 0A & \text{Property (e)} \\
(-0A + 0A) + 0A = \mathbf{0} & \text{Property (c)} \\
\mathbf{0} + 0A = \mathbf{0} & \text{Properties (b) and (e)} \\
0A = \mathbf{0} & \text{Properties (b) and (d)}
\end{array}
$$

Both proofs are valid for matrices. We, however, prefer the second proof. Since it used only the vector space properties, it will be valid in any context in which these properties hold. For example, in Section 2.2.1, we shall see that addition and scalar multiplication of functions (as used in calculus class) satisfy all the vector space properties. Thus, we can automatically state that $0f(x) = \mathbf{0}$ where $f(x)$ represents any function and $\mathbf{0}$ is the zero function. Admittedly, this is not an exciting result. (Neither, for that matter, is $0A = \mathbf{0}$.) However, it demonstrates an extremely important principal: *Anything we prove about matrices using only the vector space properties will be true in any context for which these properties hold.*

As we progress in our study of linear algebra, it will be important to keep track of exactly which facts can be proved directly from the vector space properties and which require additional structure. We do this with the concept of "vector space." We shall say that a set V is a **vector space** if it has a rule of addition and a rule of scalar multiplication defined on it so that all the vector space properties hold. By a **rule of addition** we mean a well-defined process for taking arbitrary pairs of elements A and B from V and producing a third element $A + B$ in V. (Note that the sum must lie in V.) By a **rule of scalar multiplication** we mean a well-defined process for taking arbitrary real numbers c and arbitrary elements A of V and producing a second element cA of V.

We shall usually denote the set of all $m \times n$ matrices by $M(m, n)$. Then, the set $V = M(2, 2)$ of 2×2 matrices is a vector space: any pair of 2×2 matrices may be added together or multiplied by scalars so as to produce other 2×2 matrices. Addition and scalar multiplication satisfy all the vector space properties. Similarly, the set $M(2, 3)$ of 2×3 matrices is another vector space. For each n and m, $M(n, m)$ is a vector space. The set of all possible matrices is not a vector space, at least under the usual rules of addition and scalar multiplication. This is because we cannot add matrices unless they are the same size: for example, we cannot add a 2×2 matrix to a 2×3 matrix. Thus, our "rule of addition" is not valid for all matrices.

The spaces $\mathbb{R}^2$ and $\mathbb{R}^3$ discussed in Section 1.1 are also vector spaces. (You should review the list of the vector space properties to convince yourself that they all hold for these spaces.) At the moment, the $M(m, n)$ spaces, along with $\mathbb{R}^2$ and $\mathbb{R}^3$, are the only vector spaces we know. This will change in Section 1.5 where we describe the concept of "subspace." However, if we say that something is "true for all vector spaces," we are implicitly stating that it can be proved solely on the basis of the vector space properties. Thus, the property that $0A = \mathbf{0}$ is true for all vector spaces. Another important vector space property is the following. The proof (which *must* use only the vector space properties or their consequences) is left as an exercise.

Proposition 1.   Let A be an element of a vector space $\mathcal{V}$. Then $(-1)A = -A$.   ■

Before ending this section, we need to make a comment concerning notation. Writing column vectors takes considerable text space. There is a handy space-saving notation which we shall use often. Let A be an $m \times n$ matrix. The "main diagonal" of A refers to the entries of the form a_{ii}. (Note that all these entries lie on a diagonal line starting at the upper left-hand corner of A.) If we flip A along its main diagonal, we obtain an $n \times m$ matrix, which is denoted A^t and called the **transpose** of A. Mathematically, A^t is the matrix $[b_{ij}]$ defined by the formula

$$b_{ij} = a_{ji}$$

Thus, if

$$A = \begin{bmatrix} 1 & 2 & 3 \\ 4 & 5 & 6 \end{bmatrix} \text{ then } A^t = \begin{bmatrix} 1 & 4 \\ 2 & 5 \\ 3 & 6 \end{bmatrix}$$

Notice that the columns of A become rows in A^t. Thus, $[2, 3, -4, 10]^t$ is a space-efficient way of writing the column vector

$$\begin{bmatrix} 2 \\ 3 \\ -4 \\ 10 \end{bmatrix}$$

There is a fundamental difference between mathematics and science. Science is founded on experimentation. If certain principles (such as Newton's laws of motion) seem to be valid every time experiments are done to verify them, they are accepted as "law." They will remain the law only as long as they agree with experimental evidence. Thus, Newton's laws were eventually replaced by the theory of relativity when they were found to conflict with the experiments of Michelson and Morley. Mathematics, on the other hand, is based on *proof*. No matter how many times some mathematical principle is observed to hold, we will not accept it as a "theorem" until we can produce a logical argument which shows that the principle can *never* be violated.

One reason for this insistence on proof is the wide applicability of mathematics. Linear algebra, for example, is essential to a staggering array of disciplines including (to mention just a few) engineering (all types), biology, physics, chemistry, economics, social sciences, forestry, environmental science.... We must be certain that our "laws" hold, regardless of the context in which they are applied. Beyond this, however, proofs also serve as explanations of *why* our laws are true. We cannot say that we truly understand some mathematical principle until we can prove it.

In this text, most results are completely proved, but often the proofs are presented as discussions in order to encourage the reader to read and understand them. Before stating a particular property, we usually explain why we would expect such a property to hold and its significance. If this explanation leaves little doubt that the property is always true, then our explanation is an informal, but perfectly valid, proof. In such cases, following the statement with a formal proof could seem redundant and pointless, and might diminish the reader's appreciation for the importance of the proof.

If our explanation seems to leave room for doubt, then we will typically follow the statement of the property with a "formal" proof, i.e., a precisely stated argument showing the validity of the statement. Generally, our more formal proofs are labeled "proof." There are enough formal proofs in the text that the reader should be able, by the end of the course, to become proficient in both reading and producing formal proofs.

In a more advanced text, any proposition that is not self-evident would be given a formal proof. We have not adopted this very high standard of rigor, particularly in our earlier sections, because we want the reader concentrating on understanding the ideas—not struggling with the formalities of the proofs. The reader needs to see that mathematics is not created out of thin air; we do not formulate some property as a theorem unless we have reasons for expecting it to be true and reasons for feeling that it is important. Teaching how to create mathematics is, in fact, one of our main goals.

Also, when the proof follows the statement of the theorem, it is the statement of the theorem, not the proof, which is emphasized. There is, in fact, a strong temptation to skip reading the proof altogether. This too is why we have chosen not to emphasize the statements of our theorems by highlighting them or putting boxes around them. We feel that the discussion in the text is every bit as important as the statements of the results: *why* is just as important as *how*.

EXERCISES

(1) In each case, explicitly write out the matrix A where $A = [a_{ij}]$. Also, give the third row (written as a row vector) and the second column (written as a column vector).

(a) $a_{ij} = 2i - 3j$ where $1 \le i \le 3$ and $1 \le j \le 4$

(b) $a_{ij} = i^2 j^3$ where $1 \le i \le 3$ and $1 \le j \le 2$

(c) $a_{ij} = \cos\frac{ij\pi}{3}$ where $1 \le i \le 3$ and $1 \le j \le 2$

(2) For the matrices A, B, and C below, compute (in the order indicated by the parentheses) $(A + B) + C$ and $A + (B + C)$ to illustrate that $(A + B) + C = A + (B + C)$. Also illustrate the distributive law by computing $3(A + B)$ and $3A + 3B$.

$$A = \begin{bmatrix} 1 & 1 & 2 \\ 0 & 1 & -2 \\ 2 & 0 & 1 \\ 3 & 2 & 1 \end{bmatrix} \quad B = \begin{bmatrix} 2 & 0 & 1 \\ 4 & 1 & 0 \\ 2 & 3 & 2 \\ -1 & 2 & -2 \end{bmatrix} \quad C = \begin{bmatrix} 3 & 1 & 3 \\ 4 & 2 & -2 \\ 4 & 3 & 3 \\ 2 & 4 & -1 \end{bmatrix}$$

(3) The matrices $\{A, B, C\}$ from Exercise 2 are dependent. Express one of them as a linear combination of the others. You should be able to find the constants by inspection (guessing).

(4) Let A, B, and C be as in Exercise 2. Give a fourth matrix D (reader's choice) that belongs to the span of these matrices.

(5) Each of the following sets of matrices are dependent. Demonstrate this by explicitly exhibiting one of the elements of the set as a linear combination of the others. You should be able to find the constants by inspection (guessing).

(a) $\{[1, 1, 2], [0, 0, 1], [1, 1, 4]\}$

(b) $\left\{ \begin{bmatrix} 0 & 0 \\ 1 & 0 \end{bmatrix}, \begin{bmatrix} 1 & 2 \\ 0 & 0 \end{bmatrix}, \begin{bmatrix} 1 & 0 \\ 0 & 0 \end{bmatrix}, \begin{bmatrix} 0 & 1 \\ 0 & 0 \end{bmatrix} \right\}$

(c) $\left\{ \begin{bmatrix} 1 \\ 2 \\ 3 \end{bmatrix}, \begin{bmatrix} 4 \\ 5 \\ 6 \end{bmatrix}, \begin{bmatrix} 3 \\ 3 \\ 3 \end{bmatrix}, \begin{bmatrix} 9 \\ 12 \\ 15 \end{bmatrix} \right\}$

(d) $\left\{ \begin{bmatrix} 1 & 1 \\ 2 & 3 \end{bmatrix}, \begin{bmatrix} 1 & 1 \\ 0 & 1 \end{bmatrix}, \begin{bmatrix} 1 & 2 \\ 0 & 0 \end{bmatrix}, \begin{bmatrix} 0 & -1 \\ 0 & 1 \end{bmatrix} \right\}$

(6) Verify the Remark following Example 1. That is, show that in Example 1, A_1 is not a linear combination of A_2, A_3, and A_4.

(7) What general feature of the following matrices makes it clear that they are independent?

$$\begin{bmatrix} 1 \\ 0 \\ 3 \\ 0 \end{bmatrix} \quad \begin{bmatrix} 0 \\ 0 \\ -5 \\ 1 \end{bmatrix} \quad \begin{bmatrix} 0 \\ 1 \\ 13 \\ 0 \end{bmatrix}$$

(8) Let $X = [1, -1, 0]$ and $Y = [1, 0, 0]$. Give four row vectors (reader's choice) that belong to the span of X and Y. Give an element of $M(1, 3)$ that does not belong to the span of X and Y.

(9) Let $X = [-1, 1, -1]$ and $Y = [-1, 3, 2]$.

    (a) Find an element in the span of X and Y such that each of its entries is positive.
    (b) Show that every element $[x, y, z]$ of the span of X and Y satisfies $5x + 3y - 2z = 0$.
    (c) Give an element of $M(1, 3)$ that does not belong to the span of X and Y.

(10) Let $X = [1, -1, 0]$ and $Y = [1, 0, -1]$. Are there any elements in their span with all entries positive? Explain.

(11) Let X, Y, and Z be as shown. Give four matrices (reader's choice) that belong to their span. Give a matrix that does not belong to their span.

$$X = \begin{bmatrix} 1 & 2 \\ 0 & 3 \end{bmatrix} \quad Y = \begin{bmatrix} 2 & -1 \\ 0 & 0 \end{bmatrix} \quad Z = \begin{bmatrix} 1 & 1 \\ 0 & 1 \end{bmatrix}$$

(12) Suppose that we consider the matrix $[x, y]$ as representing the point (x, y) in $\mathbb{R}^2$.

    (a) What, geometrically, would the span of the matrix $[1, 2]$ represent? Draw a picture.
    (b) What do you guess that the span of $X_1 = [1, 2]$ and $X_2 = [1, 1]$ would represent geometrically? Draw a picture to support your guess.
    (c) Do you think that it is possible to find three independent matrices in $M(1, 2)$? (Think geometrically.)
    (d) Suppose we use the row vector $[x, y, z]$ to represent the point (x, y, z) in $\mathbb{R}^3$. What, geometrically, would the span of the vectors $[1, 1, 0]$ and $[0, 1, 1]$ represent?
    (e) What, geometrically, would the span of the vectors $[1, 2, 1]$ and $[0, 1, 1]$ represent? How does this span compare with that in part (d)?
    (f) What, geometrically, would the span of the vectors $[1, 2, 1]$ and $[2, 4, 2]$ represent? Why is this answer so different from the answer to part (e)? Bring the word "dependent" into your discussion.

(13) Suppose that X and Y are elements of some vector space. Suppose that V and W both belong to the span of X and Y. Show that all linear combinations of V and W also belong to this span.

(14) The rows of the following matrix A are linearly dependent. Demonstrate this by exhibiting one row as a linear combination of the other rows.

$$\begin{bmatrix} 6 & 6 & 4 \\ 1 & 2 & 1 \\ 2 & 1 & 1 \end{bmatrix}$$

(15) Let A be as in Exercise 14. The columns of A are also linearly dependent. Demonstrate this by exhibiting one column as a linear combination of the other columns.

(16) Is it possible to find a 2×2 matrix whose rows are dependent but whose columns are independent? Prove your answer.

(17) Construct an example of your own choice of a 4×4 matrix with linearly dependent rows. Keep your answer *interesting*. Don't, for example, simply put in a zero row.

(18) Let $S = \{A, B, C, D\}$ be some set of four $m \times n$ matrices. Suppose that $D = 2A + B + 3C$ and $C = A - B$. Is $\{A, B, D\}$ dependent? Explain. Is $\{A, C, D\}$ dependent? What can you conclude (if anything) about the dependence of $\{A, B\}$?

(19) We commented that "addition and scalar multiplication of functions (as used in calculus class) satisfies all of the vector space properties." This means that the set $\mathcal{F}(\mathbb{R})$ of all real-valued functions on $\mathbb{R}$ is a vector space under the usual rules of addition and multiplication by scalars. We shall treat function spaces systematically in Sections 2.2.1 and 4.3. However, function spaces provide some very nice examples of linear dependence which can be profitably discussed now. For example, the equation

$$\sin^2 x + \cos^2 x = 1$$

expresses the constant function $y = 1$ as a linear combination of the functions $y = \sin^2 x$ and $y = \cos^2 x$, showing that the three functions $\{\sin^2 x, \cos^2 x, 1\}$ are linearly dependent.

The following sets of functions are linearly dependent in $\mathcal{F}(\mathbb{R})$. Show this by expressing one of them as a linear combination of the others. (You may need to look up the definitions of the sinh and cosh functions as well as some trig identities in a calculus book.)

(a) $\{2e^x, 3e^{-x}, \sinh x\}$
(b) $\{\sinh x, \cosh x, e^{-x}\}$
(c) $\{\cos(2x), \sin^2 x, \cos^2 x\}$
(d) $\{\cos(2x), 1, \cos^2 x\}$

(e) $\{\sin x, \sin(x + \frac{\pi}{4}), \cos(x + \frac{\pi}{4})\}$

(f) $\{(x + 3)^2, 1, x, x^2\}$

(g) $\{x^2 + 3x + 3, x + 1, 2x^2\}$

(h) $\{\ln \frac{(x^2+1)^3}{x^4+7}, \ln \sqrt{x^2 + 1}, \ln(x^4 + 7)\}$

(20) This exercise is a sequel to Exercise 19. Give two examples of functions in the span of the functions $\{1, x, x^2\}$. Describe in words what the span of these three functions is. (Some useful terminology: A function of the form $p(x) = a_0 + a_1 x + \cdots + a_n x^n$ is said to be a polynomial. If $a_n \neq 0$, the polynomial is said to have degree n.)

(21) Let

$$A = \begin{bmatrix} a & b \\ c & d \end{bmatrix}$$

(a) Use the definition of matrix addition to prove that the only 2×2 matrix B such that $A + B = A$ is the zero matrix. The point of this problem is that one should think of $A + 0 = A$ as the defining property of the zero matrix.

(b) Use the definition of matrix addition to prove that the only 2×2 matrix B such that $A + B = 0$ is

$$B = \begin{bmatrix} -a & -b \\ -c & -d \end{bmatrix}$$

The point of this problem is that one should think of $A + (-A) = 0$ as the defining property of $-A$.

(22) Prove vector space properties (b), (e), (g), (h), and (i) for $M(m, n)$.

(23) Let B, C, and X be elements of some vector space. [You may think of them as elements of $M(m, n)$ if you wish.] In the following discussion, we solved the equation $3X + B = C$ for X. At each step we used one of the vector space properties. Which property was used? [*Note:* We define $C - B = C + (-B)$.]

$$3X + B = C$$
$$(3X + B) + (-B) = C + (-B) \qquad \text{Properties (a) and (e)}$$
$$3X + (B + (-B)) = C - B \qquad \text{Definition of } C - B \text{ and Property (?)}$$
$$3X + 0 = C - B \qquad \text{Property (?)}$$
$$3X = C - B \qquad \text{Property (?)}$$
$$\tfrac{1}{3}(3X) = \tfrac{1}{3}(C - B) \qquad \text{Property (?)}$$
$$(\tfrac{1}{3}3)X = \tfrac{1}{3}(C - B) \qquad \text{Property (?)}$$
$$1X = \tfrac{1}{3}(C - B)$$
$$X = \tfrac{1}{3}(C - B) \qquad \text{Property (?)}$$

(24) Let X and Y be elements of some vector space. Prove, putting in every step, that $-(2X + 3Y) = (-2)X + (-3)Y$. You may find Proposition 1 useful.

(25) Let X, Y, and Z be elements of some vector space. [You may think of them as elements of $M(m, n)$ if you wish.] Suppose that there are scalars a, b, and c such that $aX + bY + cZ = \mathbf{0}$. Show that if $a \neq 0$ then

$$X = \left(-\frac{b}{a}\right) Y + \left(-\frac{c}{a}\right) Z$$

Do your proof in a step-by-step manner, to demonstrate the use of each vector space property needed. [*Note*: In a vector space, $X + Y + Z$ denotes $X + (Y + Z)$.]

(26) Prove that in any vector space, if $A + B = \mathbf{0}$, then $B = -A$. (Begin by adding $-A$ to both sides of the given equality.)

(27) Prove Proposition 1. [*Hint*: From Exercise 26, it suffices to prove that $A + (-1)A = \mathbf{0}$.]

 ## ON LINE

Our goal in this discussion is to plot some elements of the span of the vectors $A = [1, 1]$ and $B = [2, 3]$ using MATLAB. Before we begin, however, let us make a few general comments. When you start up MATLAB, you will see something like $>>$ followed by a blank line. If the instructions ask you to enter "$2 + 2$," then you should type $2 + 2$ on the screen behind the $>>$ prompt and then press the "enter" key. Try it!

```
>> 2+2
ans =
     4
```

Entering matrices into MATLAB is not much more complicated. Matrices begin with "[" and end with "]". Entries in rows are separated with either commas or spaces. Thus, after starting MATLAB, our matrices A and B would be entered as shown. Note that MATLAB repeats our matrix, indicating that it has understood us.

```
>> A = [1   1]
A =
     1     1
>> B = [1   3]
B =
     1     3
```

Next we construct a few elements of the span of A and B. If we enter "2*A+B", MATLAB responds

```
ans =
     3     5
```

(Note that "*" is the symbol for "times." MATLAB will complain if you simply write "2A+B".)

If we enter "(−5)∗A +7∗B", MATLAB responds

```
ans =
     2     16
```

Thus the vectors [3, 5] and [2, 16] both belong to the span.

We can get MATLAB to automatically generate elements of the span. Try entering the word "rand." This should cause MATLAB to produce a random number between 0 and 1. Enter "rand" again. You should get a different random number. It follows that entering the command "C=rand∗A+rand∗B" should produce random linear combinations of *A* and *B*. Try it!

To see more random linear combinations of these vectors, push the up-arrow key. This should return you to the previous line. Now you can simply hit "enter" to produce a new random linear combination. By repeating this process, you can produce as many random elements of the span as you wish.

Next, we will plot our linear combinations. Begin by entering the following lines. Here "figure;" creates a figure window, "hold on;" tells MATLAB to plot all points on the same graph, and "axis([-5,5,-5,5])" tells MATLAB to show the range $-5 \leq x \leq 5$ and $-5 \leq y \leq 5$ on the axes. The command "hold on" will remain in effect until we enter "hold off;"

```
>> figure;
>> hold on;
>> axis([-5,5,-5,5]);
```

A window (the "Figure window") showing a blank graph should pop up.

Points are plotted in MATLAB using the command "plot". For example, entering "plot(3,4)" will plot the point (3,4). Return to the MATLAB Command window and try plotting a few points of your own choosing. (To see your points, you will either need to return to the Figure window or move and resize the Command and Figure windows so that you can see them both at the same time. Moving between windows is accomplished by pulling down the "Window" menu.) When you are finished, clear the figure window by entering "cla;" and then enter the following line:

```
C=rand*A+rand*B; plot(C(1),C(2));
```

This will plot one point in the span. ["C(1)" is the first entry of C and "C(2)" is the second.] You can plot as many points as you wish by using the up-arrow key as before.

ON LINE EXERCISES

(1) Plot the points $[1, 1]$, $[1, -1]$, $[-1, 1]$, and $[-1, -1]$ all on the same figure. When finished clear the figure window by entering "cla;".

(2) Enter the vectors A and B from the discussion above.

    (a) Get MATLAB to compute several different linear combinations of them. [Reader's choice.]

    (b) Use C=rand*A+rand*B to create several "random" linear combinations of A and B.

    (c) Plot enough points in the span of A and B to get a discernible geometric figure. Be patient. This may require plotting over 100 points. What kind of geometric figure do they seem to form? What are the coordinates of the vertices?

        Note: If your patience runs thin, you might try entering the following three lines

```
for i=1:200
   C=rand*A+rand*B; plot(C(1),C(2));
end
```

    This causes MATLAB to execute any commands between the "for" and "end" statements 200 times.

    (d) The plot in part (c) is only part of the span. To see more of the span, enter the commands

```
for i=1:200
   C=2*rand*A+rand*B; plot(C(1),C(2),'r');
end
```

    The "r" in the plot command tells MATLAB to plot in red.

(3) Describe in words the set of points $s * A + t * B$ for $-2 \le s \le 2$ and $-2 \le t \le 2$. Create a MATLAB plot that shows this set reasonably well. Use yet another color. [Enter "help plot" to see the choice of colors.] [*Hint*: "rand" produces random numbers between 0 and 1. What would "rand-0.5" produce?]

(4) In Exercise 9 on page 25, it was stated that each element of the span of X and Y satisfies $5x + 3y - 2z = 0$.

    (a) Check this by generating a random matrix C in the span of X and Y and computing "5*C(1)+3*C(2)-2*C(3)". Repeat with another random element of the span.

    (b) Plot a few hundred elements of this span in $\mathbb{R}^3$. Before doing so, close the old Figure window. You do this by pulling down the File menu for the Figure window and selecting "Close." Next, enter "axis([-4,4,-4,4,-4,4]);" followed by "hold on;". A three-dimensional graph should pop up. You plot three-dimensional vectors C with the command "plot3(C(1),C(2),C(3));".

Describe the geometric figure so obtained. What are the coordinates of the vertices? Why is this to be expected?

1.3   SYSTEMS OF LINEAR EQUATIONS

In applications of mathematics, we are often given a set of equations such as the following.

$$\begin{aligned} x + 2y + \ z &= 1 \\ 3x + \ y + 4z &= 0 \\ 2x + 2y + 3z &= 2 \end{aligned}$$

When given such a set, we usually want to find numbers x, y, and z that make each equation valid. A set of equations in a particular collection of variables is called a **system**, and a sequence of numbers which satisfies each equation in the system is called a **solution** to the system.

In the system above, each of the equations is linear. This means that each equation is just a sum of scalars times variables, set equal to another scalar. More precisely, an equation in variables $x_1, x_2, \ldots, x_n$ will be a **linear equation** if and only if it can be expressed in the form

$$a_1 x_1 + a_2 x_2 + \cdots + a_n x_n = b$$

where a_i and b are all scalars. Thus, an equation such as $2x^2 + 3y = 7$ or $2x - 3 \sin y = 4$ would be non-linear. A system in which all the equations are linear is called a **linear system**.

Solving the system just shown is not hard. We begin by subtracting three times the first equation from the second, producing:

$$\begin{aligned} x + 2y + \ z &= 1 \\ - 5y + \ z &= -3 \\ 2x + 2y + 3z &= 2 \end{aligned}$$

Any x, y, and z that satisfy the original system will also satisfy the system above.

Conversely, notice that we can transform the above system back into the original by *adding* three times the first equation onto the second equation. Thus, any variables that satisfy the second system must also satisfy the first. Thus, both systems have exactly the same solutions. We say that the systems are equivalent. (In general, two systems that have exactly the same set of solutions are called **equivalent.**)

To continue the solution process, we next subtract twice the first equation from the third, producing:

$$x + 2y + z = 1$$
$$-5y + z = -3$$
$$-2y + z = 0$$

Note that we have eliminated all occurrences of x from the second and third equations. This system is equivalent with our second system for similar reasons that the second system was equivalent with the first. It follows that this system will have exactly the same solutions as the original system.

Next, we eliminate y from the third equation by subtracting twice the second from five times the third, again producing an equivalent system:

$$x + 2y + \ z = 1$$
$$-5y + \ z = -3$$
$$3z = 6$$

Thus, $z = 2$. Then, from the second equation, $y = 1$, and finally, from the first equation, $x = -3$.

The process we just demonstrated is called **Gaussian elimination**. The general idea is to use the first equation to eliminate all occurrences of the first variable from the equations below it. One then attempts to use the second equation to eliminate the next variable from all equations below it, and so on. In the end, the last variable is determined first (z in our example) and then the others are determined by substitution as in the example. We will describe the procedure in more detail in the next section. However, let us first consider a few more examples.

Consider the following system:

$$x + \ y + \ z = 1$$
$$4x + 3y + 5z = 7 \tag{1}$$
$$2x + \ y + 3z = 5$$

Although this system appears similar to the other, it is clear that the solution process will yield different results. In fact, a close inspection reveals that the second equation equals the third plus twice the first. Thus the second equation carries no additional information. Any numbers x, y, and z that satisfy the first and third equations will automatically satisfy the second. The second equation may be ignored. When this happens, we will say that the system is dependent. (This is an example of the concept of linear dependence discussed in Section 1.2. The set of all linear equations in a given set of variables forms a vector space. A system of linear equations is **dependent** if one equation in the system is a linear combination of other equations from the same system.

Since the second equation may be omitted, our system is really:

$$x + y + z = 1$$
$$2x + y + 3z = 5$$

We could equally well keep the first and second equations and eliminate the third, since the third is the second minus twice the first. Our system then would be

$$x + y + z = 1$$
$$4x + 3y + 5z = 7$$

Either way, though, we only have two "honest" equations to determine three unknowns. But, there is a general principle stating that *it takes at least n linear equations to uniquely determine n unknowns; a system of linear equations with more unknowns than equations either will fail to have any solutions or will have an infinite number of solutions.* (We shall prove this in the next section.) Thus, in this example, the system either will be inconsistent (have no solutions) or will have an infinite number of solutions.

Let us pretend that we didn't notice that this system was dependent. We shall apply the Gaussian elimination technique to the original system and see what happens. We begin by subtracting multiples of the first equation from the other equations so as to eliminate x from these equations. We obtain the system

$$x + y + z = 1$$
$$-y + z = 3$$
$$-y + z = 3$$

Next, we use the second equation to eliminate y from the third. Notice, however, that as we eliminate y from the third equation, we also eliminate z. We obtain the system

$$x + y + z = 1$$
$$-y + z = 3$$
$$0 = 0$$

Clearly, in this system, z can be arbitrary, as long as we let $y = z - 3$. This forces $x = 4 - 2z$. Since z is arbitrary, we are in the case of an infinite number of solutions.

We can visualize this solution geometrically. We say that a column vector $[x, y, z]^t$ is a solution to this system if the numbers x, y, and z form a solution. From the calculation above, we see that $[x, y, z]^t$ is a solution if and only if

$$\begin{bmatrix} x \\ y \\ z \end{bmatrix} = \begin{bmatrix} 4 - 2z \\ z - 3 \\ z \end{bmatrix} = \begin{bmatrix} 4 \\ -3 \\ 0 \end{bmatrix} + \begin{bmatrix} -2z \\ z \\ z \end{bmatrix} = \begin{bmatrix} 4 \\ -3 \\ 0 \end{bmatrix} + z \begin{bmatrix} -2 \\ 1 \\ 1 \end{bmatrix} \qquad (2)$$

To picture this geometrically, we think of $[x, y, z]^t$ as representing the point (x, y, z) in $\mathbb{R}^3$. Note that as z varies, $z[-2, 1, 1]^t$ describes a line through the origin. Adding the point $[4, -3, 0]^t$ to this line translates (shifts) it so that it passes through the point $[4, -3, 0]^t$. (See Figure 1.) For this reason, $[4, -3, 0]^t$ is called the "translation vector." The vector $[-2, 1, 1]^t$ is called the spanning vector because the set being translated is its span.

Formula (2) expresses our solution in what we refer to as "parametric form." In general, **parametric form** means that the solution, when expressed as a vector, is written as a constant vector plus a sum of other vectors times arbitrary parameters. In general, if we represent our solution in parametric form, then the vectors that are multiplied by arbitrary parameters are called **spanning vectors for the solution**, and the "constant vector" is the **translation vector.**

Of course, in solving our system, we could equally well have let y be arbitrary and set $z = y + 3$. This yields $x = -2 - 2y$, so

$$\begin{bmatrix} x \\ y \\ z \end{bmatrix} = \begin{bmatrix} -2 - 2y \\ y \\ y + 3 \end{bmatrix} = \begin{bmatrix} -2 \\ 0 \\ 3 \end{bmatrix} + \begin{bmatrix} -2y \\ y \\ y \end{bmatrix} = \begin{bmatrix} -2 \\ 0 \\ 3 \end{bmatrix} + y \begin{bmatrix} -2 \\ 1 \\ 1 \end{bmatrix} \qquad (3)$$

This last formula presents us with what appears to be a dilemma. We solved our system in two different ways, obtaining two seemingly different answers. Which is correct? Surely they are both correct—our mathematics seems unquestionable. Yet these expressions are so different. . . .

The answer lies in the fact that the solution set to our system is a *set*, not a *formula*. When we say that the general solution to our system is described by formula (2), we mean that each assignment of a specific value to z in this formula produces a solution to the system, and each solution arises by assigning some value to z. Thus, $z = 1$ produces the solution $[2, -2, 1]^t$ and $z = -2$ produces the solution $[8, -5, -2]^t$. We could just as well say that the general solution to our system is the set of all points of the form

$$\begin{bmatrix} 4 \\ -3 \\ 0 \end{bmatrix} + t \begin{bmatrix} -2 \\ 1 \\ 1 \end{bmatrix}$$

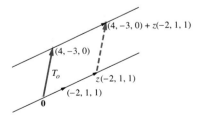

Translating a line (diagram not to scale)

Figure 1

where t is an arbitrary parameter. Every value of t produces a solution and all solutions are produced by assigning a value to t.

Similarly, formula (3) can be thought of as stating that the general solution to our system is the set of all points of the form

$$\begin{bmatrix} -2 \\ 0 \\ 3 \end{bmatrix} + s \begin{bmatrix} -2 \\ 1 \\ 1 \end{bmatrix}$$

where s is an arbitrary parameter. The relation between this representation of the solution set and the preceding one is simple: if we let $s = t - 3$ in the formula above, we obtain the preceding representation. Thus, both formulas do indeed describe the same set of points.

When we wish to stress that our solutions are sets (as above), we will usually replace arbitrary variables (such as z above) by other symbols (such as t).

Notice that in the system (1), no matter how we went about eliminating dependent equations, we were always left with exactly two equations. An analogous fact applies for more general systems of linear equations: *if one eliminates dependent equations, one at a time, until an independent system is obtained, one will always be left with the same number of equations at the end, regardless of which specific equations were eliminated.* (We shall prove this remarkable fact in Section 2.3.) *This number is called the* **rank** *of the system.* Clearly, the rank is much more important in determining the behavior of a system than the number of equations in the system.

Let us next produce an example of a system with no solutions. Consider the following system.

$$\begin{aligned} x + \ y + \ z &= 1 \\ 3x + 2y + 4z &= 4 \\ 2x + \ y + 3z &= 2 \end{aligned} \qquad (4)$$

Explicitly, consider the sum of the first and third equations:

$$3x + 2y + 4z = 3$$

Clearly, no sequence x, y, and z can satisfy both this equation and the second equation in the system. And yet, any x, y, and z that satisfy the system must satisfy both of these equations. Clearly, this system can have no solutions. We say that the equations are **inconsistent** in that there is no solution.

Let us go through the elimination process to see how this problem manifests itself. Again, we subtract multiples of the first equation from the others:

$$\begin{aligned} x + \ y + z &= 1 \\ - y + z &= 1 \\ - y + z &= 0 \end{aligned}$$

Finally, we subtract the second equation from the third:

$$x + y + z = 1$$
$$-y + z = 1$$
$$0 = -1$$

The last equation makes the obviously contradictory statement that $0 = -1$, showing that the system is indeed inconsistent.

We shall close this section with another example of dependence in systems.

$$
\begin{array}{ll}
4x + 5y + 3z + 3w = 1 & I \\
x + y + z + w = 0 & II \\
2x + 3y + z + w = 1 & III \\
5x + 7y + 3z + 3w = 2 & IV
\end{array}
\qquad (5)
$$

In this example, if we call the equations I, II, III, and IV, respectively, then

$$I = III + 2II$$
$$IV = 2III + II$$

Thus, in this system, equations II and III carry all of the information. Equations I and IV are consequences of II and III, hence can be eliminated. Or, if we preferred, we could keep I and IV and eliminate II and III, since $2IV - I = 3III$ and $2I - IV = 3II$. This is a rank 2 system: we have only two independent equations to determine four unknowns. Again, without going further, we can say that either there are no solutions or there are an infinite number of solutions.

As in the second example, this system actually has an infinite number of solutions. Again, let us suppose that we did not notice the dependence of the system. We could begin by subtracting multiples of the first equation from the others in order to eliminate x. It is easier to first interchange the first and second equations, however, since the coefficient of x in the second is one. Thus, we begin with the system:

$$x + y + z + w = 0$$
$$4x + 5y + 3z + 3w = 1$$
$$2x + 3y + z + w = 1$$
$$5x + 7y + 3z + 3w = 2$$

Subtracting multiples of the first equation from the others, we get

$$x + y + z + w = 0$$
$$y - z - w = 1$$
$$y - z - w = 1$$
$$2y - 2z - 2w = 2$$

Next we use the second equation to eliminate y from the equations below it, thereby causing the last two equations to drop out. This is to be expected since this is a rank 2 system. We obtain:

$$x + y + z + w = 0$$
$$y - z - w = 1$$

We solve the second equation by letting both z and w be arbitrary and setting $y = 1 + z + w$. We then substitute this value for y into the first equation, producing:

$$x + 2z + 2w + 1 = 0 \text{ so } x = -2z - 2w - 1$$

Expressed as a vector, our solution is written

$$[x, y, z, w]^t = [-2z - 2w - 1, 1 + z + w, z, w]^t$$

Again, we shall write this in "parametric form." We shall also set $z = r$ and $w = s$ to stress that they may be set arbitrarily. Then $[x, y, z, w]^t$ equals

$$\begin{bmatrix} -2r - 2s - 1 \\ 1 + r + s \\ r \\ s \end{bmatrix} = \begin{bmatrix} -1 \\ 1 \\ 0 \\ 0 \end{bmatrix} + \begin{bmatrix} -2r \\ r \\ r \\ 0 \end{bmatrix} + \begin{bmatrix} -2s \\ s \\ 0 \\ s \end{bmatrix}$$

$$= \begin{bmatrix} -1 \\ 1 \\ 0 \\ 0 \end{bmatrix} + r \begin{bmatrix} -2 \\ 1 \\ 1 \\ 0 \end{bmatrix} + s \begin{bmatrix} -2 \\ 1 \\ 0 \\ 1 \end{bmatrix}$$

Hence, the translation vector is $[-1, 1, 0, 0]^t$. The spanning vectors are $[-2, 1, 1, 0]^t$ and $[-2, 1, 0, 1]^t$.

Is it possible to interpret the solution just obtained geometrically? The answer is, "Yes, provided we have a good imagination." If $[x, y, z]^t$ may be thought of as representing a point in 3-space, then $[x, y, z, w]^t$ should be thought of as representing a point in four dimensional space. We let $\mathbb{R}^4$ denote the set of 4×1 matrices. [Thus, $\mathbb{R}^4 = M(4, 1)$.] *More generally, we let* $\mathbb{R}^n = M(n, 1)$.

As r and s vary, we think of $r[-2, 1, 1, 0]^t$ and $s[-2, 1, 0, 1]^t$ as describing lines in $\mathbb{R}^4$. Then

$$r[-2, 1, 1, 0]^t + s[-2, 1, 0, 1]^t \tag{6}$$

describes the result of adding each point of the first line to each point of the second line. In $\mathbb{R}^3$, this would produce a plane. (Recall that points add according to the parallelogram law.)

It is fairly natural, then, to interpret formula (6) as describing a plane in $\mathbb{R}^4$. This plane contains $[0, 0, 0, 0]^t$. (Let $r = s = 0$.) Adding $T_o = [-1, 1, 0, 0]^t$ to this plane translates it away from $\mathbf{0}$. Thus, we can interpret the solution set to our system as a plane in $\mathbb{R}^4$.

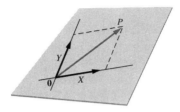

The span of two independent vectors

Figure 2

Translating a plane

Figure 3

In general, if X and Y are independent elements of any vector space whatsoever (e.g., $\mathbb{R}^4$), we will refer to their span as a "plane through $\mathbf{0}$." If T_o is another element of the same space, we refer to the set

$$T_o + \text{span } \{X, Y\}$$

as "the plane through T_o parallel to the span of X and Y."

Similarly, if X is a non-zero element of any vector space, we will refer to the span of X as a "line through $\mathbf{0}$." We refer to the set

$$T_o + \text{span } \{X\}$$

as "the line through T_o parallel to the span of X."

EXERCISES

(1) Let $X = [1, 1, 1, 1]^t$ and $Y = [1, 2, -1, 1]^t$. One of these vectors is a solution to the system below and one isn't. Which is which?

$$4x - 2y - z - w = 0$$
$$x + 3y - 2z - 2w = 0$$

(2) Let $Z = [1, 1, 2, 0]^t$ and let X be as in Exercise 1. For which scalars a and b is it true that $aZ + bX$ is a solution to the system in Exercise 1?

(3) Let X and Y be as in Exercise 1. For which scalars a and b is it true that $aX + bY$ is a solution to the system in Exercise 1?

(4) Find all solutions to the following systems. Write the solution in parametric form and give the translation vector and the spanning vectors. Give four different solutions (reader's choice) to each system. Is the solution set a line? A plane?

(a)
$$x - y + 2z - 2w = 1$$
$$2x + y \quad\quad + 3w = 4$$
$$2x + 3y + 2z \quad\quad = 6$$

(b)
$$x_1 + 2x_2 + x_3 \quad\quad = 1$$
$$x_2 + 4x_3 + 3x_4 = 2$$
$$2x_3 + 2x_4 = 4$$

(5) For each system: (i) Find all solutions (if any exist). Express your answer in parametric form and give the translation vector and the spanning vectors. State whether the solution is a line or a plane or neither. (ii) If one of the equations "disappears" during the solution process, explicitly exhibit one equation as a linear combination of the other equations.

(a)
$$x + 3y + z = 1$$
$$2x + 4y + 7z = 2$$
$$3x + 10y + 5z = 7$$

(b)
$$x + 3y + z = 1$$
$$2x + 4y + 7z = 2$$
$$4x + 10y + 9z = 4$$

(c)
$$x + 3y + z = 1$$
$$2x + 4y + 7z = 2$$
$$4x + 10y + 9z = 7$$

(d)
$$3x + 7y + 2z = 1$$
$$x - y + z = 2$$
$$5x + 5y + 4z = 5$$

Remark: The calculations may be easier if you list the equations in an order different from that given in the problem.

(e)
$$2x - 3y + 2z = 1$$
$$x - 6y + z = 2$$
$$-x - 3y - z = 1$$

(f)
$$2x + 3y - z = -2$$
$$x - y + z = 2$$
$$2x + 3y + 4z = 5$$

(g)
$$\begin{aligned} x + y + z + w &= 1 \\ 2x - 2y + z + 2w &= 3 \\ 2x + 6y + 3z + 2w &= 1 \\ 5x + -3y + 3z + 5w &= 7 \end{aligned}$$

(Here, you should find an infinite number of solutions. In fact, you should see that two of the variables can be set arbitrarily. Can you explain this in terms of the number of independent equations present?)

(h)
$$\begin{aligned} x + y + z + w &= 1 \\ 2x - 2y + z + 2w &= 3 \\ 2x + 6y + 3z + 2w &= 1 \\ 5x + -3y + 3z + 5w &= 8 \end{aligned}$$

(6) In Exercise 5b, give two different solutions to the system (reader's choice). Call your solutions X and Y. Show that $(X + Y)/2$ is a solution to the system but $X + Y$ is not. Can you find an example of a system of 3 equations in 3 unknowns having the property that the sum of two solutions is always a solution?

(7) Create an example of each of the following. Construct your examples so that none of the coefficients in your equations are zero and explain why your examples work.

   (a) A system of five equations in five unknowns which has a line as its solution set.
   (b) A system of five equations in five unknowns which has a plane as its solution set.
   (c) A system of five equations in three unknowns which has a line as its solution set.
   (d) A system of five equations in three unknowns which has a plane as its solution set.

(8) Create an example of each of the following. Construct your examples so that none of the coefficients in your equations are zero and explain why your examples work.

   (a) A system of five equations in five unknowns that has rank 2. How about one with rank 3? Rank 1?
   (b) An inconsistent system of five equations in five unknowns.

(9) The following system is obviously inconsistent. On a single sheet of graph paper, graph $y - 2x = 0$ and $y - 2x = 1$ as lines in $\mathbb{R}^2$. What geometric property of these lines do you notice? How is this relevant to the inconsistency of the system?

$$\begin{aligned} y - 2x &= 1 \\ y - 2x &= 0 \end{aligned}$$

(10) Let a be a fixed number. On a single sheet of graph paper, graph $y - x = 1$,

$y + x = 1$, and $y - 2x = a$ as lines in $\mathbb{R}^2$. Using only your graph, find all values of a for which the following system is consistent.

$$y - x = 1$$
$$y + x = 1$$
$$y - 2x = a$$

(11) Let a, b, c, and d be numbers, with at least one of a, b, and c not equal to 0. In $\mathbb{R}^3$, the solution set to the equation $ax + by + cz = d$ is a plane. Is the following statement true or false? "The solution set to a system of two linear equations in three unknowns always represents a line in three dimensions." Explain.

(12) Continuing Exercise 11, explain geometrically how the solution set of a system of three equations in three unknowns can ever represent a line. Can the solution set of a system of four equations in three unknowns represent a line?

ON LINE

(1) Solve (by hand) equation (7). You should obtain the general solution $X = [4, 0, 0]^t + s[-2, 1, 0]^t + t[-3, 0, 1]^t$ where s and t are arbitrary parameters. Then do (a)-(c).

$$x + 2y + 3z = 4 \qquad\qquad (7)$$

(a) Use the MATLAB command "rand" to create a "random" element C of the span of the vectors $V = [-2, 1, 0]^t$ and $W = [-3, 0, 1]^t$. [See the On Line portion of Section 1.2.]

    Note: When entering matrices into MATLAB, one can use semi-colons (";") to separate columns. Thus, we could enter the column vector V into MATLAB as "V=[-2;1;0]".)

(b) Let $X = [4, 0, 0]^t + C$. Check by direct substitution that X solves equation (4). [You can ask MATLAB to compute "$X(1) + 2 * X(2) + 3 * X(3)$".]

    This demonstrates that the general solution to our system is the span of the spanning vectors translated by the translation vector.

(c) Substitute C into equation (7) and describe what you get. Try substituting some other random linear combinations of V and W into this equation. What do you conjecture? Can you prove it?

(2) Repeat the above sequence (a)-(c) for the system (5) which was solved in this section. Specifically, in (a) you will create a random element C in the span of the spanning vectors $V = [-2, 1, 1, 0]^t$ and $W = [-2, 1, 0, 1]^t$. In (b) you will substitute $C + [-1, 1, 0, 0]^t$ into *each* equation in the system and in (c) you will substitute C into *each* equation in the system.

    Does the conjecture you made in Exercise (1) above still hold? Can you prove it?

1.4 GAUSSIAN ELIMINATION

One discovers very quickly that there is a large amount of writing involved in solving systems of equations. One also realizes that only the coefficients of the variables, not the names of the variables, change as we solve the system. This idea inspires the introduction of a matrix "shorthand" for systems. Consider, for example, the system (S):

$$\begin{aligned}
2x + 2y + 2z + 3w &= 4 \\
x + y + z + w &= 1 \\
2x + 3y + 4z + 5w &= 2 \\
x + 3y + 5z + 11w &= 9
\end{aligned} \tag{S}$$

We associate with it a matrix called the **augmented matrix**. This matrix consists of both the coefficients of the variables and the constants on the right side of the equation.[1]

Thus, the augmented matrix for system (S) is the matrix

$$A = \begin{bmatrix} 2 & 2 & 2 & 3 & 4 \\ 1 & 1 & 1 & 1 & 1 \\ 2 & 3 & 4 & 5 & 2 \\ 1 & 3 & 5 & 11 & 9 \end{bmatrix}$$

We can do the whole elimination procedure using the augmented matrix. Our first step in solving the system might be to interchange the first and second equations, since the coefficient of x in the second equation is 1. In terms of A, this means interchanging the first and second rows. This produces

$$\begin{bmatrix} 1 & 1 & 1 & 1 & 1 \\ 2 & 2 & 2 & 3 & 4 \\ 2 & 3 & 4 & 5 & 2 \\ 1 & 3 & 5 & 11 & 9 \end{bmatrix}$$

Next we subtract multiples of the first equation from the last three equations to eliminate x from these equations. In terms of matrices, this means subtracting multiples of the first row from the other rows. We get:

$$\begin{bmatrix} 1 & 1 & 1 & 1 & 1 \\ 0 & 0 & 0 & 1 & 2 \\ 0 & 1 & 2 & 3 & 0 \\ 0 & 2 & 4 & 10 & 8 \end{bmatrix}$$

---

[1]The matrix formed using just the coefficients of the variables, and not the constants on the right of the equalities, is called the coefficient matrix for the system. The augmented matrix is then the coefficient matrix *augmented* by the column of constants. We will discuss the coefficient matrix in the next section.

Now, we would like to use the second equation to eliminate y from the equations below it. However, the entry in the matrix above corresponding to this coefficient is zero. Thus, we exchange the second and third rows to get a non-zero coefficient in this position. We also divide both sides of the last equation by 2. We get the matrix:

$$\begin{bmatrix} 1 & 1 & 1 & 1 & 1 \\ 0 & 1 & 2 & 3 & 0 \\ 0 & 0 & 0 & 1 & 2 \\ 0 & 1 & 2 & 5 & 4 \end{bmatrix}$$

Subtracting the second row from the fourth eliminates y from the last two equations:

$$\begin{bmatrix} 1 & 1 & 1 & 1 & 1 \\ 0 & 1 & 2 & 3 & 0 \\ 0 & 0 & 0 & 1 & 2 \\ 0 & 0 & 0 & 2 & 4 \end{bmatrix}$$

Notice that in eliminating y from the last two equations, we also eliminated z. Since z is gone, we turn to w. We subtract twice row 3 from row 4, to eliminate w from the fourth equation. We get

$$\begin{bmatrix} 1 & 1 & 1 & 1 & 1 \\ 0 & 1 & 2 & 3 & 0 \\ 0 & 0 & 0 & 1 & 2 \\ 0 & 0 & 0 & 0 & 0 \end{bmatrix} \tag{1}$$

Now, the basic elimination process stops. Subtracting any given row from any row below it would only make the system more complicated. Fortunately, however, we don't need to reduce further. The matrix above corresponds to the system

$$\begin{aligned} x + y + z + w &= 1 \\ y + 2z + 3w &= 0 \\ w &= 2 \end{aligned}$$

The last equation yields $w = 2$. Substitution into the third equation yields

$$y = -2z - 3w = -6 - 2z$$

Finally, the first yields

$$x = 1 - y - z - w = 1 - (-6 - 2z) - z - 2 = 5 + z$$

In parametric form, the solution becomes

$$\begin{bmatrix} x \\ y \\ z \\ w \end{bmatrix} = \begin{bmatrix} 5 + z \\ -6 - 2z \\ z \\ 2 \end{bmatrix} = \begin{bmatrix} 5 \\ -6 \\ 0 \\ 2 \end{bmatrix} + z \begin{bmatrix} 1 \\ -2 \\ 1 \\ 0 \end{bmatrix}$$

There are a few important observations to be drawn from this calculation. First, at each step we performed an operation of one of the following types on the augmented matrix:

(I) Interchange two rows.

(II) Add a multiple of one row onto another.

(III) Multiply one row by a non-zero constant.

These operations are called **elementary row operations**. If we apply an elementary row operation to the augmented matrix, we obtain the augmented matrix for a new system. Every row of the new matrix is a linear combination of rows of the original matrix. Hence, every equation in the new system is a linear combination of equations from the original system. It follows that every solution of the original system is also a solution of the new one.

Conversely, every solution of the new system is also a solution of the original. This is due to the reversibility of elementary row operations. For example, we could undo the effect of multiplying some row by 2 by dividing the new row by 2. We could undo the effect of adding one particular row onto another by subtracting it. Thus, we could, if desired, transform the new system back into the original. It follows that *elementary row operations do not change the solution sets for the corresponding systems.* In particular, matrix (1) describes a system that has exactly the same set of solutions as system (S).

Matrix (1) is in what is called **echelon form**, which is recognizable by the "step-like" arrangement of the zeros in the lower left corner. It is defined by stipulating that the first non-zero entry in any non-zero row occurs to the right of the first such entry in the row directly above it. We also require that all zero rows be grouped together at the bottom. Thus, any matrices of the following forms are in echelon form if the "#" entries are non-zero.

$$\begin{bmatrix} \# & * & * & * & * \\ 0 & \# & * & * & * \\ 0 & 0 & 0 & \# & * \\ 0 & 0 & 0 & 0 & 0 \end{bmatrix} \qquad \begin{bmatrix} 0 & \# & * & * & * & * & * \\ 0 & 0 & 0 & \# & * & * & * \\ 0 & 0 & 0 & 0 & \# & * & * \\ 0 & 0 & 0 & 0 & 0 & 0 & 0 \end{bmatrix} \qquad (2)$$

▶ EXAMPLE 1: Which of the following matrices are in echelon form?

$$\begin{bmatrix} 1 & 0 \\ 0 & 1 \end{bmatrix} \qquad \begin{bmatrix} 1 & 3 & 0 \\ 0 & 0 & 1 \\ 0 & 0 & 0 \end{bmatrix} \qquad \begin{bmatrix} 2 & 0 & 2 & 3 \\ 0 & 1 & 4 & 5 \\ 0 & 0 & 4 & 1 \end{bmatrix}$$

$$\begin{bmatrix} 1 & 2 & 1 & 3 \\ 1 & 0 & 1 & 4 \\ 0 & 0 & 1 & 5 \end{bmatrix} \qquad \begin{bmatrix} 1 & 2 & 0 & 1 \\ 0 & 1 & 2 & 1 \\ 0 & 0 & 0 & 1 \\ 0 & 0 & 1 & 2 \\ 0 & 0 & 0 & 0 \end{bmatrix}$$

Solution: The first three matrices are in echelon form. The last two matrices are not in echelon form. In the fourth matrix, the first non-zero entry of row 2 is directly below the first non-zero entry of row 1. In the last matrix, rows 3 and 4 must be interchanged to get an echelon form. ◄

The solution technique just described produces matrices in echelon form. This is because a matrix that is not in echelon form may always be reduced further—if any step is more than one row high, we can reduce it to a height of 1 by subtracting multiples of one of the rows in this step from the others and/or by switching some of these rows. For example, in the last of the five matrices above, we can interchange the third and fourth rows to reduce the step size; and in the next to last matrix, we can subtract the first row from the second.

Two matrices are said to be **row equivalent** if one can be transformed into the other using elementary row operations. The foregoing comments explain the following important theorem:

Theorem 1. *Every matrix is row equivalent to a matrix in echelon form.* ■

Once echelon form has been reached, the solutions may be computed as follows. First, we write down the system of equations described by the echelon matrix. The variables that correspond to the corners of the steps—the # entries in formula (2)—are called **pivot** variables. The other variables may be set arbitrarily. (These are called **free variables**.) We then solve the last equation for the last pivot variable. We substitute this into the second-to-last equation and solve for the second-to-last pivot variable, and so on, until all of the pivot variables have been determined. (This process is called **back substitution**.)

Actually, the solution process is somewhat simpler if we carry the elimination process beyond echelon form. In matrix (1), we may subtract multiples of the third row from the rows *above* it to eliminate w from these equations. We get:

$$\begin{bmatrix} 1 & 1 & 1 & 0 & -1 \\ 0 & 1 & 2 & 0 & -6 \\ 0 & 0 & 0 & 1 & 2 \\ 0 & 0 & 0 & 0 & 0 \end{bmatrix}$$

Next, we subtract the second row from the first, getting:

$$\begin{bmatrix} 1 & 0 & -1 & 0 & 5 \\ 0 & 1 & 2 & 0 & -6 \\ 0 & 0 & 0 & 1 & 2 \\ 0 & 0 & 0 & 0 & 0 \end{bmatrix}$$

This matrix is in **reduced echelon form** (or "reduced form" for short). This means that the pivot entries are 1 and the entries above and below the pivots are 0. The first two matrices in Example 1 are in reduced form, the third is not: it has non-zero entries above the pivot in the third row. Also, the pivots in the first and third rows are not 1.

The advantage of reduced form is that the answer may be obtained directly, without back substitution. Thus, the first row of the preceding matrix tells us that $x = 5 + z$, the second tells us that $y = -6 - 2z$, and the third says $w = 2$.

Example 2 demonstrates most of the "wrinkles" that can occur in the elimination process.

▶ **EXAMPLE 2:**   Bring the following matrix into echelon and reduced echelon forms. What is the general solution for the system with augmented matrix A?

$$A = \begin{bmatrix} 1 & -1 & 1 & 3 & 0 & 3 & 6 \\ 2 & -2 & 2 & 6 & 0 & 1 & 7 \\ -1 & 1 & 1 & -1 & -2 & 0 & 1 \\ 4 & -4 & 1 & 9 & 3 & 0 & 6 \end{bmatrix}$$

Solution: We begin by eliminating the first variable from the equations below the first equation:

$$\begin{bmatrix} 1 & -1 & 1 & 3 & 0 & 3 & 6 \\ 0 & 0 & 0 & 0 & 0 & -5 & -5 \\ 0 & 0 & 2 & 2 & -2 & 3 & 7 \\ 0 & 0 & -3 & -3 & 3 & -12 & -18 \end{bmatrix} \qquad \begin{array}{l} R_2 \rightarrow R_2 - 2R_1 \\ R_3 \rightarrow R_3 + R_1 \\ R_4 \rightarrow R_4 - 4R_1 \end{array}$$

$$\begin{bmatrix} 1 & -1 & 1 & 3 & 0 & 3 & 6 \\ 0 & 0 & 1 & 1 & -1 & 4 & 6 \\ 0 & 0 & 2 & 2 & -2 & 3 & 7 \\ 0 & 0 & 0 & 0 & 0 & -5 & -5 \end{bmatrix} \qquad \begin{array}{l} R_4 \rightarrow R_4/(-3) \\ R_2 \leftrightarrow R_4 \end{array}$$

$$\begin{bmatrix} 1 & -1 & 1 & 3 & 0 & 3 & 6 \\ 0 & 0 & 1 & 1 & -1 & 4 & 6 \\ 0 & 0 & 0 & 0 & 0 & -5 & -5 \\ 0 & 0 & 0 & 0 & 0 & -5 & -5 \end{bmatrix} \qquad \begin{array}{l} R_3 \rightarrow R_3 - 2R_2 \end{array}$$

$$\begin{bmatrix} 1 & -1 & 1 & 3 & 0 & 3 & 6 \\ 0 & 0 & 1 & 1 & -1 & 4 & 6 \\ 0 & 0 & 0 & 0 & 0 & 1 & 1 \\ 0 & 0 & 0 & 0 & 0 & 0 & 0 \end{bmatrix} \qquad \begin{array}{l} R_2 \rightarrow R_4 - R3 \\ R_3 \rightarrow R_3/(-5) \end{array}$$

This is echelon form. In producing reduced form, it is usually most efficient to begin with the last non-zero row and work from the bottom up. Thus, we subtract multiples of row 3 from the rows above it, yielding:

$$
\begin{bmatrix}
1 & -1 & 1 & 3 & 0 & 0 & 3 \\
0 & 0 & 1 & 1 & -1 & 0 & 2 \\
0 & 0 & 0 & 0 & 0 & 1 & 1 \\
0 & 0 & 0 & 0 & 0 & 0 & 0
\end{bmatrix}
\qquad
\begin{aligned}
R_2 &\rightarrow R_2 - 4R_3 \\
R_1 &\rightarrow R_1 - 3R_3
\end{aligned}
$$

Next, we subtract row 2 from row 1:

$$
\begin{bmatrix}
1 & -1 & 0 & 2 & 1 & 0 & 1 \\
0 & 0 & 1 & 1 & -1 & 0 & 2 \\
0 & 0 & 0 & 0 & 0 & 1 & 1 \\
0 & 0 & 0 & 0 & 0 & 0 & 0
\end{bmatrix}
\qquad
R_2 \rightarrow R_1 - R_2
$$

This is the reduced form. If we interpret this matrix as an augmented matrix, it describes the system

$$
\begin{aligned}
x_1 - x_2 \quad + 2x_4 + x_5 \quad &= 1 \\
x_3 + \ x_4 - x_5 \quad &= 2 \\
x_6 &= 1
\end{aligned}
$$

The pivot variables are x_1, x_3, and x_6 and the free variables are x_2, x_4, and x_5. The general solution is $x_6 = 1$, $x_3 = 2 - x_4 + x_5$, and $x_1 = 1 + x_2 - 2x_4 - x_5$. Let us set $x_2 = s$, $x_4 = t$ and $x_5 = u$. Then, in parametric form, the solution is

$$
\begin{bmatrix} x_1 \\ x_2 \\ x_3 \\ x_4 \\ x_5 \\ x_6 \end{bmatrix}
=
\begin{bmatrix} 1 + s - 2t - u \\ s \\ 2 - t + u \\ t \\ u \\ 1 \end{bmatrix}
=
\begin{bmatrix} 1 \\ 0 \\ 2 \\ 0 \\ 0 \\ 1 \end{bmatrix}
+ s \begin{bmatrix} 1 \\ 1 \\ 0 \\ 0 \\ 0 \\ 0 \end{bmatrix}
+ t \begin{bmatrix} -2 \\ 0 \\ -1 \\ 1 \\ 0 \\ 0 \end{bmatrix}
+ u \begin{bmatrix} -1 \\ 0 \\ 1 \\ 0 \\ 1 \\ 0 \end{bmatrix}
$$

◀

We know, of course, that not all systems have solutions. In the following pair of matrices, the matrix on the left is the augmented matrix for a system with no solutions, since the last non-zero row describes the equation $0 = 1$. The matrix on the right is the augmented matrix for a system that does have solutions. The last equation says $x_4 = 1$.

A system can be inconsistent only if its augmented matrix has an echelon form whose last row describes the equation $0 = 1$.

$$\begin{bmatrix} 1 & 2 & -1 & 4 & 5 \\ 0 & 0 & 1 & 2 & 1 \\ 0 & 0 & 0 & 0 & 1 \end{bmatrix} \qquad \begin{bmatrix} 1 & 2 & -1 & 4 & 5 \\ 0 & 0 & 1 & 2 & 1 \\ 0 & 0 & 0 & 1 & 1 \end{bmatrix}$$

▶ **EXAMPLE 3:** Find conditions on a, b, c, and d for the following system to be consistent.

$$\begin{aligned} x + y + 2z + w &= a \\ 3x - 4y + z + w &= b \\ 4x - 3y + 3z + 2w &= c \\ 5x - 2y + 5z + 3w &= d \end{aligned}$$

Solution: We reduce the augmented matrix:

$$\begin{bmatrix} 1 & 1 & 2 & 1 & a \\ 3 & -4 & 1 & 1 & b \\ 4 & -3 & 3 & 2 & c \\ 5 & -2 & 5 & 3 & d \end{bmatrix}$$

$$\begin{bmatrix} 1 & 1 & 2 & 1 & a \\ 0 & -7 & -5 & -2 & b - 3a \\ 0 & -7 & -5 & -2 & c - 4a \\ 0 & -7 & -5 & -2 & d - 5a \end{bmatrix} \qquad \begin{aligned} R_2 &\rightarrow R_2 - 3R_3 \\ R_3 &\rightarrow R_3 - 4R_1 \\ R_4 &\rightarrow R_4 - 5R_1 \end{aligned}$$

$$\begin{bmatrix} 1 & 1 & 2 & 1 & a \\ 0 & -7 & -5 & -2 & b - 3a \\ 0 & 0 & 0 & 0 & c - a - b \\ 0 & 0 & 0 & 0 & d - 2a - b \end{bmatrix} \qquad \begin{aligned} R_3 &\rightarrow R_3 + R_2 \\ R_4 &\rightarrow R_4 + R_2 \end{aligned}$$

To avoid inconsistencies, we require $c - a - b = 0$ and $d - 2a - b = 0$.   ◀

Now that we have an efficient way of solving systems, we can deal much more effectively with some of the issues discussed in Section 1.2, as the next example demonstrates.

▶ **EXAMPLE 4:** Decide whether or not the vector B belongs to the span of the vectors X_1, X_2, and X_3 below.

$$B = [13, -16, 1]^t \quad X_1 = [2, -3, 1]^t \quad X_2 = [-1, 1, 1]^t \quad X_3 = [-3, 4, 0]$$

Solution: The vector B will belong to the span if there exist constants x, y, and z such that

$$x \begin{bmatrix} 2 \\ -3 \\ 1 \end{bmatrix} + y \begin{bmatrix} -1 \\ 1 \\ 1 \end{bmatrix} + z \begin{bmatrix} -3 \\ 4 \\ 0 \end{bmatrix} = \begin{bmatrix} 13 \\ -16 \\ 1 \end{bmatrix}$$

We simplify the left side of this equality and equate coefficients, obtaining the system

$$\begin{aligned} 2x - y - 3z &= 13 \\ -3x + y + 4z &= -16 \\ x + y \quad\;\; &= 1 \end{aligned}$$

The augmented matrix for this system is

$$\begin{bmatrix} 2 & -1 & -3 & 13 \\ -3 & 1 & 4 & -16 \\ 1 & 1 & 0 & 1 \end{bmatrix}$$

We now reduce this matrix, obtaining

$$\begin{bmatrix} 1 & 0 & -1 & 0 \\ 0 & 1 & 1 & 0 \\ 0 & 0 & 0 & 1 \end{bmatrix}$$

This matrix represents an inconsistent system, showing that B *is not* in the span of the X_i. ◀

Theorem 1 has some very important theoretical consequences. One of the most important is the following:

More Unknowns Theorem. *A system of linear equations with more unknowns than equations will either fail to have any solutions or will have an infinite number of solutions.* ∎

To prove this, consider an echelon form of the augmented matrix. It might look something like the following matrix:

$$\begin{bmatrix} 0 & 1 & * & * & * & * & * \\ 0 & 0 & 0 & 1 & * & * & * \\ 0 & 0 & 0 & 0 & 1 & * & * \end{bmatrix}$$

There is at most one pivot variable per equation. If there are fewer equations than variables, some of the variables must be free variables. As long as the system is not inconsistent, these variables may be set arbitrarily, proving the theorem.

EXERCISES

In these exercises, if you are asked for a general solution, the answer should be expressed in "parametric form" as in the text. Indicate the spanning and translation vectors.

(1) Use elementary row operations to reduce each of these matrices to *reduced* echelon form.

(a)
$$\begin{bmatrix} 2 & 4 & 3 & 0 & 6 \\ 0 & 1 & 1 & 1 & 1 \\ 0 & 0 & 0 & 2 & 4 \end{bmatrix}$$

(b)
$$\begin{bmatrix} 1 & 2 & 3 & 4 \\ 0 & 5 & 6 & 7 \\ 0 & 0 & 9 & 10 \\ 0 & 0 & 0 & 13 \end{bmatrix}$$

(c)
$$\begin{bmatrix} a & b & c & d \\ 0 & e & f & g \\ 0 & 0 & h & i \\ 0 & 0 & 0 & j \end{bmatrix}$$

where a, e, h, and j are all non-zero.

(2) Suppose that the matrices in Exercise 1 are the augmented matrices for a system of equations. In each case, write the system down and find all solutions (if any) to the system.

(3) Which of the following matrices are in echelon form? Which are in reduced echelon form?

(a)
$$\begin{bmatrix} 1 & 0 & 2 & 4 & 0 & 1 \\ 1 & 1 & 0 & 4 & 3 & 2 \\ 0 & 1 & 0 & 4 & 3 & 2 \\ 0 & 0 & 0 & 2 & 1 & 3 \end{bmatrix}$$

(b)
$$\begin{bmatrix} 0 & 1 & 2 & 2 & 4 \\ 0 & 0 & 1 & 2 & 4 \\ 0 & 0 & 0 & 0 & 1 \end{bmatrix}$$

(c)
$$\begin{bmatrix} 0 & 0 & 2 & 4 & 1 \\ 3 & 1 & 2 & 6 & 0 \\ 1 & 1 & 1 & 1 & 1 \\ 0 & 1 & 2 & -1 & 1 \end{bmatrix}$$

(4) Bring each of the matrices in Exercise 3 that are not already in echelon form to echelon form. Interpret each matrix as the augmented matrix for a system of equations. Give the system and give the general solution for each system.

(5) Find the reduced echelon form of each of the following matrices.

(a)
$$\begin{bmatrix} 2 & 7 & -5 & -3 & 13 \\ 1 & 0 & 1 & 4 & 3 \\ 1 & 3 & -2 & -2 & 6 \end{bmatrix}$$

(b)
$$\begin{bmatrix} 1 & 1 & 1 & 1 & 1 \\ 2 & 2 & 1 & 1 & 1 \\ 1 & 0 & 2 & 3 & 2 \\ 4 & 3 & 2 & 1 & 0 \end{bmatrix}$$

(c)
$$\begin{bmatrix} 3 & 9 & 13 \\ 2 & 7 & 9 \end{bmatrix}$$

(d)
$$\begin{bmatrix} 2 & 1 & 3 & 4 & 0 & -1 \\ -2 & -1 & -3 & -4 & 5 & 6 \\ 4 & 2 & 7 & 6 & 1 & -1 \end{bmatrix}$$

(e)
$$\begin{bmatrix} 5 & 4 \\ 1 & 2 \end{bmatrix}$$

(f)
$$\begin{bmatrix} a & b \\ c & d \end{bmatrix} \quad \text{where } ad - bc \neq 0$$

(6) Find conditions on a, b, and c for which the following system has solutions.

$$3x + 2y - z = a$$
$$x + y + 2z = b$$
$$5x + 4y + 3z = c$$

(7) You are given a vector B and vectors X_i. Decide whether B is in the span of the X_i by attempting to solve the equation $B = x_1 X_1 + x_2 X_2 + x_3 X_3$.
   (a) $B = [3, 2, 1]^t$, $X_1 = [1, 0, -1]^t$, $X_2 = [1, 2, 1]^t$, and $X_3 = [1, 1, 1]^t$
   (b) $B = [a, b, c]^t$, $X_1 = [1, 0, -1]^t$, $X_2 = [1, 2, 1]^t$, and $X_3 = [1, 1, 1]^t$
   (c) $B = [3, 2, 1]^t$, $X_1 = [1, 0, -1]^t$, $X_2 = [1, 2, 1]^t$, and $X_3 = [3, 4, 1]^t$

(8) In a vector space, we say that a particular set of vectors "spans the space" if every vector in the space may be written as a linear combination of the given vectors. Show that the vectors $X = [1, 2]^t$ and $Y = [1, -2]^t$ span $\mathbb{R}^2$. Specifically, show that the equation $[a, b]^t = xX + yY$ is solvable, regardless of a and b.

(9) Show that the vectors $X_1 = [3, 1, 5]^t$, $X_2 = [2, 1, 4]^t$, and $X_3 = [-1, 2, 3]^t$ *do not* span $\mathbb{R}^3$ by finding a vector that cannot be expressed as a linear combination of them.

(10) We said in Section 1.3 that the rank of a system of linear equations is the number of equations left after dependent equations have been eliminated, one at a time. We shall prove in Section 2.3 that the rank equals the number of non-zero rows in the reduced form of the augmented matrix. Use this statement to prove that a consistent, rank 3, system of five equations in five unknowns will always have an infinite number of solutions. (Think about the number of free variables in the reduced form of the augmented matrix.)

(11) Write out the possible reduced echelon forms of a 2×2 matrix.

(12) Write out the possible reduced echelon forms of the general 3×3 matrix.

(13) A system is said to be homogeneous if the constants on the right sides of the equations are 0. Why is a homogeneous system always consistent? What does the more unknowns theorem tell you about a homogeneous system that has fewer equations than unknowns?

(14) If you begin with a 5×3 matrix and row reduce until echelon form is obtained, is it possible that all the rows of the echelon form are non-zero? If not, what is the smallest number of zero rows you can get? Explain. Give (if possible) examples of 5×3 matrices in echelon form that represent (a) an inconsistent system, (b) a system with exactly one solution, (c) a system with exactly two solutions, and (d) a system with an infinite number of solutions.

(15) In any linear system, the last column of the augmented matrix is called the "vector of constants," while the matrix obtained from the augmented matrix by deleting the last column is called the "coefficient" matrix. For example, in Exercise 6, the vector of constants is $[a, b, c]^t$, which is the vector formed from the constants on the right sides of the equations in this system and the coefficient matrix is

$$\begin{bmatrix} 3 & 2 & -1 \\ 1 & 1 & 2 \\ 5 & 4 & 3 \end{bmatrix}$$

In applications, it often happens that the coefficient matrix remains fixed while the vector of constants changes periodically. We shall say that the coefficient matrix is "invertible" if there is one and only one solution to the system, regardless of the value of the vector of constants. [2]

(a) Is the coefficient matrix for the system in the Exercise 6 invertible? Explain.

---

[2] This exercise is intended as a "preview" of the concept of invertibility which is discussed in depth in Chapter 3.

(b) Suppose a given system has three equations in three unknowns with an invertible coefficient matrix. Describe the row-reduced form of the augmented matrix as explicitly as possible.

(c) Can a system with two equations and three unknowns have an invertible coefficient matrix? Explain in terms of the row-reduced form of the system.

(d) Can a system with four equations and three unknowns have an invertible coefficient matrix? Explain in terms of the row-reduced form of the system.

(e) Can a system with a non–square coefficient matrix have an invertible coefficient matrix? Explain in terms of the row-reduced form of the system. (A "square matrix" is one that has the same number of rows as columns.)

(16) Suppose that $ad - bc = 0$. Show that the rows of the matrix A are linearly dependent.

$$A = \begin{bmatrix} a & b \\ c & d \end{bmatrix}$$

(17) In the following system, show that for any choice of x and y, there is one (and only one) choice of z and w that solves the system. Write the general solution in parametric form using x and y as free variables.

$$\begin{aligned} x \quad + \quad z + w &= 0 \\ y + 2z + w &= 0 \end{aligned}$$

(18) Can x and y be set arbitrarily in this system? (If you aren't sure, try a few values.)

$$\begin{aligned} x \quad + 2z + 2w &= 0 \\ y + \quad z + \quad w &= 0 \end{aligned}$$

ON LINE

MATLAB can row reduce many matrices almost instantaneously. One begins by entering the matrix into MATLAB. We mentioned in the last section that rows can be separated by semi-colons. One can also separate rows by putting them on different lines. Thus, we can enter a matrix A by stating that

```
>> A = [  1   -1   1    3    0    6
          2   -2   2    6    0    7
         -1    1   1   -1   -2    1
          4   -4   1    9    3    6 ]
```

To row reduce A we simply enter

```
>> rref(A)
```

We obtain the row reduced form almost instantaneously:

```
ans  =
    1   -1    0    2    1    0
    0    0    1    1   -1    0
    0    0    0    0    0    1
    0    0    0    0    0    0
```

(1) Enter "format long" and then compute "$(1/99) * 99 - 1$" in MATLAB. You should get zero. Next compute "$(1/999) * 999 - 1$" and "$(1/9999) * 9999 - 1$." Continue increasing the number of 9s until MATLAB does not produce 0. How many 9s does it require? Why do you not get zero?

   This calculation demonstrates a very important point: MATLAB can interpret as non-zero numbers some answers that should be zero. If this happens with a pivot in a matrix, MATLAB may produce totally incorrect reduced forms. MATLAB guards against such errors by automatically setting all sufficiently small numbers to zero. How small is "sufficiently small" is determined by a number called "eps," which, by default, is set to around 10^{-17}. MATLAB will also issue a warning if it suspects that its answer is unreliable.

   In some calculations (especially calculations using measured data), the default value of eps will be too small. For this reason, MATLAB allows you to include a tolerance in rref. If you wanted to set all numbers less than 0.0001 to zero in the reduction process, you would enter "rref(A,.0001)".

(2) Use "rref " to find all solutions to the system in Exercise 5g in Section 1.3.

(3) We said that the rank of a system is the number of equations left after dependent equations have been eliminated. We commented (but certainly did not prove) that this number does not depend on which equations were kept and which eliminated. It seems natural to suppose that the rows in the row-reduced form of a matrix represent independent equations, hence that the rank should also be computable as the number of non-zero rows in the row-reduced form of the matrix. We can check this conjecture experimentally using the MATLAB command "rank(A)", which computes the rank of a given matrix.

   Use MATLAB to create four 4×5 matrices A, B, C, and D, having respective ranks of 1, 2, 3, and 4. Design your matrices so that none of their entries are zero. Compute their rank by (a) using the rref command and (b) using the rank command.

   Note: Try executing the following sequence of commands and see what happens. This could save you some time!

```
>> M = [ 1 2 1 5 3
         2 1 1 4 3 ]
>> M(3,:) = 2*M(2,:) + M(1,:)
```

This works because, in MATLAB, "M(i,:)" represents the ith row of M. (Similarly, "M(:,j)" represents the jth column.) The equation "M(3,:)=2*M(2,:)+ M(1,:)" both creates a third row for M and sets it equal to twice the second plus the first row.

Note: If you choose "nasty" enough coefficients, then you may need to include a tolerance in the rank command in order to get the answer to agree with what you expect. This is done just as with the "rref" command.

(4) For each of the four matrices from Exercise 3, use "rref" to row reduce the *transpose*. The MATLAB symbol for the transpose of A is "A'". How does the rank of each matrix compare with that of its transpose?

(5) Let $X = [1, 2, -5, 4, 3]'$, $Y = [6, 1, -8, 2, 10]'$, and $Z = [-5, 12 - 19, 24, 1]'$.

   (a) Determine which of the vectors U and V is in the span of X, Y, and Z by using "rref" to solve a system of equations as in Exercise 7 on page 51.
   $$U = [-5, 23, -41, 46, 8]', V = [22, 0, -22, 0, 34]'$$

   (b) Imagine as the head of an engineering group, you supervise a computer technician who knows *absolutely nothing* about linear algebra, other than how to enter matrices and commands into MATLAB. You need to get this technician to do problems similar to part (a). Specifically, you will be giving him or her an initial set of three vectors X, Y, and Z from $\mathbb{R}^n$. You will then provide an additional vector U, and you want the technician to determine whether U is in the span of X, Y, and Z.

   Write a brief set of instructions that will tell your technician how to do this job. Be as explicit as possible. Remember that the technician cannot do linear algebra! You must provide instructions on constructing the necessary matrices, telling what to do with them and how to interpret the answers. The final "output" to you should be a simple "Yes" or "No." You don't want to see matrices!

   (c) One of your assistant engineers comments that it would be easier for the technician in part (b) to use MATLAB's "rank" command rather than "rref." What does your assistant have in mind?

1.4.1 Network Flow

An interesting context in which dependent systems of equations can arise is in the study of traffic flow. Figure 1a is a map of the downtown of a city. Each street is one way in its respective direction. The numbers represent the average number of cars per minute that enter or leave a given street at 3:30 P.M. The variables also represent average numbers of cars per minute. Of course, barring accidents, the total number of cars entering any intersection must equal the total number leaving. Thus, from the intersection of East and North, we see that $x + y = 50$. Continuing clockwise around the square we get the system

$$
\begin{aligned}
x + y &= 50 \\
y + z &= 80 \\
z + w &= 50 \\
x + w &= 20
\end{aligned}
$$

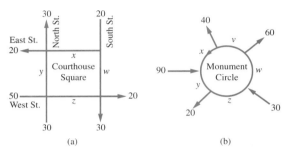

(a)   (b)

Figure 1

In the exercises which follow, you will analyze this system, as well as the system that describes diagram (b). The type of analysis done in these exercises can be applied in any context where some quantity is flowing through prescribed channels. In circuit theory, the statement that the amount of current entering a node equals the amount leaving is called Kirchoff's current law.

EXERCISES

(1) Find all solutions to the system that describes Figure 1a. Use w as your free variable. Then answer the following:

   (a) Suppose that over the month of December the traffic on South Street in front of the courthouse at 3:30 P.M. ranged from 6 to 8 cars per minute. Determine which of the streets on Courthouse Square had the greatest volume of traffic. What were the maximum and minimum levels of traffic flow on this street?

   (b) Suppose that it is observed that in June, the traffic flow past the courthouse is heaviest on North Street. Prove that $10 < w \leq 20$.

(2) Figure 1(b) is a map of a one way traffic circle. Five one-way streets feed in or out of the circle as indicated. The numbers represent the average number of cars per minute that enter or leave on the given street at 3:30 P.M. The variables also represent average numbers of cars per minute. As in Exercise 1, write the equations that determine the traffic flow and find the general solution. (Use v as the free variable.) Which segment of the circle has the least flow? Which has the most?

   Note: In this exercise, some of the variables will need to be subtracted. Thus, for example, the node with 60 cars leaving produces the equation $60 = w - v$.

(3) In Figure 2 below, you see a map of a system of one-way streets in a town. The numbers represent the average number of cars per hour which enter or leave on the given street at 3:30 P.M. The variables also represent average numbers of cars per hour. The section of North Street between First and Second is under construction. Assuming that as few cars as possible use this block, what levels of traffic flow do you expect on the other blocks? [*Hint:* It is easiest if you order your variables in such a way that s becomes a free variable. Note that all of the variables must have non-negative values.]

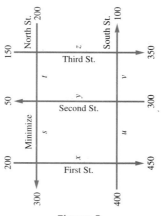

Figure 2

1.5   COLUMN SPACE AND NULLSPACE

The two most fundamental questions concerning any system of equations are (a) Is it solvable? and (b) If it is solvable, what is the nature of the solution set? In this section we discuss both issues. Example 1 addresses the first question.

▶ EXAMPLE 1:   Describe geometrically the set of vectors $B = [b_1, b_2, b_3]^t$ for which the system below is solvable.

$$\begin{aligned} x \quad\;\; + 2z &= b_1 \\ x + y + 3z &= b_2 \\ y + \;\; z &= b_3 \end{aligned}$$

Solution: The system may be written as a single matrix equation:

$$\begin{bmatrix} b_1 \\ b_2 \\ b_3 \end{bmatrix} = \begin{bmatrix} x \quad + 2z \\ x + y + 3z \\ y + z \end{bmatrix} = x\begin{bmatrix} 1 \\ 1 \\ 0 \end{bmatrix} + y\begin{bmatrix} 0 \\ 1 \\ 1 \end{bmatrix} + z\begin{bmatrix} 2 \\ 3 \\ 1 \end{bmatrix} \quad (1)$$

We denote the three columns on the right by A_1, A_2, and A_3, respectively, so that our system becomes

$$B = xA_1 + yA_2 + zA_3$$

We conclude that our system is solvable if and only if B is in the span of the A_i.

To describe this span geometrically, note that $A_3 = 2A_1 + A_2$. Hence, the foregoing equation for B is equivalent with

$$B = xA_1 + yA_2 + z(2A_1 + A_2) = (x + 2z)A_1 + (y + z)A_2 = x'A_1 + y'A_2$$

where $x' = x + 2z$ and $y' = y + z$. This describes the plane spanned by the vectors A_1 and A_2 (see Figure 1). We conclude that our system is solvable if and only if B belongs to this plane.

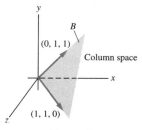

Figure 1

The ideas we used to analyze Example 1 generalize readily to other systems. The columns on the right in equation (1) are the columns of the matrix

$$A = \begin{bmatrix} 1 & 0 & 2 \\ 1 & 1 & 3 \\ 0 & 1 & 1 \end{bmatrix}$$

This matrix is referred to as the "coefficient matrix" for the system because its entries are the coefficients of the unknowns in the system. By definition, the **coefficient matrix** for a system is the augmented matrix with its last column deleted.

If A is the coefficient matrix for some linear system, then, as in Example 1, the corresponding system may be written

$$B = x_1 A_1 + x_2 A_2 + \cdots + x_n A_n \qquad (2)$$

where the A_i are the columns of A, $[x_1, x_2, \ldots, x_n]^t$ is the vector of unknowns, and B is the column vector formed from the constants. Thus the system is solvable if and only if B belongs to the span of the columns of A. In general, the span of the columns of a matrix is called the **column space** of the matrix. Hence, we arrive at the following theorem.

Theorem 1. *A linear system is solvable if and only if the vector of constants belongs to the column space of the coefficient matrix.* ∎

Thinking in terms of the column space can help us to understand why some systems are solvable only if the vector of constants satisfies certain constraints.

▶ EXAMPLE 2: Find a system of two equations in three unknowns such that the system is solvable if and only if the vector of constants lies on the line spanned by the vector $[1, 4]^t$.

Solution: The coefficient matrix for our system will be 2×3. From the preceding comments, the span of its columns must equal the line spanned by the vector $[1, 4]^t$. This will happen only if each column is a multiple of this vector. Thus, we could, for example, let the coefficient matrix be

$$A = \begin{bmatrix} 1 & 2 & 3 \\ 4 & 8 & 12 \end{bmatrix}$$

The corresponding system would be

$$x + 2y + 3z = a$$
$$4x + 8y + 12z = b$$

◀

The column space in Example 1 turned out to be a plane through the origin as a result of an algebraic property of the column space. To explain this, note that planes through the origin in $\mathbb{R}^3$ have the property that if we take any two vectors X and Y lying in a given plane, then all linear combinations of X and Y will also lie in this plane (see Figure 2). We say that planes through the origin are "closed under linear combinations." Figure 3 shows that lines through the origin are closed under linear combinations. In general, any subset of a vector space that is closed under linear combinations is called a subspace.

Definition. *A nonempty subset W of some vector space is a subspace if it is "closed under linear combinations." This means that for every X and Y in W, all linear combinations of X and Y also belong to W.* ■

Planes and lines that do not contain the origin, and curves that are not "straight," will not be closed under linear combinations (see Figures 2 and 3). In fact, we will prove in Section 2.2 that only certain subsets of $\mathbb{R}^3$ can be subspaces: (a) $\mathbb{R}^3$ itself, (b) a plane through the origin, (c) a line through the origin, or (d) the set consisting of just the zero vector.

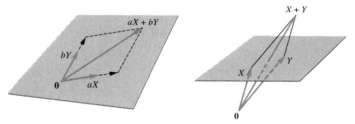

Planes which contain **0** are subspaces.   Planes not containing **0** are not subspaces.

Figure 2

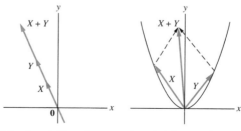

A line through **0** is a subspace   A parabola is not a subspace

Figure 3

We shall show momentarily that the column space of any $m \times n$ matrix is a subspace (of $\mathbb{R}^m$). This, together with the preceding comments, proves that the column space of a non-zero $3 \times n$ matrix must either be all of $\mathbb{R}^3$, a plane through the origin, or a line through the origin. In general we think of subspaces of $\mathbb{R}^m$ as being higher dimensional analogues of lines and planes through the origin.

To see that the column space is closed under linear combinations, let A be an $m \times n$ matrix and let X and Y belong to the column space of A. Then there are constants s_i and t_i such that

$$\begin{aligned} X &= s_1 A_1 + s_2 A_2 + \cdots + s_n A_n \\ Y &= t_1 A_1 + t_2 A_2 + \cdots + t_n A_n \end{aligned} \tag{3}$$

where A_i are the columns of A.

Then, as the reader may check, for c and d any scalars,

$$cX + dY = (cs_1 + dt_1)A_1 + (cs_2 + dt_2)A_2 + \ldots + (cs_n + dt_n)A_n \tag{4}$$

Thus, any linear combination of X and Y is again in the span of the A_i, proving that the column space is indeed closed under linear combinations.

Actually, the same reasoning applies to any set that is representable as a span. We say that a subset $\mathcal{W}$ of a vector space is **spanned by** elements $X_1, X_2, \ldots, X_n$ from the vector space if $\mathcal{W}$ equals the span of these elements. Note that this means, in particular, that the X_i must belong to $\mathcal{W}$. The following proposition is proved using formulas similar to (3) and (4).

Proposition 1. *Suppose that $\mathcal{W}$ is a subset of a vector space spanned by elements $X_1, X_2, \ldots, X_n$. Then $\mathcal{W}$ is a subspace of the given vector space.* ∎

The process of forming linear combinations of the columns of the coefficient matrix is so important that it has its own notation. Let $A = [A_1, A_2, \ldots, A_n]$ be an $m \times n$ matrix where the A_i are the columns of A, and let $X = [x_1, x_2, \ldots, x_n]^t$ be a column of length n. We define

$$AX = x_1 A_1 + x_2 A_2 + \cdots + x_n A_n \tag{M1}$$

We refer to AX as the **product** of the matrix A with the column X. Thus, for example,

$$\begin{bmatrix} 3 & 2 & -1 \\ 1 & 1 & 2 \end{bmatrix} \begin{bmatrix} 1 \\ 2 \\ 3 \end{bmatrix} = 1\begin{bmatrix} 3 \\ 1 \end{bmatrix} + 2\begin{bmatrix} 2 \\ 1 \end{bmatrix} + 3\begin{bmatrix} -1 \\ 2 \end{bmatrix} = \begin{bmatrix} 3+4-3 \\ 1+2+6 \end{bmatrix} = \begin{bmatrix} 4 \\ 9 \end{bmatrix}$$

and

$$\begin{bmatrix} 4 & -5 \\ 2 & 1 \\ -3 & 1 \end{bmatrix} \begin{bmatrix} 2 \\ 3 \end{bmatrix} = 2\begin{bmatrix} 4 \\ 2 \\ -3 \end{bmatrix} + 3\begin{bmatrix} -5 \\ 1 \\ 1 \end{bmatrix} = \begin{bmatrix} -7 \\ 7 \\ -3 \end{bmatrix}$$

Formula (2) says that any system of linear equations may be written in product form as

$$B = AX$$

where A is the coefficient matrix, X is the vector of unknowns, and B is the matrix of constants. Thus, for example, the system from Example 1, written as a product, becomes

$$\begin{bmatrix} b_1 \\ b_2 \\ b_3 \end{bmatrix} = \begin{bmatrix} 1 & 0 & 2 \\ 1 & 1 & 3 \\ 0 & 1 & 1 \end{bmatrix} \begin{bmatrix} x \\ y \\ z \end{bmatrix}$$

Formulas (3) and (4) above make an important statement about matrix products. Formula (3) says exactly that $X = AS$ and $Y = AT$ where $S = [s_1, s_2, \ldots, s_n]^t$ and $T = [t_1, t_2, \ldots, t_n]^t$. Formula (4) says that

$$cAS + dAT = A(cS + dT)$$

This property is often split into two properties, which together are referred to as the linearity properties.

Linearity Properties.   *Let A be an* $m \times n$ *matrix and let X and Y be* $n \times 1$ *matrices. Let a be a scalar. Then*

$$A(X + Y) = AX + AY \quad \textit{(distributive law)}$$
$$A(aX) = aAX \quad \textit{(scalar law)}$$

Next, let us address the question of describing the solution set to a system. Consider, for example, the system

$$\begin{aligned} x + y + z &= 1 \\ 4x + 3y + 5z &= 7 \\ 2x + y + 3z &= 5 \end{aligned} \tag{5}$$

This is system (1) from Section 1.3. From formula (2) in Section 1.3, the general solution is

$$X = [4, -3, 0]^t + z[-2, 1, 1]^t \tag{6}$$

The translation vector $T_o = [4, -3, 0]^t$ is a solution to the system: it is the solution obtained by letting the free variable equal zero. Thus, if we set $x = 4$, $y = -3$, and $z = 0$ in the system, we find that each equation is satisfied.

This is not the case for the spanning vector. If we set $x = -2$, $y = 1$, and $z = 1$, the system is not satisfied. In fact, the spanning vector satisfies the system

$$
\begin{aligned}
x + y + z &= 0 \\
4x + 3y + 5z &= 0 \\
2x + y + 3z &= 0
\end{aligned}
$$

Recall that a system is homogeneous if all the constants on one side of the equations are zero. We refer to the second system as the homogeneous system corresponding to the first system. Thus, in this example, the spanning vector satisfies the homogeneous system corresponding to our original system.

It is not difficult to see why this happens. Let A be the coefficient matrix for this system and let X be as in formula (6). Then, since both X and T_o satisfy the nonhomogeneous system,

$$
\begin{aligned}
AX &= B \\
AT_o &= B
\end{aligned}
$$

where $B = [1, 7, 5]^t$. Subtracting these equations, and using the linearity properties, we see that

$$
A(X - T_o) = \mathbf{0} \tag{7}
$$

It follows immediately that for all z, $z[-2, 1, 1]^t$ satisfies the homogeneous system since $X - T_o = z[-2, 1, 1]^t$.

Actually, more is true. Let Z be any solution to the homogeneous system. Then

$$
A(T_o + Z) = AT_o + AZ = B + \mathbf{0} = B \tag{8}
$$

Thus, $T_o + Z$ is a solution to the nonhomogeneous system. Since the general solution is given by formula (6), it follows that $Z = z[-2, 1, 1]^t$ for some z. Thus, the term "$z[-2, 1, 1]^t$" in formula (4) represents the general solution of the homogeneous system.

Most of what we said carries over to more general systems. Suppose that we are studying a system $AX = B$ and that X and T are solutions of this system. Then the argument leading up to formula (7) shows that $X = T + Z$, where Z is a solution of the corresponding homogeneous system. Conversely, the argument leading up to formula (8) shows that if Z is a solution to the homogeneous system, then $X = T + Z$ solves the nonhomogeneous system. It follows that *the general solution to the nonhomogeneous system may be expressed as "T plus the general solution of the corresponding homogeneous system."* These comments are important enough to be called a theorem:

Translation Theorem.   *Let T be any solution to the system*

$$
AX = B
$$

Then Y satisfies this system if and only if $Y = T + Z$ where Z satisfies $AZ = \mathbf{0}$.   ■

Notice that in this theorem, T need not be the translation vector found by row reduction. In fact, T can be *any arbitrary solution to the system*. (It is often called the "particular solution.")

The translation theorem has a geometric interpretation. If A is a matrix, then the set of vectors X such that $AX = 0$ is called the **nullspace** of A. Thus, the nullspace is just the solution set of the homogeneous system that has A as its coefficient matrix. It is an important observation that the nullspace is a subspace.

Theorem 2.   *The nullspace of an $m \times n$ matrix A is a subspace of $\mathbb{R}^n$.*

Proof: Note first that the zero vector in $\mathbb{R}^n$ is an element of the nullspace since $A0 = 0$. Hence the nullspace is nonempty.

Next, suppose that X and Y satisfy $AX = AY = 0$. Then, for all scalars a and b

$$A(aX + bY) = A(aX) + A(bY) = aAX + bAY = a0 + b0 = 0$$

Hence, the nullspace is closed under linear combinations, proving that it is a subspace. ∎

Thus, in $\mathbb{R}^3$, the nullspace will either be all of $\mathbb{R}^3$, a plane through the origin, a line through the origin, or the set consisting of the zero vector. The translation theorem may be paraphrased as saying that the general solution to $AX = B$ is "T plus the nullspace" where T is any particular solution. We depict the addition of T to the nullspace as shifting the nullspace away from the origin (see Figure 4).

It turns out that the nullspace is extremely important. Indeed, it is referred to as one of the three fundamental spaces associated with a matrix. (The other two are the column space, which was defined earlier, and the row space, which we will study in Section 2.3.) Computing the nullspace is something that we already know how to do. All we need remember is that the nullspace is the solution set to $AX = 0$.

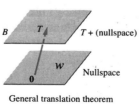

General translation theorem

Figure 4

▶ EXAMPLE 3: Find the nullspace to the matrix A:

$$A = \begin{bmatrix} 1 & 2 & 4 & 1 \\ 1 & 1 & 3 & 2 \\ 2 & 3 & 7 & 3 \end{bmatrix}$$

Solution: The system $AX = 0$ has the augmented matrix

$$\begin{bmatrix} 1 & 2 & 4 & 1 & 0 \\ 1 & 1 & 3 & 2 & 0 \\ 2 & 3 & 7 & 3 & 0 \end{bmatrix}$$

The row-reduced form of this matrix is

$$\begin{bmatrix} 1 & 0 & 2 & 3 & 0 \\ 0 & 1 & 1 & -1 & 0 \\ 0 & 0 & 0 & 0 & 0 \end{bmatrix}$$

The general solution is

$$\begin{bmatrix} x \\ y \\ z \\ w \end{bmatrix} = z \begin{bmatrix} -2 \\ -1 \\ 1 \\ 0 \end{bmatrix} + w \begin{bmatrix} -3 \\ 1 \\ 0 \\ 1 \end{bmatrix}$$

Thus, the nullspace is the span of the two column vectors above. ◀

Remark: Notice that the nullspace in Example 3 turned out to be a span. This is due to its being a subspace. It is a theorem (which will be discussed in some of the exercises for Section 2.2) that all subspaces of $\mathbb{R}^n$ are representable as the span of some finite number of vectors.

We finish this section with an important comment concerning subspaces. The property of being a subspace can be broken down into three statements called the subspace properties:

Subspace Properties.

(1) For all X and Y in $\mathcal{W}$, $X + Y$ is in $\mathcal{W}$.

(2) For all X in $\mathcal{W}$ and all scalars a, aX belongs to $\mathcal{W}$.

(3) $\mathcal{W}$ is nonempty. ■

Recall that in Section 1.2 we said that a "rule of addition" for a set $\mathcal{W}$ is a way of adding pairs of elements of $\mathcal{W}$ that always produces elements of $\mathcal{W}$, and a "rule of scalar multiplication" is a way of multiplying elements of $\mathcal{W}$ by scalars so as to produce elements of $\mathcal{W}$. Statement 1 says that subspaces have a rule of addition, and statement 2 says that they have a rule of scalar multiplication. In fact, we have the following proposition:

Proposition 2. *A subspace of a vector space* $\mathcal{V}$ *is a vector space under the addition and scalar multiplication of* $\mathcal{V}$.

Proof: Let $\mathcal{W}$ be a subspace of a vector space $\mathcal{V}$. To prove that $\mathcal{W}$ is a vector space, we should prove that the addition and scalar multiplication for $\mathcal{W}$ satisfy all the vector space properties [properties (a)–(j) from Section 1.2]. Most of these properties, however, are automatic by virtue of being already valid for $\mathcal{V}$. For example, addition in $\mathcal{W}$ is certainly commutative because addition in $\mathcal{V}$ is commutative. In fact, a close inspection of the list of vector space properties shows that the only properties that require proof are

(a) $\mathbf{0}$ belongs to $\mathcal{W}$.

(b) If A belongs to $\mathcal{W}$, then $-A$ also belongs to $\mathcal{W}$.

Both these follow from statement 2. [For (a), we let X be any element of $\mathcal{W}$ and let $a = 0$. For (b), we let $X = A$ and $a = -1$.] ∎

Summary: This section introduced a large number of new (and important) ideas. It is perhaps worth summarizing the main points:

(a) The column space for a matrix A is the span of the columns of A. The column space can also be described as the set of vectors B for which the system $AX = B$ is solvable.

(b) The nullspace for a matrix A is the set of X for which $AX = \mathbf{0}$. The nullspace describes the solutions to the system $AX = B$ in the sense that the general solution to this system is a translate of the nullspace.

(c) Both the column space and the nullspace are subspaces. Specifically, this means that they are "closed under linear combinations."

EXERCISES

(1) For each matrix A and each vector X, compute AX.

(a)
$$A = \begin{bmatrix} 1 & 0 & -3 & 2 \\ 2 & -2 & 1 & 1 \\ 3 & 2 & -2 & -3 \end{bmatrix} \quad X = \begin{bmatrix} 1 \\ 2 \\ 3 \\ 4 \end{bmatrix}$$

(b)

$$A = \begin{bmatrix} -5 & 17 \\ 4 & 2 \\ 3 & 1 \\ 5 & -5 \end{bmatrix} \quad X = \begin{bmatrix} 2 \\ 1 \end{bmatrix}$$

(c)

$$A = \begin{bmatrix} 1 & 2 & 3 \\ 4 & 5 & 6 \end{bmatrix} \quad X = \begin{bmatrix} x_1 \\ x_2 \\ x_3 \end{bmatrix}$$

(2) Write each of the systems in Exercises 4a, 4b, 5f, and 5g from Section 1.3 in the form $AX = B$.

(3) Give an example of a nonhomogeneous system of equations that has the matrix A from Exercise 1a as its coefficient matrix. Repeat for parts (b) and (c).

(4) Find the nullspace for each of the matrices in Exercise 1. Express each answer as a span.

(5) Find the nullspace for each of the matrices in Exercise 5, Section 1.4. Express each answer as a span.

(6) Create a 2×3 matrix A such that the equation $AX = B$ is solvable if and only if B belongs to the line in $\mathbb{R}^2$ spanned by the vector $[1, 2]^t$. Choose your matrix so that no two of its entries are equal. As a check on your work, choose a specific B not on this line and show (by row reduction) that the system $AX = B$ is inconsistent.

(7) Create a 3×4 matrix A such that the equation $AX = B$ is solvable if and only if B belongs to the line in $\mathbb{R}^3$ spanned by the vector $[1, 2, 3]^t$. Choose your matrix so that no two of its entries are equal. As a check on your work, choose a specific B on this line and show (by row reduction) that the system $AX = B$ is consistent.

(8) Create a 3×4 matrix A such that the equation $AX = B$ is solvable if and only if B belongs to the plane in $\mathbb{R}^3$ spanned by the vectors $[1, 0, 1]^t$ and $[1, 1, 1]^t$. Choose your matrix so that none of its entries are 0. As a check on your work, choose a specific B not on this plane and show (by row reduction) that the system $AX = B$ is inconsistent.

(9) Can a homogeneous system be inconsistent? Explain. What can you say about the number of solutions to a homogeneous system that has more unknowns than equations?

(10) Find the nullspace for matrix A. Compare your answer with the answer to Example 2 in Section 1.4. What theorem does this exercise demonstrate?

$$A = \begin{bmatrix} 1 & -1 & 1 & 3 & 0 & 3 \\ 2 & -2 & 2 & 6 & 0 & 1 \\ -1 & 1 & 1 & -1 & -2 & 0 \\ 4 & -4 & 1 & 9 & 3 & 0 \end{bmatrix}$$

(11) I claim that the general solution to Example 2 in Section 1.4 may be expressed as

$$\begin{bmatrix} -1 \\ 1 \\ 2 \\ 1 \\ 1 \\ 1 \end{bmatrix} + s \begin{bmatrix} 1 \\ 1 \\ 0 \\ 0 \\ 0 \\ 0 \end{bmatrix} + t \begin{bmatrix} -2 \\ 0 \\ -1 \\ 1 \\ 0 \\ 0 \end{bmatrix} + u \begin{bmatrix} -1 \\ 0 \\ 1 \\ 0 \\ 1 \\ 0 \end{bmatrix}$$

where s, t, and u are arbitrary parameters. Am I right? Explain your answer on the basis of the translation theorem.

(12) I claim that the general solution to Example 2 in Section 1.4 may be expressed as

$$\begin{bmatrix} -1 \\ 1 \\ 2 \\ 1 \\ 1 \\ 1 \end{bmatrix} + s \begin{bmatrix} -1 \\ 1 \\ -1 \\ 1 \\ 0 \\ 0 \end{bmatrix} + t \begin{bmatrix} -2 \\ 0 \\ -1 \\ 1 \\ 0 \\ 0 \end{bmatrix} + u \begin{bmatrix} -1 \\ 0 \\ 1 \\ 0 \\ 1 \\ 0 \end{bmatrix}$$

where s, t, and u are arbitrary parameters. Am I right? Explain your answer on the basis of the translation theorem.

(13) "The vectors $X_1 = [1, -1, 1]'$ and $X_2 = [1, 0, 1]'$ span the same plane as $Y_1 = [4, -2, 4]'$ and $Y_2 = [0, -1, 0]'$." True or false? Explain.

(14) "The vectors $X_1 = [1, 2, 1, 1]'$, $X_2 = [1, 1, 1, 1]'$, $X_3 = [1, 0, 1, 2]'$ span the same subspace of $\mathbb{R}^4$ as $Y_1 = [2, 3, 2, 2]'$, $Y_2 = [0, 1, 0, 0]'$, and $Y_3 = [1, 1, 1, 1]'$." True or false? Explain.

(15) (a) Find an equation in three unknowns whose solution space is the subspace of $\mathbb{R}^3$ spanned by the vectors $[1, 2, 1]'$ and $[1, 0, -3]'$. [*Hint*: Is the equation homogeneous?]
(b) Find a system of three equations in three unknowns whose solution space is the subspace of $\mathbb{R}^3$ spanned by the vectors $[1, 2, 1]'$ and $[1, 0, -3]'$.

(16) Find a system of three equations in three unknowns that has the set of vectors of the form $[1, 1, 1'] + a[1, 2, 1]' + b[1, 0, -3]'$ as its general solution.

(17) As CEO of an engineering firm, you have two groups of engineers working on solving the same linear system. Group I tells you that the solution is

$$[1, 0, 0]^t + s[-3, 1, 1]^t + t[-1, 0, 1]^t$$

where s and t are arbitrary parameters. Group II tells you that the solution is

$$[1, -1, 1]^t + s[-4, 1, 2]^t + t[-3, 1, 1]^t$$

Are the answers consistent? Explain.

(18) Let $\{X, Y, Z, W\}$ be four non-zero elements in some vector space. Suppose that $Z = 3X$ and $W = -2Y$. Prove that span $\{X, Y, Z, W\}$ = span $\{X, Y\}$. Under what further conditions would span $\{X, Y, Z, W\}$ = span $\{Y, W\}$?

(19) Let $\{X, Y, Z\}$ be three elements in some vector space. Suppose that $Z = 2X + 3Y$.

(a) Prove that span $\{X, Y, Z\}$ = span $\{X, Y\}$.
(b) Under the same hypothesis, prove that span $\{X, Y, Z\}$ = span $\{X, Z\}$.

(20) We commented that subspaces satisfy the subspace properties (conditions 1, 2, and 3 on page 65). Explain why any subset of a vector space that satisfies these properties must be a subspace.

(21) In these exercises, you are given some set $\mathcal{W}$ of matrices. You should (i) give a non-zero matrix that belongs to $\mathcal{W}$, (ii) give a matrix (of the same size as the elements of $\mathcal{W}$) that does not belong to $\mathcal{W}$, (iii) use the subspace properties to prove that $\mathcal{W}$ is a subspace, and (iv) express $\mathcal{W}$ as a span. We present the solution to part (a) as an example of what is required.

(a) $\mathcal{W}$ is the set of all matrices of the form

$$\begin{bmatrix} a + b + 3c & 2a - b \\ a & 2a + b + 4c \end{bmatrix}$$

where a, b, and c range over all real numbers.

Solution:

(i) To find an element of $\mathcal{W}$, we simply assign values to the variables. For example, $a = 1, b = 0, c = -1$ yields the matrix

$$\begin{bmatrix} -2 & 2 \\ 1 & -2 \end{bmatrix}$$

This is one element of $\mathcal{W}$. Other assignments produce other elements.

(ii) To find a matrix that is not in $\mathcal{W}$, we simply pick some "random" matrix and check whether it is in $\mathcal{W}$. For example, let us try a matrix all of whose entries are 1. This will be in $\mathcal{W}$ only if there are constants a, b, and c such that

$$\begin{bmatrix} 1 & 1 \\ 1 & 1 \end{bmatrix} = \begin{bmatrix} a + b + 3c & 2a - b \\ a & 2a + b + 4c \end{bmatrix}$$

This yields the system

$$a + b + 3c = 1$$
$$2a - b \quad\quad = 1$$
$$a \quad\quad\quad\quad = 1$$
$$2a + b + 4c = 1$$

We solve this system by means of row reduction, finding it inconsistent; hence our matrix does not belong to $\mathcal{W}$.

(iii) To prove that $\mathcal{W}$ is a subspace, it suffices to prove the subspace properties. (Note that we already know that $\mathcal{W}$ is nonempty from solution (i).)

To prove property 1, let X and Y be elements of $\mathcal{W}$. Then

$$X = \begin{bmatrix} a + b + 3c & 2a - b \\ a & 2a + b + 4c \end{bmatrix}$$

where a, b, and c are scalars and

$$Y = \begin{bmatrix} a' + b' + 3c' & 2a' - b' \\ a' & 2a' + b' + 4c' \end{bmatrix}$$

where a', b', and c' are scalars. We must show that $X + Y$ has the form

$$\begin{bmatrix} a'' + b'' + 3c'' & 2a'' - b'' \\ a'' & 2a'' + b'' + 4c'' \end{bmatrix}$$

for some scalars a'', b'', and c''. However, direct computation shows that $X + Y$ equals

$$\begin{bmatrix} (a + a') + (b + b') + 3(c + c') & 2(a + a') - (b + b') \\ a + a' & 2(a + a') + (b + b') + 4(c + c') \end{bmatrix}$$

This is of the desired form, where $a'' = a + a'$, $b'' = b + b'$, and $c'' = c + c'$.

To prove property 2, suppose that k is a scalar. Then kX equals

$$\begin{bmatrix} ka + kb + 3kc & 2ka - kb \\ ka & 2ka + kb + 4kc \end{bmatrix} = \begin{bmatrix} a' + b' + 3c' & 2a' - b' \\ a' & 2a' + b' + 4c' \end{bmatrix}$$

where $a' = ka$, $b' = kb$, and $c' = kc$. This proves that kX also belongs to $\mathcal{W}$.

(iv) To find elements that span $\mathcal{W}$, we write the general element of $\mathcal{W}$ as

$$\begin{bmatrix} a+b+3c & 2a-b \\ a & 2a+b+4c \end{bmatrix} = \begin{bmatrix} a & 2a \\ a & 2a \end{bmatrix} + \begin{bmatrix} b & -b \\ 0 & b \end{bmatrix} + \begin{bmatrix} 3c & 0 \\ 0 & 4c \end{bmatrix}$$

$$= a\begin{bmatrix} 1 & 2 \\ 1 & 2 \end{bmatrix} + b\begin{bmatrix} 1 & -1 \\ 0 & 1 \end{bmatrix} + c\begin{bmatrix} 3 & 0 \\ 0 & 4 \end{bmatrix}$$

This shows that W is the span of the matrices

$$A = \begin{bmatrix} 1 & 2 \\ 1 & 2 \end{bmatrix} \quad B = \begin{bmatrix} 1 & -1 \\ 0 & 1 \end{bmatrix} \text{ and } C = \begin{bmatrix} 3 & 0 \\ 0 & 4 \end{bmatrix}$$

Remark: Actually, we could have done the preceding exercise with much less work, had we not been required to "use the subspace properties." Since spans are subspaces, to show that W is a subspace, all we need show is that it is a span. This is what we did in part (iv). Knowing that W was a subspace helped us only in that it told us to look for spanning elements.

(b) W is the set of all elements of $\mathbb{R}^2$ of the form $[x, -x]^t$ where x ranges over all real numbers.

(c) W is the set of all elements of $M(2, 2)$ of the form

$$\begin{bmatrix} a & a+b+c \\ b & c \end{bmatrix}$$

where a, b, and c range over all real numbers.

(d) W is the set of points in $\mathbb{R}^3$ of the form

$$[s - 2t + u, -3s + t - u, -2s - t]^t$$

where s, t, and u range over all real numbers.

(22) Let W denote the set of all elements A of $M(2, 2)$ such that the sum of all of the entries of A is zero.

(a) Give an example of two non-zero elements A and B in W.
(b) For your A and B, show that $3A + 4B \in W$.
(c) Show that W is a subspace of $M(2, 2)$.

(23) Let W be the set of all points $[x, y]^t$ in $\mathbb{R}^2$ such that both $x \geq 0$ and $y \geq 0$. Draw a graph representing W. Is W a subspace of $\mathbb{R}^2$? Explain.

(24) Graph the set W of points $[n, m]^t$ in $\mathbb{R}^2$ such that both n and m are integers. Is W a subspace of $\mathbb{R}^2$? Explain.

(25) Which lines in $\mathbb{R}^2$ form subspaces and which don't? Explain.

(26) Let W be the set of all points in $\mathbb{R}^3$ of the form $[x, y, z]^t$ where $x^2 = y + z$. Find two specific points (reader's choice) X and Y in W such that $X + Y$ is not in W. Are there any non-zero points X and Y in W such that $X + Y$ also belongs to W? Explain.

(27) A square matrix A is upper triangular if all of the entries below the main diagonal are zero. Thus, for example, the following matrix A is upper triangular. Prove that the set T of all 3×3 upper triangular matrices is a subspace of $M(3, 3)$.

$$A = \begin{bmatrix} 1 & 2 & 3 \\ 0 & 4 & 5 \\ 0 & 0 & 6 \end{bmatrix}$$

(28) An upper triangular matrix B is unipotent if its diagonal entries are all equal to 1. Thus, for example, these matrices A and B are unipotent.

$$A = \begin{bmatrix} 1 & 1 & -2 \\ 0 & 1 & 0 \\ 0 & 0 & 1 \end{bmatrix} \quad B = \begin{bmatrix} 1 & 2 & 3 \\ 0 & 1 & 5 \\ 0 & 0 & 1 \end{bmatrix}$$

(a) Find non-zero scalars a and b such that $aA + bB$ is unipotent.
(b) Is the set of all unipotent matrices a subspace of $M(3, 3)$? Explain.

(29) Consider a system $AX = B$ where $B \neq 0$. Let X and Y satisfy this system. Find all constants a and b such that $aX + bY$ also satisfies this system.

(30) Suppose that S and T are subspaces of some vector space V. Is $S \cap T$ also a subspace of V? Prove your answer. (Recall that $S \cap T$ denotes the set of all elements that belong to both S and T.)

(31) Suppose that S and T are subspaces of $\mathbb{R}^3$. Under what circumstances will $S \cup T$ be a subspace of $\mathbb{R}^3$? What if S and T are subspaces of $\mathbb{R}^n$? Prove your answer. (Recall that $S \cup T$ denotes the set of all elements that belong to either S or T or both.)

(32) Let S and T be subspaces of some vector space V. By $S + T$ we mean the set of all vectors of the form $X + Y$ where $X \in S$ and $Y \in T$. Prove that $S + T$ is a subspace of V.

(33) Suppose that $V = \mathbb{R}^3$, S is a line through 0 and T is a plane through 0. What can you say about $S + T$? How is it different from $S \cup T$? (See Exercise 32 for the definition of $S + T$.)

ON LINE

MATLAB is very good at matrix multiplication. For example:

```
>> A = [   1   2   1   3
          -5   7   2   2
          13   4   4   3 ]
```

```
   A =
        1       2      1      3
       -5       7      2      2
       13       4      4      3
>> X = [1;3;-2;4]
   X =
        1
        3
       -2
        4
>> B = A * X
   B =
       17
       20
       29
```

ON LINE EXERCISES

(1) Check the value of AX computed above by asking MATLAB to compute $A_1 + 3A_2 - 2A_3 + 4A_4$ where A_i is the ith column of A. (Recall that MATLAB for A_i is "A(:,i)".)

(2) For the matrices A and B above, find the general solution to the system $AX = B$ using "rref". Express your solution in parametric form. For which value of the parameter do you obtain X?

   Note: A neat way of producing the augmented matrix is to enter "C=[A,B]".

(3) Here is another way to find the general solution to the system in Exercise 2. Try this (with X as before):

```
>> format short
>> Z=null(A)
>> A*(X+20*Z)
```

What null does is find a spanning vector for the nullspace. This example demonstrates the translation theorem in that it shows that the general solution is any one solution plus the nullspace. There is a catch though; null only gives *approximate* answers. To see this, try

```
>>format long
>> A*Z
```

The term "1.0e-15" means that the given vector should be multiplied by 10^{-15}. Small, but not 0.

(4) Let

$$A = \begin{bmatrix} 17 & -6 & 13 & 27 & 64 & 19 \\ 4 & -6 & -33 & 25 & 7 & 9 \\ 55 & -24 & 6 & 106 & 199 & 66 \\ 89 & -36 & 32 & 160 & 327 & 104 \end{bmatrix} \text{ and } B = \begin{bmatrix} 17 \\ 4 \\ 55 \\ 89 \end{bmatrix}$$

(a) Compute the rank of A. How many free variables will the system $AX = 0$ have?

(b) How many spanning vectors will the nullspace for A have?

(c) Use the "W=null(A)" command to find a spanning set for the nullspace of A. (The columns of W will be the desired spanning set.)

(d) Find (by inspection) a vector X_0 such that $AX_0 = B$.

(e) Let the columns of W be denoted by W_i. Check that $X_0 + 2W_1 - 3W_2 + 7W_3 - 9W_4$ is, at least approximately, a solution to $AX = B$.

(f) I claim that the general solution to $AX = B$ is $X_0 + WY$ where $Y \in \mathbb{R}^4$. Explain.

(g) Use "rref" to find a basis for the nullspace of A by solving $AX = 0$. Your answer will appear to be totally different from the answer that null gives. Demonstrate that the two answers are really equivalent by showing that the second column of W is a linear combination of your basis and that your third basis element is a linear combination of the columns of W.

   Note: You may need to use "rref([A,B],tol)" where the tolerance "tol" is larger than the default value. Try to use the smallest possible tolerance.

1.5.1 PREDATOR-PREY PROBLEMS
LINEAR DYNAMICAL SYSTEMS

Requires a Computer

Rabbits are pleasant little animals which, when left to their own devices, like nothing better than eating and producing more rabbits. In the presence of sufficient food and lacking natural competitors, we might guess that each spring the number of rabbits would increase by a fixed percentage, say, 10%. Thus, if the initial rabbit population is R_o, next spring there will be $R_1 = (1.1)R_o$ rabbits. In two years, the rabbit population will be $R_2 = (1.1)^2 R_o$. The number of rabbits after n years will be

$$R_n = (1.1)^n R_o$$

The rabbit population grows without bounds. Let us assume that R_n is measured in units of hundreds of rabbits.

Suppose now that we introduce a small number of mountain lions into our environment to keep the rabbit population under control. The rate of growth of the mountain

lions will depend on both the number of mountain lions and the number of rabbits. We shall assume that the growth of the mountain lion population is governed by the equation

$$M_{n+1} = (0.4)R_n + (0.7)M_n \qquad (1)$$

[Note that if $R_n = 0$, then equation (1) says that the mountain lion population *decreases* by 30% each year. The mountain lions need the rabbits as food to survive.]

On the other hand, each year, the number of rabbits will be reduced by the number of rabbits eaten. We will assume that this is proportional to the number of mountain lions. Specifically we assume that

$$R_{n+1} = (1.1)R_n - (0.1)M_n \qquad (2)$$

We combine these two equations into a single matrix equation as

$$\begin{bmatrix} R_{n+1} \\ M_{n+1} \end{bmatrix} = \begin{bmatrix} (1.1)R_n - (0.1)M_n \\ (0.4)R_n + (0.7)M_n \end{bmatrix} = \begin{bmatrix} 1.1 \\ 0.4 \end{bmatrix} R_n + \begin{bmatrix} -0.1 \\ 0.7 \end{bmatrix} M_n = \begin{bmatrix} 1.1 & -0.1 \\ 0.4 & 0.7 \end{bmatrix} \begin{bmatrix} R_n \\ M_n \end{bmatrix}$$

It follows that each year's population is determined by multiplying the preceding year's population by the 2×2 matrix above! Specifically, if T is this 2×2 matrix and we set $P_n = [R_n, M_n]^t$, then

$$P_{n+1} = T P_n$$

Thus, for example, if our initial population is 10×10^2 rabbits and 5×10^2 mountain lions, then $P_1 = [10, 5]^t$. In year 2, the population is

$$P_2 = \begin{bmatrix} 1.1 & -0.1 \\ 0.4 & 0.7 \end{bmatrix} \begin{bmatrix} 10 \\ 5 \end{bmatrix} = \begin{bmatrix} 10.5 \\ 7.5 \end{bmatrix}$$

In the third year, we have

$$P_3 = \begin{bmatrix} 1.1 & -0.1 \\ 0.4 & 0.7 \end{bmatrix} \begin{bmatrix} 10.5 \\ 7.5 \end{bmatrix} = \begin{bmatrix} 10.80 \\ 9.45 \end{bmatrix}$$

It seems that the rabbit population is growing slowly and the mountain lion population much more rapidly. We can see this clearly using MATLAB. Enter the following sequence of commands. The last commands should be entered *all on one line* as shown. (We omit MATLAB's responses.) The "hold on" command will cause a Figure window to pop up. You have to return to the Command window to enter the next commands.

```
>> hold on
>> T=[1.1 -.1; .4 .7];
>> P1=[10;5]
>> P2=T*P1;R=[P1,P2];plot(R(1,:),R(2,:));P1=P2;
```

After you have entered the last line, MATLAB will plot a line segment from P1 to P2. Resize both the Figure and Command windows so that you can see them both at the same time.

To plot the population in the third year, go back to the Command window, press the up-arrow key and then enter. Every time you press up-arrow followed by enter, you will compute and plot another value for the population.

EXERCISES

(1) Explain the function of each of the commands in the last line of the computer code just quoted.

(2) For the example of rabbits and mountain lions, plot the first five population vectors.

(3) Plot enough population vectors to get a good idea of the long-term population distribution of rabbits and mountain lions. (It should take 30 or more points.) Approximately what is the maximum mountain lion population? What is the maximum rabbit population? What does the model predict for the future of the rabbits and the mountain lions?

(4) Suppose that we begin with 10,000 rabbits and 200 mountain lions. What is the mountain lion population in year 2? Is this rate of growth reasonable?

(5) The -0.1 entry in T is called the "kill rate." Try decreasing this to $= -0.4$. Begin with P1=[10,5]. Plot a large number of points (50 or so). You should get a very "pretty" graph! Your pretty graph, however, is physically impossible! Explain.

(6) Exercises 4 and 5 suggest some problems with our model which, in fact, is seriously flawed. In this exercise, you are to discuss some of these flaws. For your discussion consider the following:

(a) Suppose that we initially have 10×10^2 mountain lions in our population. According to equation (2), how many rabbits get eaten by mountain lions in the first year? (Be careful. The answer is not 100. Remember that eaten rabbits don't reproduce.) Note that this number is *independent of the number of rabbits currently in the population*. Do you think many mountain lions go to bed hungry? Explain.

(b) According to equation (1), each increase of the rabbit population by 10 rabbits allows four more mountain lions to survive to the next generation. Does this suggest a food surplus or a food shortage for the mountain lions? Explain. Is this consistent with the answer to part (a)?

Remark: The reader might wonder why we discuss a "flawed" model. One answer is that all models are to some extent flawed. In real-world problems, there may be too many variables, or unknown variables or unknown relationships. A model builder must decide which factors are significant and which are not and how they relate to each other. A model builder also must be able to analyze the model produced and be aware that all results it generates are suspect until the model has proven its value. The rabbit/mountain lion model is valuable as an example of this kind of analysis. Actually, this model is similar to one that does produce more credible results.

Specifically, we can produce a more accurate model by replacing equations (1) and (2) with equations of the form

$$R_{n+1} = (1.1)R_n - aR_n M_n$$
$$M_{n+1} = bR_n W_n + (0.7)M_n$$

where a and b are constants to be determined experimentally. The first equation says that the number of rabbits eaten is proportional to *both* the number of rabbits and the number of mountain lions. Thus, if there are few rabbits, they will be hard to find, and few will be eaten.

The second equation says that having additional rabbits in the population increases the *percentage* of mountain lions that survive. Equation (1), on the other hand, essentially allows us to change rabbits into mountain lions! We did not analyze the more accurate model because it is a nonlinear model and this is a text on *linear* algebra. The graphical technique we used could, however, have been used on the nonlinear model as well.

1 CHAPTER SUMMARY

The central topic of this chapter was systems of linear equations, which we solved using *Gaussian Elimination* (Section 1.4). Specifically, we used *elementary row operations* to reduce the *augmented matrix* for the system until either *echelon form* or *reduced echelon form* was obtained. The variables corresponding to the corners of the steps in the echelon form are the *pivot variables* and the other variables are the *free* variables. We obtain the *general solution* by setting the free variables arbitrarily and solving for the pivot variables.

The idea that we can produce all solutions by setting the free variables arbitrarily is crucial. This is the basis for the description of the solution set in *parametric form* as well as the *more unknowns* theorem, which tells us that we need at least n equations to uniquely determine n unknowns. Note that there is one *spanning vector* for each free variable. The *translation vector* is obtained by setting all of the free variables equal to zero.

This chapter also discussed the nature of the solution sets. We saw in Section 1.5 that the system could be written in terms of *matrix multiplication* as $AX = B$ where A is the *coefficient matrix* for the system and B is the *matrix of constants*. Two fundamental questions were (1) for A given, describe the set of B for which the system is solvable and (2) for A and B given, describe the set of X for which $AX = B$ is solvable. The answer to (1) is that $AX = B$ is solvable if and only if B belongs to the *column space* of A, which is by definition the span of the columns of A. (The *span* of a set of vectors is the set of all their linear combinations. An element X is a linear combination of elements X_i if there are scalars c_i such that $X = c_1 X_1 + c_2 X_2 + \ldots + c_n X_n$.) The answer to (2) involves the *nullspace* of A, which is the solution space to the equation $AX = \mathbf{0}$. The *translation theorem* says that the general solution to $AX = B$ is obtained by translating the nullspace of A by any vector T which satisfies $AT = B$.

It is important that both the column space and nullspace are *subspaces*. Algebraically, this means that they are *closed under linear combinations*. Geometrically, this means that we may think of them as higher dimensional analogues of lines and planes through the origin. Since subspaces are vector spaces, the fact that the nullspace and column space are subspaces will allow us to use the theory that we develop in Chapter 2 to study these spaces.

Another fundamental concept was *rank* (Section 1.3). We saw that a system of, say, five linear equations could be equivalent to a system of two linear equations if, for example, the last three equations were linear combinations of the first two. (Two systems are *equivalent* if they have the same solution set.) The rank of a system of linear equations is, in this chapter, defined to be the number of equations left after eliminating dependent equations, one at a time, until an independent system is obtained. It is a fundamental result (to be proved in Chapter 2) that the rank is independent of which equations were kept and which eliminated.

The concept of rank is based on the concept of linear independence. A system of equations is dependent if one equation is a linear combination of other equations from the same system. In general, a set of elements from some vector space is dependent if one of the elements is a linear combination of other elements from the same set.

By a vector space, we mean any set of elements which can be added and multiplied by scalars so as to satisfy the *vector space properties* from Section 1.2. Our main example is the set $M(m, n)$ of $m \times n$ matrices. However, when discussing dependent equations, we used the fact that the set of linear equations in a particular set of variables is a vector space, and we also noted that nullspaces and column spaces are vector spaces. The significance of the concept of vector space is that anything that we prove using only the vector space properties will automatically be true for any vector space. In particular, it will apply to nullspaces, column spaces, and systems of equations, as well as to $m \times n$ matrices. We will see more examples of vector spaces later.

CHAPTER 2

LINEAR INDEPENDENCE AND DIMENSION

2.1 THE TEST FOR LINEAR INDEPENDENCE

One important application of the Gaussian technique in linear algebra is deciding whether a given set of elements of $M(m, n)$ is independent. In principle, this task could be quite tedious. For example, to show that $\{A_1, A_2, A_3\}$ is independent, we must show that:

(1) A_1 is not a linear combination of A_2 and A_3.

(2) A_2 is not a linear combination of A_1 and A_3.

(3) A_3 is not a linear combination of A_1 and A_2.

Fortunately, this can all be checked in a single step. Consider three dependent matrices, A_1, A_2, and A_3. Suppose

$$A_1 = d_2 A_2 + d_3 A_3$$

We may write this as

$$0 = (-1)A_1 + d_2 A_2 + d_3 A_3$$

Thus, there are constants k_1, k_2, and k_3, with $k_1 \neq 0$ such that

$$0 = k_1 A_1 + k_2 A_2 + k_3 A_3$$

On the other hand, suppose we didn't know that the A_is were dependent, but we did have constants k_1, k_2, and k_3 with $k_1 \neq 0$, such that

$$0 = k_1 A_1 + k_2 A_2 + k_3 A_3$$

Then, we may write

$$-k_1 A_1 = k_2 A_2 + k_3 A_3$$

Hence

$$A_1 = \left(\frac{-k_2}{k_1} \right) A_2 + \left(\frac{-k_3}{k_1} \right) A_3$$

Thus A_1 is a combination of A_2 and A_3. If $k_2 \neq 0$, then A_2 would be a combination of A_1 and A_3. If $k_3 \neq 0$, then A_3 would be a combination of A_1 and A_2. As long as one of the $k_i \neq 0$, then one of the A_is is a combination of the others.

The same reasoning, applied to n elements instead of three, proves the following theorem, which is called the "test for independence." This property is taken as the definition of independence in many linear algebra texts. Notice that in the calculations above, we never explicitly used the fact that we were dealing with matrices; we only used the vector space properties [properties (a)–(j) from Section 1.2]. Hence we will state our result as a vector space theorem.

Test for Independence. Let $\{A_1, A_2, \ldots, A_n\}$ be a set of n elements of a vector space V. Consider the equation

$$k_1 A_1 + \ldots + k_n A_n = 0$$

If the only solution to this equation is $k_1 = k_2 = \cdots = k_n = 0$, *then the elements* $A_1, A_2, \ldots, A_n$ *are independent.*

On the other hand, if there is a solution with $k_l \neq 0$ *for some l, then the elements are dependent and* A_l *is a linear combination of the others.*

Proof: Suppose first that we are given

$$k_1 A_1 + \ldots + k_n A_n = 0$$

with $k_l \neq 0$. Then we may write

$$k_1 A_1 + \ldots + k_{l-1} A_{l-1} + k_{l+1} A_{l+1} + \ldots + k_n A_n = -k_l A_l$$

Since $k_l \neq 0$, we may multiply both sides of the equation by $(-k_l)^{-1}$, solving for A_l and proving dependence.

Conversely, if A_l is a linear combination of the other A_j, then, there are constants k_i such that

$$k_i A_1 + k_2 A_2 + \cdots + k_{e-1} A_{e-1} + k_{e+1} A_{e+1} + \cdots + k_n A_n = A_l$$

This may be expressed in the form

$$k_1 A_1 + \cdots + k_n A_n = \mathbf{0}$$

where $k_l = -1$. This finishes the proof.   ∎

▶ EXAMPLE 1: Prove that the row vectors $A_1 = [1, 1, 2]$, $A_2 = [1, 2, 1]$ and $A_3 = [0, 1, 1]$ are independent.

Solution: Consider the equation

$$k_1 A_1 + k_2 A_2 + k_3 A_3 = \mathbf{0}$$
$$k_1 [1, 1, 2] + k_2 [1, 2, 1] + k_3 [0, 1, 1] = [0, 0, 0]$$
$$[k_1 + k_2, k_1 + 2k_2 + k_3, 2k_1 + k_2 + k_3] = [0, 0, 0]$$

Equating corresponding entries, we get the system

$$k_1 + k_2 = 0$$
$$k_1 + 2k_2 + k_3 = 0$$
$$2k_1 + k_2 + k_3 = 0$$

The augmented matrix is

$$\begin{bmatrix} 1 & 1 & 0 & 0 \\ 1 & 2 & 1 & 0 \\ 2 & 1 & 1 & 0 \end{bmatrix}$$

After row reduction, we obtain

$$\begin{bmatrix} 1 & 1 & 0 & 0 \\ 0 & 1 & 1 & 0 \\ 0 & 0 & 2 & 0 \end{bmatrix}$$

From this, it follows that $k_3 = 0$, $k_2 = 0$ and $k_1 = 0$, proving independence.   ◀

▶ EXAMPLE 2:   Check the following column vectors for independence using the test.

$$A = \begin{bmatrix} 1 \\ 3 \\ 1 \end{bmatrix} \quad B = \begin{bmatrix} 1 \\ 0 \\ 1 \end{bmatrix} \quad C = \begin{bmatrix} 0 \\ 1 \\ 0 \end{bmatrix}$$

Solution: Before we begin, let us first note that the test will yield "dependent" as an answer, since

$$A = B + 3C$$

If we had not been required to use the test, we could stop here.

For the test, consider the equation

$$x \begin{bmatrix} 1 \\ 3 \\ 1 \end{bmatrix} + y \begin{bmatrix} 1 \\ 0 \\ 1 \end{bmatrix} + z \begin{bmatrix} 0 \\ 1 \\ 0 \end{bmatrix} = \mathbf{0} \tag{1}$$

This is the same as

$$\begin{bmatrix} x + y \\ 3x + z \\ x + y \end{bmatrix} = \begin{bmatrix} 0 \\ 0 \\ 0 \end{bmatrix}$$

Thus

$$\begin{aligned} x + y \phantom{{}+ z} &= 0 \\ 3x \phantom{{}+ y} + z &= 0 \\ x + y \phantom{{}+ z} &= 0 \end{aligned}$$

The augmented matrix for this system is:

$$\begin{bmatrix} 1 & 1 & 0 & 0 \\ 3 & 0 & 1 & 0 \\ 1 & 1 & 0 & 0 \end{bmatrix}$$

We row-reduce, obtaining:

$$\begin{bmatrix} 1 & 1 & 0 & 0 \\ 0 & -3 & 1 & 0 \\ 0 & 0 & 0 & 0 \end{bmatrix}$$

We see that z is arbitrary. In particular, z may be chosen non-zero, showing that C is dependent on A and B. In fact, if we chose some specific value for z, say $z = 1$, we find $y = \frac{1}{3}$, and $x = -\frac{1}{3}$. Formula (1) then says that

$$-\frac{A}{3} + \frac{B}{3} + C = \mathbf{0}$$

Hence, $C = \frac{A}{3} - \frac{B}{3}$. This is easily seen to be equivalent with the relation observed above.

◄

▶ EXAMPLE 3:   Test the following matrices for dependence. If dependence is found, use your results to express one as a linear combination of the others.

$$\begin{bmatrix} 1 & 2 \\ 1 & 3 \end{bmatrix} \quad \begin{bmatrix} 1 & 2 \\ 2 & 4 \end{bmatrix} \quad \begin{bmatrix} -1 & -2 \\ -3 & -5 \end{bmatrix} \quad \begin{bmatrix} -1 & -2 \\ 0 & -2 \end{bmatrix}$$

Solution: Let us denote these matrices by A, B, C, and D, respectively. Consider the equation

$$xA + yB + zC + wD = \mathbf{0} \tag{2}$$

Substituting for A, B, C, and D and combining, we have

$$x\begin{bmatrix} 1 & 2 \\ 1 & 3 \end{bmatrix} + y\begin{bmatrix} 1 & 2 \\ 2 & 4 \end{bmatrix} + z\begin{bmatrix} -1 & -2 \\ -3 & -5 \end{bmatrix} + w\begin{bmatrix} -1 & -2 \\ 0 & -2 \end{bmatrix} = \begin{bmatrix} 0 & 0 \\ 0 & 0 \end{bmatrix}$$

$$\begin{bmatrix} x + y - z - w & 2x + 2y - 2z - 2w \\ x + 2y - 3z & 3x + 4y - 5z - 2w \end{bmatrix} = \begin{bmatrix} 0 & 0 \\ 0 & 0 \end{bmatrix}$$

This yields the system

$$\begin{aligned} x + y - z - w &= 0 \\ 2x + 2y - 2z - 2w &= 0 \\ x + 2y - 3z &= 0 \\ 3x + 4y - 5z - 2w &= 0 \end{aligned}$$

The reduced augmented matrix for this system is

$$\begin{bmatrix} 1 & 0 & 1 & -2 & 0 \\ 0 & 1 & -2 & 1 & 0 \\ 0 & 0 & 0 & 0 & 0 \\ 0 & 0 & 0 & 0 & 0 \end{bmatrix}$$

The variables z and w are free and $y = 2z - w$ and $x = -z + 2w$. Clearly, the test yields dependency, since we may, for example, choose any $z \neq 0$. Since z is the coefficient of C in formula (2), it follows that C must be a combination of the other vectors. To explicitly realize C as a combination, we must assign specific values to z and w. We could use any values as long as $z \neq 0$. However, a convenient assignment is $z = -1$ and $w = 0$. This yields $x = 1$ and $y = -2$. From formula (2), we see that

$$(1)A + (-2)B + (-1)C + 0D = \mathbf{0}$$

Hence

$$A - 2B = C$$

What if, instead, we had set $z = 0$ and $w = -1$? Then $x = -2$, $y = 1$ and thus

$$(-2)A + (1)B + 0C + (-1)D = \mathbf{0}$$

Therefore

$$-2A + B = D$$

It follows that both D and C depend on A and B. The elements A and B are independent since neither is a multiple of the other. ◄

Example 3 demonstrates a general technique for finding dependencies. Let $S = \{A_1, A_2, \ldots, A_k\}$ be some dependent set of elements in $M(m, n)$. Consider the "dependency equation":

$$x_1 A_1 + x_2 A_2 + \cdots + x_k A_k = \mathbf{0} \tag{3}$$

This equation results in a homogeneous system of equations in the variables x_i, which can be solved. Some of the x_i will be free variables and some will be pivot variables. Let us call A_i either a free element or a pivot element, depending on whether its coefficient x_i is free or pivot. Each time we set one of the free variables equal to -1, while setting the others equal to 0, we express the corresponding free element A_i as a linear combination of the pivot elements. Thus, *all the free elements depend on the pivot elements*.

The pivot elements form an independent set. To see this, imagine a linear combination of the pivot elements that is equal to zero. This is the same as having a solution to equation (3) in which all the free variables are set equal to zero. But the pivot variables are computed in terms of the free variables. In a homogeneous system, if the free variables are set to zero, then the pivot variables must also equal zero. Thus, the independence of the pivot elements follows from the test for independence. The pivot elements are in fact a maximal independent subset of S, in the sense that not only are they independent, but any larger subset of S which contains all of them is dependent.

One important application of this principle is to the case in which the A_i are the columns of some matrix A.

▶ EXAMPLE 4:   For the following matrix A, find a maximal independent set of columns and express the other columns as linear combinations of these columns.

$$A = \begin{bmatrix} 1 & 2 & -1 & 3 \\ 2 & 2 & -4 & 4 \\ 1 & 3 & 0 & 4 \end{bmatrix}$$

Solution: The dependency equation for the columns is

$$x_1 \begin{bmatrix} 1 \\ 2 \\ 1 \end{bmatrix} + x_2 \begin{bmatrix} 2 \\ 2 \\ 3 \end{bmatrix} + x_3 \begin{bmatrix} -1 \\ -4 \\ 4 \end{bmatrix} + x_4 \begin{bmatrix} 3 \\ 4 \\ 4 \end{bmatrix} = \begin{bmatrix} 0 \\ 0 \\ 0 \end{bmatrix}$$

This is the same as the matrix equation $AX = \mathbf{0}$. Hence the augmented matrix for this system is

$$\begin{bmatrix} 1 & 2 & -1 & 3 & 0 \\ 2 & 2 & -4 & 4 & 0 \\ 1 & 3 & 0 & 4 & 0 \end{bmatrix}$$

We row-reduce, finding the reduced form to be

$$\begin{bmatrix} 1 & 0 & -3 & 1 & 0 \\ 0 & 1 & 1 & 1 & 0 \\ 0 & 0 & 0 & 0 & 0 \end{bmatrix}$$

The pivot variables are x_1 and x_2. Thus, the pivot columns are $A_1 = [1, 2, 1]'$ and $A_2 = [2, 2, 3]'$. They are automatically independent of each other and form a maximal independent set of columns.

Since, x_3 and x_4 are free variables, A_3 and A_4 should be linear combinations of A_1 and A_2. In fact, the general solution to this system is $x_1 = 3x_3 - x_4$ and $x_2 = -x_3 - x_4$. Setting $x_3 = -1$ and $x_4 = 0$ yields $x_1 = -3$ and $x_2 = 1$; hence

$$-3A_1 + A_2 = A_3$$

Similarly, setting $x_3 = 0$ and $x_4 = -1$ yields

$$A_1 + A_2 = A_4$$ ◀

In general, if A is some matrix, then the dependency equation for the columns of A results in the system $AX = \mathbf{0}$. The free columns are exactly the columns that correspond to the free variables in this system. They will all be linear combinations of the pivot columns. The pivot columns will form a maximal independent set of columns in A. We summarize this discussion in the following theorem.

Theorem 1.   *The pivot columns for a matrix A are the columns corresponding to the pivot variables in the system $AX = \mathbf{0}$. The pivot columns for A are independent and the nonpivot columns are linear combinations of them.*   ■

EXERCISES

(1) Test the given matrices for dependence using the test for independence. For each dependent set, use the method of Example 3 to express one of the matrices as a combination of the other matrices in the set.

(a)
$$\begin{bmatrix} 1 \\ 2 \\ 1 \\ 2 \end{bmatrix} \quad \begin{bmatrix} 2 \\ 3 \\ -1 \\ 0 \end{bmatrix} \quad \begin{bmatrix} 1 \\ 0 \\ 1 \\ 0 \end{bmatrix}$$

(b)
$$[1, 2, 1] \quad [3, -1, 2] \quad [7, -7, 4]$$

(c)
$$\begin{bmatrix} 2 & 3 \\ 0 & 1 \end{bmatrix} \quad \begin{bmatrix} 1 & 3 \\ 0 & 0 \end{bmatrix} \quad \begin{bmatrix} 17 & 0 \\ 9 & 1 \end{bmatrix} \quad \begin{bmatrix} 0 & 5 \\ 0 & 6 \end{bmatrix}$$

(d)
$$\begin{bmatrix} 2 & -2 \\ 3 & -1 \\ 0 & 2 \end{bmatrix} \quad \begin{bmatrix} 4 & -1 \\ 2 & 3 \\ 1 & 0 \end{bmatrix} \quad \begin{bmatrix} 8 & -2 \\ 4 & 6 \\ 2 & 0 \end{bmatrix}$$

(e)
$$\begin{bmatrix} 1 & 2 \\ 3 & 2 \end{bmatrix} \quad \begin{bmatrix} 2 & 1 \\ 3 & 4 \end{bmatrix} \quad \begin{bmatrix} 5 & -2 \\ 3 & 10 \end{bmatrix}$$

(f)
$$\begin{bmatrix} 1 \\ 2 \\ 3 \\ 1 \end{bmatrix} \quad \begin{bmatrix} 2 \\ 0 \\ 1 \\ 0 \end{bmatrix} \quad \begin{bmatrix} 1 \\ 1 \\ 1 \\ 1 \end{bmatrix}$$

(g) $[4, 2, -1], [3, 3, 2], [1, 0, 1]$

(2) Consider the accompanying matrix. Use the test for independence to test the rows of this matrix for dependence. (You will find that they are dependent. In fact, you will discover that the dependency equation has two free variables.) Suppose that this matrix is the augmented matrix for a system of equations. What is the rank of this system? (Recall that the rank is the number of equations left after dependent equations have been discarded, one at a time.) Which equations could be discarded? Give as complete an answer as possible.

$$\begin{bmatrix} 1 & 0 & 1 & 1 \\ 2 & 1 & 3 & 0 \\ 3 & 3 & 6 & -3 \\ 4 & 1 & 5 & 2 \end{bmatrix}$$

(3) Let A be the matrix from Exercise 2. Use the test for independence to test the columns of this matrix for dependence. Find a maximal independent set of columns. Express the other columns as linear combinations of these columns.

(4) Let A be an $m \times n$ matrix. Is there a relationship between the maximal number of independent columns and the number of nonzero rows in an echelon form of A? Do you expect a relationship between the maximal number of independent columns and the maximal number of independent rows?

(5) Use the test for independence to prove that the rows of the following 3×6 matrix are linearly independent.

$$A = \begin{bmatrix} 1 & a & 0 & b & 0 & c \\ 0 & 0 & 1 & d & 0 & e \\ 0 & 0 & 0 & 0 & 1 & f \end{bmatrix}$$

(6) Use the test for independence to prove that the rows of the following 3×6 matrix are linearly independent.

$$A = \begin{bmatrix} 1 & a & b & c & d & e \\ 0 & 0 & 1 & f & g & h \\ 0 & 0 & 0 & 0 & 1 & k \end{bmatrix}$$

(7) In Exercises 5 and 6, find a maximal independent set of columns. Express the other columns as linear combinations of the elements of this set.

(8) Suppose that X_1, X_2, and X_3 form a linearly independent set of elements in some vector space. Let $Y_1 = X_1 + 2X_2 - X_3$, $Y_2 = X_1 + X_2$, and $Y_3 = 7X_2$. Prove that Y_1, Y_2, and Y_3 also forms a linearly independent set.

(9) Prove that the columns of a matrix are independent if and only if the nullspace consists of only the zero vector.

(10) Prove that any set of three matrices in $M(2, 1)$ is dependent.

(11) Prove that any set of five matrices in $M(2, 2)$ is dependent.

(12) What is the largest number of independent matrices one can have in $M(3, 2)$?

(13) In Exercise 19, Section 2.1, we commented that the identity

$$\sin^2 x + \cos^2 x = 1$$

can be thought of as expressing the function $y = 1$ as a linear combination of the functions $y = \cos^2 x$ and $y = \sin^2 x$. As we shall see in Section 2.2.1, it is also important to know when certain functions *cannot* be expressed as a linear combination of other functions. For example, we might want to know if there is an identity of the form

$$k_1 e^x + k_2 e^{-x} = 1$$

which is true for all x, where k_1 and k_2 are fixed scalars.

It turns out that there is no such identity. We can prove this using the test for independence. Specifically, it will be shown in Section 2.2.1 that the set of all infinitely differentiable functions on $\mathbb{R}$ forms a vector space. If there were such an identity, then the set $\{e^x, e^{-x}, 1\}$ would be dependent in this space. From the test for independence, there would exist scalars a, b, and c, not all zero, such that

$$ae^x + be^{-x} + c\,1 = \mathbf{0} \tag{4}$$

where "$\mathbf{0}$" is the constant function $y = 0$. Since this identity must hold for all x, we can set $x = 0$, concluding that

$$a + b + c = 0$$

One equation is not sufficient to determine 3 unknowns. However, we can differentiate both sides of equation (4) with respect to x, producing

$$ae^x - be^{-x} = \mathbf{0} \tag{5}$$

Setting $x = 0$ produces

$$a - b = 0$$

Next, we differentiate equation (5)

$$ae^x + be^{-x} = \mathbf{0}$$

and set $x = 0$

$$a + b = 0$$

The equations $a + b = 0$ and $a - b = 0$ prove that $a = b = 0$. The equation $a + b + c = 0$ then shows that $c = 0$. Hence, the given functions are independent, showing that no such identity is possible.

Show that the following sets of infinitely differentiable functions are linearly independent. Your proof will involve differentiating the dependency equation several times and evaluating at a convenient value as before.

(a) $\{e^x, e^{2x}, e^{3x}\}$ (b) $\{1, e^x, e^{2x}, e^{3x}\}$

(c) $\{1, x, x^2\}$ (d) $\{1, x, x^2, x^3, x^4\}$

(e) $\{\ln x, x \ln x\}$ (f) $\{\ln x, x \ln x, x^2 \ln x\}$

(14) We know that the functions $\{\cos^2 x, \sin^2 x, 1\}$ are dependent. Attempt to prove that they are *independent* using the technique demonstrated in Exercise 13. How does the technique break down?

ON LINE

(1) Let

$$A = \begin{bmatrix} 1 & 2 & -3 \\ 4 & 5 & -1 \\ 3 & 2 & 1 \\ 1 & 1 & 1 \end{bmatrix}$$

Use MATLAB to find the row-reduced form of A. How can you tell just from this reduced form that the columns of A are independent? Relate your answer to Theorem 1.

(2) Let A be a matrix with more rows than columns. State a general rule for using "rref(A)" to decide whether the columns of A are independent. Demonstrate your condition by (a) producing a 5×4 matrix A with no non-zero coefficients which has independent columns and computing "rref(A)" and (b) producing a 5×4 matrix A that has no non-zero coefficients but does have dependent columns and computing "rref(A)." Prove your condition using Theorem 1.

(3) Let

$$A = \begin{bmatrix} -1 & 2 & 6 & -8 & -14 & 3 \\ 2 & 4 & 1 & -8 & 5 & -1 \\ -3 & 1 & 4 & -9 & -10 & 0 \\ 3 & -2 & -1 & 12 & -1 & 4 \\ 5 & 7 & 11 & -11 & -19 & 9 \end{bmatrix}$$

Use "rref(A)" to find the pivot columns of A. Write them out explicitly as columns. Then express the other columns of A as linear combinations of the pivot columns as in Example 4. You should discover that the first three columns of A are the pivot columns.

(4) In MATLAB, enter "B=[A(:,4),A(:,1),A(:,2),A(:,3),A(:,5),A(:,6)]" where A is the matrix from Exercise 3. The columns of B are just those of A, listed in a different order. Compute "rref(B)" to find the pivot columns. Do you obtain a different set of pivot columns? Use your answer to express the other columns as linear combinations of the pivot columns. Could you have derived these expressions from those in Exercise 3? If so, how?

(5) Find a matrix C whose columns are just those of A listed in a different order, such that the column of C that equals "A(:,5)" and the column that equals "A(:,1)" are both pivot columns. Is it possible to find such a C where "A(:,2)" is a pivot column as well? If so, find an example. If not, explain why it is not possible.

2.2   DIMENSION

The term "dimension" has slightly different meanings in mathematics and in ordinary speech. Roughly, a mathematician would say that the dimension of some set is the smallest number of variables necessary to describe the set. Thus, to a mathematician, space is three-dimensional because the position of any point may be described using exactly three coordinates. These coordinates may be freely chosen, and different choices yield different points. To a mathematician, a problem such as describing the flight of a cannon ball is at least six-dimensional; one must know the location of the cannon (x, y, and z coordinates), the direction the barrel was pointed (elevation and compass bearing), and the initial speed of the cannon ball. Thus, from this point of view, the physical reality of six dimensions is clear.

This concept of dimension is consistent with the notion that a plane should be two-dimensional. Recall that in Section 1.3 we said that in $\mathbb{R}^3$, two independent elements X_1 and X_2 span a plane through **0**. This means that every element of the plane may be written as $xX_1 + yX_2$ for real numbers x and y. Thus, it takes two numbers (x and y) to describe the general element of the plane. (We consider X_1 and X_2 as fixed.) No fewer than two vectors can span a plane. The span of a single vector will be a line, not a plane. These comments suggest the following definition:

Definition. *The dimension of a vector space* V *is the smallest number of elements necessary to span* V. ∎

It is comforting that according to this definition, $\mathbb{R}^2$ is two-dimensional. To prove this, we must show that

(1) $\mathbb{R}^2$ may be spanned by two elements.

(2) $\mathbb{R}^2$ cannot be spanned by one element.

Item 2 is simple, since any single element in $\mathbb{R}^2$ will span only a line. For item 1, we write the general element of $\mathbb{R}^2$ as

$$\begin{bmatrix} x \\ y \end{bmatrix} = \begin{bmatrix} x \\ 0 \end{bmatrix} + \begin{bmatrix} 0 \\ y \end{bmatrix} = x \begin{bmatrix} 1 \\ 0 \end{bmatrix} + y \begin{bmatrix} 0 \\ 1 \end{bmatrix}$$

The two column vectors on the right are denoted I_1 and I_2, respectively. Our equation says exactly that I_1 and I_2 span $\mathbb{R}^2$, as desired.

The dimension of a space puts limits on the number of independent elements that can exist in the space. For example it appears that any three vectors in $\mathbb{R}^2$ will always be dependent (see Figure 1).

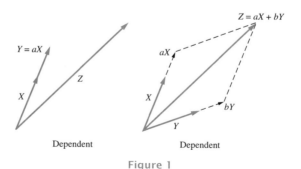

Figure 1

In $\mathbb{R}^3$, the set of linear combinations of a pair of independent vectors lies in the plane they determine. Thus, three non-coplanar vectors will be independent, as shown in Figure 2.

Although the picture is hard to draw, one feels that four vectors in $\mathbb{R}^3$ must always be dependent.

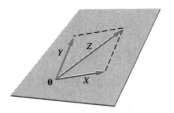 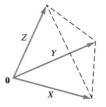

3 Coplanar vectors are dependent!   3 Non-coplanar vectors are independent!

Figure 2

What we observed for $\mathbb{R}^2$ and for $\mathbb{R}^3$ suggests the following theorem, which is one of the most fundamental theorems in all of linear algebra.

Theorem 1. *In an n-dimensional vector space, there can be at most n independent elements.*

Proof: We may restate this as saying that in an n-dimensional space, any set of more than n elements must be dependent. The proof is most easily understood in the case of three elements in a two-dimensional space. We will first discuss the proof in this case and then we will describe how to modify the argument to cover the general case.

Thus, we assume initially that V is a two-dimensional vector space. Then there are two elements X_1 and X_2 which span V. Let Y_1, Y_2, and Y_3 be any three elements of V. We must prove that the Y_i are dependent. This will be true if there are constants x_1, x_2, and x_3 such that

$$x_1 Y_1 + x_2 Y_2 + x_3 Y_3 = \mathbf{0} \tag{1}$$

with at least one $x_i \neq 0$.

Since $\{X_1, X_2\}$ spans V, each Y_i may be written in terms of the X_i:

$$Y_1 = a_{11} X_1 + a_{21} X_2$$
$$Y_2 = a_{12} X_1 + a_{22} X_2$$
$$Y_3 = a_{13} X_1 + a_{23} X_2$$

We substitute these expressions for Y_i into formula (1) and factor out X_1 and X_2, obtaining an expression of the form

$$c_1 X_1 + c_2 X_2 = \mathbf{0}$$

where

$$c_1 = a_{11} x_1 + a_{12} x_2 + a_{13} x_3$$
$$c_2 = a_{21} x_1 + a_{22} x_2 + a_{23} x_3$$

We conclude that equation (1) will be satisfied if

$$a_{11}x_1 + a_{12}x_2 + a_{13}x_3 = 0$$
$$a_{21}x_1 + a_{22}x_2 + a_{23}x_3 = 0$$

Now, observe that this system consists of two homogeneous equations in three unknowns. A homogeneous system cannot be inconsistent. (Why?) Thus, from the more unknowns theorem, it has an infinite number of solutions and, in particular, it has one with at least one of the $x_i \neq 0$. This proves dependence.

In the general (n-dimensional) case, V will be spanned by n elements $X_1, X_2, \ldots, X_n$ and we will be given m elements $Y_1, Y_2, \ldots, Y_m$ where $m > n$. Now the dependency equation for the Y_i is

$$x_1 Y_1 + x_2 Y_2 + \ldots x_m Y_m = \mathbf{0} \qquad (2)$$

We wish to prove that there are non-zero x_i, which make this equation valid.

Just as before, we express each of the Y_i as a linear combination of the X_i:

$$Y_i = a_{i1} X_1 + a_{i2} X_2 + \ldots + a_{in} X_n$$

As before, we substitute each of these expressions into the dependency equation [equation (2)] and factor out the X_i. Now we obtain an expression of the form

$$c_1 X_1 + c_2 X_2 + \ldots + c_n X_n = \mathbf{0}$$

where now

$$c_i = a_{i1}x_1 + a_{i2}x_2 + \ldots + a_{im}x_m$$

Formula (2) will be satisfied if each of the $c_i = 0$. However, setting the $c_i = 0$ produces a system of n, homogeneous, linear equations in the m unknowns $x_1, \ldots, x_m$. As in the two-dimensional case, the crucial observations are that the system is homogeneous and that there are fewer equations than unknowns. (Recall that $n < m$.) Thus, the system has an infinite number of solutions and, in particular, it has one with at least one of the $x_i \neq 0$. This proves dependence. ∎

Remark: The above argument is more or less the same as the argument required for Exercises 10, 11, and 12 in Section 2.1.

Since subspaces are vector spaces, it is meaningful to discuss their dimension.

▶ EXAMPLE 1:   Show that the set of matrices of the form

$$\begin{bmatrix} a+b+3c & 2a-b \\ 0 & 2a+b+4c \end{bmatrix}$$

where a, b, and c range over all real numbers, is a subspace of $M(2, 2)$, and find its dimension.

Solution: We begin by finding a spanning set for $\mathcal{W}$. We write:

$$\begin{bmatrix} a+b+3c & 2a-b \\ 0 & 2a+b+4c \end{bmatrix} = \begin{bmatrix} a & 2a \\ 0 & 2a \end{bmatrix} + \begin{bmatrix} b & -b \\ 0 & b \end{bmatrix} + \begin{bmatrix} 3c & 0 \\ 0 & 4c \end{bmatrix}$$

$$= a\begin{bmatrix} 1 & 2 \\ 0 & 2 \end{bmatrix} + b\begin{bmatrix} 1 & -1 \\ 0 & 1 \end{bmatrix} + c\begin{bmatrix} 3 & 0 \\ 0 & 4 \end{bmatrix}$$

This shows that $\mathcal{W}$ is the span of the matrices

$$A = \begin{bmatrix} 1 & 2 \\ 0 & 2 \end{bmatrix} \quad B = \begin{bmatrix} 1 & -1 \\ 0 & 1 \end{bmatrix} \text{ and } C = \begin{bmatrix} 3 & 0 \\ 0 & 4 \end{bmatrix}$$

In particular, since spans are subspaces, $\mathcal{W}$ is a subspace.

The presence of three spanning elements suggests that $\mathcal{W}$ could be three-dimensional. However, inspection reveals that our spanning set is dependent. Specifically

$$C = A + 2B$$

Hence, the general element X of $\mathcal{W}$ may be written

$$X = aA + bB + cC = aA + bB + c(A + 2B) = (a+c)A + (b+2c)B = a'A + b'B$$

where $a' = a + c$ and $b' = b + 2c$. This expresses X as a linear combination of A and B. It follows that A and B span $\mathcal{W}$. Thus, the dimension is at most 2. In fact, the dimension must be 2, since A and B are independent (neither is a multiple of the other) and (from Theorem 1) there cannot exist two independent elements in a one-dimensional space. ◄

This example makes two extremely important points. First, if a space is spanned by a dependent set of elements, we may always find a smaller spanning set by eliminating a dependent element. Thus, if we span a set using the fewest possible elements, those elements *must* be independent. Since the dimension is the smallest number of elements needed to span the space, it follows that:

Proposition 1. *If a vector space is n-dimensional, then any set of n elements that spans the space must be independent.* ∎

The second point made by our example is that if we can span a space by n-independent elements, then that space must be n dimensional. To see this, note that the dimension is at most n, since there are n elements that span it. If the dimension were less than n, then (from Theorem 1) we couldn't have n independent elements. Thus, we have the following proposition:

Proposition 2.   *If a vector space can be spanned by n independent elements, then it is n-dimensional.* ∎

Since independent elements span a space as efficiently as possible, the most important spanning sets are those that are independent. These, in fact, are so important that we give them a special name:

Definition.   *A basis for a vector space $\mathcal{V}$ is a set of linearly independent elements which spans $\mathcal{V}$.* ∎

Thus, for example, the matrices A and B from Example 1 together form a basis for $\mathcal{W}$.

It follows from Proposition 1 that every n-dimensional vector space has a basis. Proposition 2 says that the number of elements in any basis is the dimension of the space. Thus, we arrive at the following fundamental result.

Dimension Theorem.   *A vector space $\mathcal{V}$ is n-dimensional if and only if it has a basis containing n elements. In this case, all bases contain n elements.* ∎

▶ **EXAMPLE 2:**   Show that $\mathbb{R}^n$ is n-dimensional.

Solution: We begin by finding a spanning set. Given any $X = [x_1, x_2, \ldots, x_n]^t$, we write

$$\begin{bmatrix} x_1 \\ x_2 \\ x_3 \\ \vdots \\ x_n \end{bmatrix} = x_1 \begin{bmatrix} 1 \\ 0 \\ 0 \\ \vdots \\ 0 \end{bmatrix} + x_2 \begin{bmatrix} 0 \\ 1 \\ 0 \\ \vdots \\ 0 \end{bmatrix} + x_3 \begin{bmatrix} 0 \\ 0 \\ 1 \\ \vdots \\ 0 \end{bmatrix} + \ldots + x_n \begin{bmatrix} 0 \\ 0 \\ 0 \\ \vdots \\ 1 \end{bmatrix}$$

The jth column vector on the right is usually denoted I_j. Our formula says that

$$X = x_1 I_1 + x_2 I_2 + x_3 I_3 + \cdots + x_n I_n \tag{3}$$

Hence, the I_j span $\mathbb{R}^n$.

These vectors are also independent. To see this, note that each has a 1 in a position where all the others have a 0. Specifically, I_j has a 1 in the jth position while every other I_k has a zero in this position. It follows that I_j cannot be a linear combination of the other I_k since any such linear combination would have a zero in the jth position. Thus the I_j form a basis for $\mathbb{R}^n$, showing that $\mathbb{R}^n$ is indeed n-dimensional.   ◄

The basis formed by the vectors I_j in Example 2 is particularly important. It is referred to as the **standard basis** for $\mathbb{R}^n$. The reason for the notation "I_j" is that I_j is the jth column of the $n \times n$ **identity matrix** I. This is the matrix with ones on its main diagonal and all its other entries zero. Thus, for example, the 3×3 identity matrix is

$$I = \begin{bmatrix} 1 & 0 & 0 \\ 0 & 1 & 0 \\ 0 & 0 & 1 \end{bmatrix}$$

and

$$I_1 = \begin{bmatrix} 1 \\ 0 \\ 0 \end{bmatrix} \qquad I_2 = \begin{bmatrix} 0 \\ 1 \\ 0 \end{bmatrix} \qquad I_3 = \begin{bmatrix} 0 \\ 0 \\ 1 \end{bmatrix}$$

In $\mathbb{R}^3$, the vectors I_1, I_2, and I_3 are usually denoted by $\mathbf{i}$, $\mathbf{j}$, and $\mathbf{k}$, respectively. Formula (3), in this case, is just the familiar expression

$$[x, y, z]^t = x\mathbf{i} + y\mathbf{j} + z\mathbf{k}$$

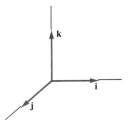

The standard basis in 3-space

Given a dependent spanning set, we can always find a basis by eliminating dependent elements. Often this involves using the test for independence, as the next example demonstrates.

▶ **EXAMPLE 3:** Find a basis for the subspace of $\mathbb{R}^4$ spanned by the vectors $V_1 = [1, 2, 0, 3]^t$, $V_2 = [2, 1, 1, 1]^t$, $V_3 = [1, 4, 3, -3]^t$, $V_4 = [3, 15, 8, -4]^t$, $V_5 = [3, -11, -9, 18]^t$. What is the dimension of this subspace?

Solution: By definition, the subspace in question is spanned by V_1 through V_5. We know, however, that five vectors in $\mathbb{R}^4$ cannot be independent. At least one of the V_i is a combination of the others. To determine which of the V_i are combinations of the others, we proceed as in Example 3, Section 2.1. Explicitly, we consider the dependency equation

$$x_1\begin{bmatrix}1\\2\\0\\3\end{bmatrix}+x_2\begin{bmatrix}2\\1\\1\\1\end{bmatrix}+x_3\begin{bmatrix}1\\4\\3\\-3\end{bmatrix}+x_4\begin{bmatrix}3\\15\\8\\-4\end{bmatrix}+x_5\begin{bmatrix}3\\-11\\-9\\18\end{bmatrix}=\begin{bmatrix}0\\0\\0\\0\end{bmatrix}\quad(4)$$

This yields a homogeneous system of equations with augmented matrix:

$$\begin{bmatrix}1&2&1&3&3&0\\2&1&4&15&-11&0\\0&1&3&8&-9&0\\3&1&-3&-4&18&0\end{bmatrix}$$

We reduce, obtaining:

$$\begin{bmatrix}1&0&0&2&1&0\\0&1&0&-1&3&0\\0&0&1&3&-4&0\\0&0&0&0&0&0\end{bmatrix}$$

We see that x_4 and x_5 are free and $x_3 = -3x_4 + 4x_5$, $x_2 = x_4 - 3x_5$ and $x_1 = -2x_4 - x_5$. Letting $x_5 = -1$ and $x_4 = 0$, yields $x_3 = -4$, $x_2 = 3$, and $x_1 = 1$. Hence, from formula (4)

$$V_1 + 3V_2 - 4V_3 = V_5$$

Similarly, setting $x_4 = 1$ and $x_5 = 0$ yields

$$2V_1 - V_2 + 3V_3 = V_4$$

Since V_4 and V_5 are linear combinations of V_1, V_2, and V_3, we see that our subspace may be spanned by V_1, V_2, and V_3. The discussion following Example 3 in Section 2.1 proves that these vectors are independent. Thus, our subspace is three-dimensional and these three vectors form a basis.　◀

Proving that a set of elements is a basis for a given vector space involves proving that the set both spans the space and is independent. Actually, if we already know the dimension of the space, this work can be cut in half. To explain this statement, consider Figure 3, where we have indicated two independent vectors X and Y along with a third vector Z.

It appears that Z will be a linear combination of X and Y. It seems, then, that any two independent elements in $\mathbb{R}^2$ will span $\mathbb{R}^2$. This, in fact, follows directly from Theorem 1. Any three elements of $\mathbb{R}^2$ must be dependent. Hence the set $\{X, Y, Z\}$ is

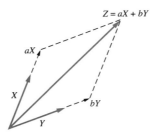

Two independent vectors span R^2

Figure 3

dependent. On the other hand, $\{X, Y\}$ is independent. It seems logical to suppose that $\{X, Y, Z\}$ is dependent *because* Z is a linear combination of X and Y. This is true and is the basis of the proof of the following proposition.

Proposition 3.   *In an n-dimensional vector space* V, *any set of n linearly independent elements spans* V.

Proof: Let $S = \{X_1, \ldots, X_n\}$ be a set of n linearly independent elements in V. Let $W \in V$. Then, from Theorem 1, $\{W, X_1, \ldots, X_n\}$ is dependent. Thus, there are scalars c_i, not all zero, such that

$$c_0 W + c_1 X_1 + \cdots + c_n X_n = \mathbf{0}$$

If $c_0 = 0$, then

$$c_1 X_1 + \cdots + c_n X_n = \mathbf{0}$$

which contradicts the independence of the X_i. Hence, $c_0 \neq 0$. We may therefore solve for W in terms of the X_i:

$$W = -\frac{c_1}{c_0} X_1 - \cdots - \frac{c_n}{c_0} X_n$$

Thus the X_i span V. ∎

The information from Propositions 1 and 3 can be summarized in the following statement, which is also one of the most important theorems in linear algebra.

Theorem 2.   *In an n-dimensional space* V, *any set of n elements that spans* V *must be independent and any set of n independent elements must span* V. ∎

Theorem 2 can, at times, save a considerable amount of work, as the following examples show.

▶ EXAMPLE 4:   Show that the following vectors form a basis for $\mathbb{R}^4$.

$$X_1 = \begin{bmatrix} 1 \\ 2 \\ 1 \\ -1 \end{bmatrix} \quad X_2 = \begin{bmatrix} 2 \\ 0 \\ 2 \\ 0 \end{bmatrix} \quad X_3 = \begin{bmatrix} 1 \\ 1 \\ 1 \\ 1 \end{bmatrix} \quad X_4 = \begin{bmatrix} 0 \\ 1 \\ 1 \\ 0 \end{bmatrix}$$

Solution: To show that our set is a basis, we must show that the set both spans $\mathbb{R}^4$ and is independent. However, we know from Example 2 that $\mathbb{R}^4$ is four-dimensional. From Theorem 2, any four independent vectors in $\mathbb{R}^4$ will automatically span $\mathbb{R}^4$. Hence, we need only show independence. For this, we proceed as usual, setting up the dependency equation and solving. We leave the details to the reader.   ◀

▶ EXAMPLE 5:   Show that the following vectors do not span $\mathbb{R}^4$.

$$X_1 = \begin{bmatrix} 1 \\ 2 \\ 1 \\ -1 \end{bmatrix} \quad X_2 = \begin{bmatrix} 2 \\ 0 \\ 2 \\ 0 \end{bmatrix} \quad X_3 = \begin{bmatrix} 1 \\ 1 \\ 1 \\ 1 \end{bmatrix} \quad X_4 = \begin{bmatrix} 3 \\ 2 \\ 3 \\ -1 \end{bmatrix}$$

Solution: According to Theorem 2, if four vectors span a four-dimensional space, they must be independent. Hence, if we show that the given vectors are dependent, it follows that they do not span. We could set up the dependency equation and solve it, finding that there are indeed nontrivial solutions. However, we may also just note that $X_4 = X_1 + X_2$.   ◀

Theorem 2 also helps us describe certain spaces.

▶ EXAMPLE 6:   Suppose that $\mathcal{W}$ is a two dimensional subspace of $\mathbb{R}^3$ that contains the vectors $X_1 = [1, 2, -3]^t$ and $X_2 = [3, -7, 2]^t$. Does $\mathcal{W}$ contain $Y = [1, 1, 1]^t$?

Solution: Since $\mathcal{W}$ is two-dimensional, any two independent vectors in $\mathcal{W}$ will span $\mathcal{W}$. Thus, X_1 and X_2 span $\mathcal{W}$. The vector Y will belong to $\mathcal{W}$ if and only if there are scalars x and y such that $xX_1 + yX_2 = Y$. Substituting for Y, X_1, and X_2, we see that the vector equation is equivalent with the system

$$\begin{aligned} x + 3y &= 1 \\ 2x - 7y &= 1 \\ -3x + 2y &= 1 \end{aligned}$$

We row-reduce the augmented matrix, obtaining

$$\begin{bmatrix} 1 & 0 & 0 \\ 0 & 1 & 0 \\ 0 & 0 & 1 \end{bmatrix}$$

Thus, the system is inconsistent, showing that Y is not an element of $\mathcal{W}$.   ◄

Remark: We close this section with a rather sobering comment. All the spaces considered in this section were finite-dimensional in the sense that they could be spanned by a finite number of elements. There do, however, exist spaces so big that they cannot be spanned by any finite number of elements. Such spaces are called "infinite dimensional." The simplest example is the space $\mathbb{R}^\infty$. This is by definition the set of all "vectors" of the form

$$[x_1, x_2, \ldots, x_n, \ldots$$

where $\{x_n\}$ is an infinite sequence. Thus, for example, both

$$X = [1, 2, 3, 4, \ldots, n, \ldots$$

and

$$Y = [1, \frac{1}{2}, \frac{1}{3}, \frac{1}{4}, \ldots, \frac{1}{n}, \ldots$$

represent elements of $\mathbb{R}^\infty$.

We can add elements of $\mathbb{R}^\infty$ and multiply them by scalars just as we do for elements of $\mathbb{R}^n$. Thus, for X and Y as above

$$X + Y = [2, 2\frac{1}{2}, 3\frac{1}{3}, 4\frac{1}{4}, \ldots, n + \frac{1}{n}, \ldots$$

Similarly,

$$2X = [2, 4, 6, 8, 10, \ldots, 2n, \ldots$$

It is easily seen that $\mathbb{R}^\infty$ satisfies all the vector space properties listed in Section 1.2. Thus, $\mathbb{R}^\infty$ is a vector space.

Let I_j be the element of $\mathbb{R}^\infty$ that has a one in the jth position and zeros in all other positions. Thus, for example

$$I_3 = [0, 0, 1, 0, 0, \ldots, 0, \ldots$$

It is clear that each of the I_j is independent of the other I_k, since each I_j has a one in a position where all the others have a zero. Thus $\mathbb{R}^\infty$ has an infinite set of independent vectors. This proves that $\mathbb{R}^\infty$ is infinite-dimensional because (by Theorem 1) in an n-dimensional space we can have at most n independent elements.

EXERCISES

(1) The vectors $A_1 = [1, 2]^t$ and $A_2 = [2, -3]^t$ are (obviously) independent in $\mathbb{R}^2$. According to Theorem 2, they must therefore span $\mathbb{R}^2$. As an example of this, choose some random *non-zero* vector B in $\mathbb{R}^2$ and find constants x and y such that $B = xA_1 + yA_2$. Please do not choose either A_1 or A_2 as B. This is not "random"!

(2) Redo Exercise 1 with another choice of B. This is what spanning the space means: no matter which B you choose, the constants always exist.

(3) Let $A_1 = [1, 2]^t$ and $A_2 = [2, 4]^t$ in $\mathbb{R}^2$. Find a vector B in $\mathbb{R}^2$ for which there are no constants x and y, such that $B = xA_1 + yA_2$. [*Hint*: Draw a picture.] Find another vector C such that x and y do exist. What geometric condition must C satisfy?

(4) "The vectors $A_1 = [1, 2]^t$, $A_2 = [2, -3]^t$, and $A_3 = [2, 4]^t$ span $\mathbb{R}^2$." True or false? Explain.

(5) Does the subspace W in Example 6 contain $Z = [9, -8, -5]^t$?

(6) Prove that the given sets are subspaces of $\mathbb{R}^n$ for the appropriate n. Find spanning sets for these spaces and find at least two different bases for each space. Give the dimension of each space.
  (a) $S = \{[a + b + 2c, 2a + b + 3c, a + b + 2c, a + 2b + 3c]^t | a, b, c \in \mathbb{R}\}$
  (b) $S = \{[a + 2c, 2a + b + 3c, a + b + c]^t | a, b, c \in \mathbb{R}\}$
  (c) $S = \{[a + b + 2c, 2a + b + 3c, a + b + c]^t | a, b, c \in \mathbb{R}\}$

(7) Show that the following set W of matrices is a subspace of $M(2, 2)$ and find a basis. What is the dimension of this subspace? [*Warning*: The dimension is *not* 3.]
$$S = \left\{ \begin{bmatrix} a + b + 2c + 3d & a + b + 2c + 3d \\ 2a + 2c + 4d & a + b + 2c + 3d \end{bmatrix} | a, b, c, d \in \mathbb{R} \right\}$$

(8) Show that $M(2, 2)$ is four-dimensional. For your proof, reason as in Exercise 7, recalling that the general element of $M(2, 2)$ may be written
$$\begin{bmatrix} a & b \\ c & d \end{bmatrix}$$

(9) What is the dimension of $M(2, 3)$? Prove your answer.

(10) What is the dimension of $M(m, n)$? Prove your answer.

(11) What is the dimension of the space of all upper triangular 3×3 matrices? Prove your answer. (Recall that a matrix is upper triangular if its only non-zero entries lie on or above the main diagonal.)

(12) Find a basis for the subspace of $M(2, 2)$ spanned by the following matrices. What is the dimension of this subspace? [*Hint*: These matrices were considered in Example 3 of Section 2.1.]

$$X_1 = \begin{bmatrix} 1 & 2 \\ 1 & 3 \end{bmatrix} \quad X_2 = \begin{bmatrix} 1 & 2 \\ 2 & 4 \end{bmatrix} \quad X_3 = \begin{bmatrix} -1 & -2 \\ -3 & -5 \end{bmatrix} \quad X_4 = \begin{bmatrix} -1 & -2 \\ 0 & -2 \end{bmatrix}$$

(13) The vectors in parts (a)–(c) form a spanning set for some subspace of $M(m, n)$ for some m and n. Use the technique of Example 3 (from this section) to find a basis for this subspace. What is the dimension of the subspace?

(a) $[5, -3, 2, 4], [-2, 1, -1, -2], [4, -3, 1, 2], [-5, 1, -4, -8]$

(b)
$$\begin{bmatrix} 1 \\ 1 \\ 1 \\ 1 \\ 1 \end{bmatrix} \begin{bmatrix} 1 \\ 5 \\ 6 \\ -15 \\ 0 \end{bmatrix} \begin{bmatrix} 2 \\ -2 \\ 1 \\ 4 \\ 5 \end{bmatrix} \begin{bmatrix} 3 \\ -1 \\ 4 \\ -2 \\ 7 \end{bmatrix}$$

(c)
$$\begin{bmatrix} 1 & 2 \\ -2 & 3 \end{bmatrix} \begin{bmatrix} 2 & 5 \\ -1 & -1 \end{bmatrix} \begin{bmatrix} 1 & 2 \\ 0 & -1 \end{bmatrix}$$

(14) Find a basis for the nullspace for each of the following matrices. What is the dimension of the nullspace? [*Hint*: Find a set of vectors that spans the nullspace.] Do you see a relationship between the rank and the dimension of the nullspace?[1]

(a)
$$\begin{bmatrix} 1 & 2 & 1 \\ 2 & 4 & 2 \end{bmatrix}$$

(b)
$$\begin{bmatrix} 1 & 3 & 1 & 1 \\ 2 & 1 & 2 & 1 \\ 4 & 7 & 4 & 3 \end{bmatrix}$$

(c)
$$\begin{bmatrix} 1 & -5 & 3 & 4 \\ 2 & -10 & 6 & 8 \\ 3 & -15 & 9 & 12 \end{bmatrix}$$

---

[1] Note to instructor: This exercise serves as a preview of Section 2.3, where the relation between the rank and the dimension of the nullspace is discussed in depth.

(15) Prove that these vectors form a basis for $\mathbb{R}^3$:

$$X_1 = [2, -3, 1]^t \quad X_2 = [1, 2, 3]^t \quad X_3 = [7, -2, 4]^t$$

(16) The most efficient way to do Exercise 15 is to prove that the given vectors are independent and then use Theorem 2 to conclude that they span $\mathbb{R}^3$. To see how much work this saves, prove directly that the given vectors span $\mathbb{R}^3$ without using dimension theory.

(17) The following sequence of questions studies forming bases that contain a given set of vectors.

(a) Let $X_1 = [-3, 4, 1]^t$ and $X_2 = [1, 1, 1]^t$. Find a third vector X_3 (reader's choice) such that X_1, X_2, and X_3 form a basis for $\mathbb{R}^3$.

(b) Suppose, in general, that I give you two independent vectors X_1 and X_2 in $\mathbb{R}^3$. Is it always possible to find X_3 such that X_1, X_2, and X_3 form a basis for $\mathbb{R}^3$? Explain.

(c) Let $X_1 = [1, 0, 1, 0]^t$ and $X_2 = [1, 2, 3, 4]^t$. Find a vector X_3 (reader's choice) in $\mathbb{R}^4$ that is independent of X_1 and X_2. Find a fourth vector X_4 such that X_1, X_2, X_3, and X_4 form a basis for $\mathbb{R}^4$.

(d) Suppose, in general, that I give you two independent vectors X_1 and X_2 in $\mathbb{R}^4$. Is it always possible to find X_3 and X_4 such that X_1, X_2, X_3, and X_4 form a basis for $\mathbb{R}^4$? Explain.

(e) Suppose, in general, that I give you k independent elements $X_1, X_2, \ldots, X_k$ in an n-dimensional space. Is it always possible to find $X_{k+1}, X_{k+2}, \ldots, X_n$ such that $X_1, X_2, \ldots, X_n$ form a basis? Explain.

(18) A certain 5×4 matrix A is known to have a three-dimensional nullspace. It is also known that the three elements X_1, X_2, and X_3 below satisfy $AX = 0$.

$$X_1 = [3, -2, 1, -1]^t \quad X_2 = [2, 2, 1, 2]^t \quad X_3 = [-1, 4, 2, 3]^t$$

(a) Do the given elements span the nullspace of A? Explain.

(b) Is the vector $[1, 1, 1, 1]^t$ in the nullspace of A? Explain.

(c) Do the vectors below satisfy $AX = 0$? Do they span the nullspace?

$$Y_1 = [3, -2, 1, -1]^t \quad Y_2 = [2, 2, 1, 2]^t \quad Y_3 = [1, -4, 0, -3]^t$$

(19) Suppose that $\mathcal{W}$ is a vector space spanned by three elements $\{A, B, C\}$. Suppose that $\{A, B\}$ forms a basis for $\mathcal{W}$. Does it follow that $\{A, B, C\}$ is dependent? Explain. Suppose that $\{A, C\}$ is also a basis of $\mathcal{W}$. Does it follow that $\{B, C\}$ is a basis of $\mathcal{W}$? Explain. (Think about a plane in $\mathbb{R}^3$.)

(20) Suppose that the elements A and B form a basis for a vector space $\mathcal{W}$. Prove that the elements $X = 2A + 3B$ and $Y = 3A - 5B$ also form a basis for $\mathcal{W}$.

(21) In our proof of Theorem 1, we began with the case of three elements in a two-dimensional space. Give a similar proof of Theorem 1 for the case of 4 elements Y_1, Y_2, Y_3, and Y_4 in a three-dimensional space. You are not allowed to use the expression ". . . ."

(22) We have seen that in $\mathbb{R}^3$, every line through $\mathbf{0}$ and every plane through $\mathbf{0}$ is a subspace of $\mathbb{R}^3$. Two further examples of subspaces are $\mathbb{R}^3$ itself and the set consisting of the zero vector only. Prove that every subspace of $\mathbb{R}^3$ is one of these four types. For your proof, let $\mathcal{W}$ be a non-zero subspace of $\mathbb{R}^3$. Of course, $\mathcal{W}$ contains an infinite number of elements. However, $\mathcal{W}$ can contain at most three *independent* elements. (Why?) If $\mathcal{W}$ contains three independent elements, then what is $\mathcal{W}$? Suppose the maximal number of independent elements in $\mathcal{W}$ is two. Prove that then any two independent elements in $\mathcal{W}$ span $\mathcal{W}$. What is $\mathcal{W}$ in this case? Suppose $\mathcal{W}$ contains only one independent element. What is $\mathcal{W}$ in this case? What if $\mathcal{W}$ contains no independent elements?

(23) Let $\mathcal{W}$ be a subspace of a finite-dimensional vector space $\mathcal{V}$. Prove that $\mathcal{W}$ has a finite spanning set. For the proof, it suffices to assume that $\mathcal{W} \neq \{\mathbf{0}\}$. (Why?) In $\mathcal{W}$, let $S = \{X_1, X_2, \ldots, X_k\}$ contain as many linearly independent elements as possible. (Why is there a limit on the number of independent elements in $\mathcal{W}$?) Prove that S spans $\mathcal{W}$.

(24) Let $\mathcal{W}$ be a subspace of an n-dimensional vector space $\mathcal{V}$. Prove that $\dim \mathcal{W} \leq n$. Prove that if $\dim \mathcal{W} = n$, then $\mathcal{W} = \mathcal{V}$.

(25) Let $\{X_1, X_2, \ldots, X_n\}$ be a linearly independent subset of some vector space $\mathcal{V}$. Suppose that there are scalars c_i and d_i such that $c_1 X_1 + c_2 X_2 + \cdots + c_n X_n = d_1 X_1 + d_2 X_2 + \cdots + d_n X_n$. Prove that $c_i = d_i$. Thus, elements of $\mathcal{V}$ are representable in at most one way as a linear combination of the X_i. This statement is called the "uniqueness theorem." [*Hint*: For the proof, note that $X = Y$ can also be expressed as $X - Y = \mathbf{0}$.]

(26) Prove that the uniqueness theorem from Exercise 25 is false for spanning sets that are not bases. That is, suppose that $\{X_1, X_2, \ldots, X_n\}$ spans a vector space $\mathcal{V}$ and is dependent. Prove that every element $X \in \mathcal{V}$ has at least two different expressions as a linear combination of the spanning vectors. It might help to consider first the case of $X = \mathbf{0}$.

(27) In $\mathbb{R}^\infty$, let $\mathcal{W}$ be the set of vectors $[x_1, x_2, \ldots, x_n, \ldots$ where

$$\lim_{n \to \infty} x_n = 0$$

(a) Give an example of an element $X \in \mathbb{R}^\infty$ that belongs to $\mathcal{W}$ and an element $Y \in \mathbb{R}^\infty$ that does not.

(b) Prove that $\mathcal{W}$ is a subspace of $\mathbb{R}^\infty$.

(c) Prove that $\mathcal{W}$ is infinite-dimensional.

(d) Exhibit a three-dimensional subspace of $\mathbb{R}^{\infty}$.

(28) We have commented in numerous earlier exercises that the space $C^{\infty}(\mathbb{R})$ of in-finitely differentiable functions on $\mathbb{R}$ is a vector space. A polynomial function is a function $p(x)$ which is expressible in the form

$$p(x) = a_n x^n + a_{n-1} x^{n-1} + \cdots + a_0$$

where the a_i are scalars. If $a_n \neq 0$, the polynomial is said to have degree n. We let P_n denote the space of all polynomials of degree less than or equal to n. Thus, for example, P_2 is the set of all functions of the form

$$p(x) = a_2 x^2 + a_1 x + a_0$$

where a_i are scalars. In this exercise, you will study the dimension of P_2.

(a) Prove that P_2 is a subspace of $C^{\infty}(\mathbb{R})$. The significance of this is that it proves that P_2 is a vector space and hence the theory developed in this section is applicable to P_2.

(b) Explain why the functions 1, x, and x^2 span P_2.

(c) Use the technique of Exercise 13, Section 2.1, to prove that the 3 functions from part (b) form an independent set. How does this prove that P_2 is three-dimensional?

(d) Prove that the functions 1, $x - 1$, and $(x - 1)^2$ belong to P_2 and form a linearly independent set. (Use the technique of Exercise 13, Section 2.1.) What theorem from this section allows you to conclude that if $p(x)$ belongs to P_2, then there exist scalars a, b, and c such that

$$p(x) = a + b(x - 1) + c(x - 1)^2$$

Remark: The equality just stated is equivalent with

$$\frac{p(x)}{(x - 1)^3} = \frac{a}{(x - 1)^3} + \frac{b}{(x - 1)^2} + \frac{c}{x - 1}$$

which is the partial fractions decomposition of the rational function on the left side of this equality. In general, the existence of partial fractions decompositions can be proved using dimension theory. Exercise 29 is another example of this.

(29) Show that the functions $x - 1$, $(x - 1)(x - 2)$, and $(x - 2)^2$ form an indepen-dent set in P_2. [*Hint:* Rather than differentiating the dependency equation, try substituting $x = 0$, $x = 1$, and $x = 2$ into it.]

What theorem from this section, together with part (c) of Exercise 28, allows you to conclude that if $p(x)$ belongs to P_2, then there exist scalars a, b, and c such that

$$p(x) = a(x - 1) + b(x - 1)(x - 2) + c(x - 2)^2$$

Divide both sides of this equality by $(x - 1)(x - 2)^2$ to prove the existence of the partial fractions decomposition for $p(x)/((x - 1)(x - 2)^2)$

(30) Generalize Exercise 28 to P_3.

(31) What is the dimension of P_n? Prove your answer.

ON LINE

Let

$$A = \begin{bmatrix} -1 & 2 & 6 & -8 & -14 & 3 \\ 2 & 4 & 1 & -8 & 5 & -1 \\ -3 & 1 & 4 & -9 & -10 & 0 \\ 3 & -2 & -1 & 12 & -1 & 4 \\ 5 & 7 & 11 & -11 & -19 & 9 \end{bmatrix}$$

Please use A for the following exercises.

ON LINE EXERCISES

(1) Use the "rank" command in MATLAB to find the rank of A. *Using only the value of the rank*, explain why statements (a) and (b) are true. Then do the rest of the exercises.

(a) The reduced form for the augmented matrix for the system $AX = \mathbf{0}$ has three free variables. (Recall that in an earlier On Line section it was noted that the rank is the number of non-zero rows in the reduced form.)

(b) The nullspace of A has dimension at most 3. [*Hint*: How many spanning vectors are there in the general solution to $AX = \mathbf{0}$?]

(2) Show that each of the following vectors satisfies $AX = \mathbf{0}$.

$$X_1 = [-5, 13, -10, 2, -3, 1]^t$$
$$X_2 = [3, -6, 11, 1, 2, -5]^t$$
$$X_3 = [-4, 7, 9, 5, 1, -6]^t$$

(3) Prove that X_1, X_2, and X_3 are linearly independent by computing the rank of the matrix $X = [X_1, X_2, X_3]$. (Recall that the maximal number of linearly independent columns equals the rank.)

(4) How does it follow that the dimension of the nullspace of A is 3? How does it follow that the X_i constitute a basis for the nullspace?

(5) Use "Y=null(A)" to find a basis for the nullspace of A. Express each column of Y as a linear combination of the X_i.

> Remark: An efficient way of doing this is as follows. In MATLAB, set "B=[X,Y]" where X is as in Exercise 3. Then $B = [X_1, X_2, X_3, Y_1, Y_2, Y_3]$ where the Y_i are the columns of Y. This matrix should have rank 3. (Why?) If you reduce B, the first three columns will be the pivot columns. According to Theorem 1 in Section 2.1, the last three columns will then be linear combinations of the first three. In fact, the technique of Example 4 in Section 2.1 shows how to find the explicit expressions for the Y_i in terms of the X_i.
> [*Note*: You may need to increase the tolerance in "rref" to get B to have rank 3.]

(6) In Exercise 5, what made us so sure that the first three columns would be the pivot columns? Why, for example, couldn't the pivot columns be columns 1, 3, and 4? [*Hint*: Think about what this would say about the reduced form of $[X_1, X_2, X_3]$.]

(7) Express each of the X_i as a linear combination of the Y_i. This shows that the Y_i also span the nullspace of A.

(8) Find (by inspection) a vector T that solves the equation $AX = [6, 1, 4, -1, 11]'$.

(9) Generate a 1×3 random matrix $[r, s, t]$ and let $Z = T + r * X_1 + s * X_2 + t * X_3$ where T is as in Exercise 8. Compute AZ. Explain why you get what you get. Find constants u, v, and w such that $Z = T + u * Y_1 + v * Y_2 + w * Y_3$. What theorem does this demonstrate?

2.2.1 DIFFERENTIAL EQUATIONS

Let us begin with an example. Let y be the function $y(x) = e^{2x}$. We let y' be the derivative of y. Then, as the reader may easily verify,

$$y' = 2y$$

This is an important equation. It is a "differential" equation in that it relates a function and its derivative. If x represents time, then this equation describes a quantity that is always increasing twice as fast as the amount currently present. For example, a rapidly growing population might approximate such a formula.

The particular y given above is not the only function that satisfies our differential equation. The function Cy also satisfies this equation for any constant C. Thus, $g(x) = 3e^{2x}$ satisfies $g' = 2g$. It turns out that the only solutions are, in fact, Cy for C some constant where $y(x) = e^{2x}$.

This (we hope) will remind you of the concept of spanning. If Y is an element of some vector space, then the set of all multiples of Y is the space spanned by Y. We can say that the solution space to our differential equation is the space spanned by the single solution $y(x) = e^{2x}$. But what vector space are we discussing? The answer depends on what kind of functions one wishes to allow as solutions to the equation. In this section, we will be fairly restrictive, limiting our discussion to solutions that are defined for all real numbers x and can be differentiated any number of times. The space of all such functions is denoted $C^\infty(\mathbb{R})$. Thus, $y(x) = e^x$ and $y(x) = \sin x$ are elements of $C^\infty(\mathbb{R})$. The function $y(x) = 1/x$ is not because it is not defined for all real numbers. Neither is the function $y(x) = |x|$ because its derivative does not exist at 0. *In this section, "solution" always refers to a solution in $C^\infty(\mathbb{R})$.*

The space $C^\infty(\mathbb{R})$ is a vector space. The operations of addition of functions and multiplication of functions by scalars are the same as those studied in calculus class. Abstractly, they are defined by

$$(f + g)(x) = f(x) + g(x) \text{ and } (cf)(x) = cf(x)$$

The zero function (denoted $\mathbf{0}$) is the constant function $y = 0$. The sum of any two C^∞ functions is still C^∞ as is any scalar multiple.

It is very simple to prove that addition of functions satisfies the vector space properties listed in Section 1.2. For example, to prove the commutative law for addition of functions, let f and g be two C^∞ functions. Then, for all $x \in \mathbb{R}$,

$$(f + g)(x) = f(x) + g(x) = g(x) + f(x) = (g + f)(x)$$

Since $(f + g)(x) = (g + f)(x)$ for all x, it follows that $f + g = g + f$, as claimed.

The proofs of the other vector space properties are left as exercises (see Exercise 2, page 116).

Let us consider another differential equation. Suppose we wish to find all functions y in $C^\infty(\mathbb{R})$ such that

$$y'' + y' - 2y = \mathbf{0} \tag{1}$$

Based on the example above, we guess that there may be a solution of the form $y(x) = e^{rx}$ where r is some constant. Then, to determine r, we substitute our guess into formula (1). We find that for our guess to work, we must have

$$0 = r^2 e^{rx} + r e^{rx} - 2e^{rx} = (r^2 + r - 2)e^{rx}$$

for all x. This will be true if and only if

$$0 = r^2 + r - 2 = (r - 1)(r + 2)$$

Thus, we obtain two values of r: $r = 1$ and $r = -2$. This yields solutions $y_1(x) = e^x$ and $y_2(x) = e^{-2x}$. These are the only solutions of the form $y(x) = e^{rx}$.

Are these the only solutions? Definitely not! Suppose that y and z represent any two solutions. Then

$$y'' + y' - 2y = 0$$
$$z'' + z' - 2z = 0$$

If we add these two equations together, we obtain

$$(y + z)'' + (y + z)' - 2(y + z) = 0$$

Thus, $y + z$ is also a solution.

Similarly, if we multiply both sides of formula (1) by a scalar c, we conclude that

$$(cy)'' + (cy)' - 2cy = 0$$

showing that cy is a solution. It follows that any linear combination of y_1 and y_2 will also be a solution. Thus, for example, the function

$$y_3(x) = 2e^x - 7e^{-2x}$$

is another solution. In linear algebraic terms, the set of all solutions is a subspace of $C^\infty(\mathbb{R})$.

We try again: Are the linear combinations of y_1 and y_2 the only solutions of formula (1)? The answer to this question comes from a theorem in the subject of differential equations. The proof is based on the existence and uniqueness theorem for differential equations.

Theorem 1.   *Let a, b, and c be constants with $a \neq 0$. Then the set of all solutions to the following differential equation is a two-dimensional subspace of $C^\infty(\mathbb{R})$.*

$$ay'' + by' + cy = 0$$

According to this theorem, the set of solutions to formula (1) is two-dimensional. We know that in a two-dimensional space, any two linearly independent elements will span the space. If our solutions y_1 and y_2 can be shown to be independent, they will span the solution space. This means exactly that any solution is a linear combination of y_1 and y_2.

So, are y_1 and y_2 independent? Suppose that there is a scalar C such that $y_1 = Cy_2$. This means that for all x,

$$e^x = Ce^{-2x}$$

This is the same as $e^{3x} = C$. This equation says that the function $y(x) = e^{3x}$ is a constant function! Clearly, no such C exists. Similarly, there is no constant C such that $y_2 = Cy_1$.

Thus, y_1 and y_2 are indeed independent, proving (from Theorem 1) that the general solution to formula (1) is

$$y(x) = C_1 e^x + C_2 e^{-2x}$$

Let us consider another example.

▶ EXAMPLE 1:   Find the general solution to the following differential equation:

$$y'' - 2y' + y = 0 \qquad (2)$$

Solution: As before, we try a solution of the form $y(x) = e^{rx}$. We substitute into the differential equation, discovering that this will be a solution if and only if $r^2 - 2r + 1 = 0$. This factors as $(r - 1)^2 = 0$, so $r = 1$. Hence we obtain the solution $y_1(x) = e^x$. Since the solution space is a subspace, we know that any multiple of y_1 is also a solution.

This cannot, however, describe the general solution. The set of multiples of y_1 defines a one-dimensional subspace of $C^\infty(\mathbb{R})$. According to Theorem 1, the solution space must be two-dimensional. There must be another, independent, solution y_2. Furthermore, y_2 cannot be of the form $y(x) = e^{rx}$ since the only solution of this form is y_1.

How do we find y_2? If this were a text in differential equations, we would derive the answer. Since this is a text in linear algebra, we will simply tell you that the answer is $y_2(x) = xe^x$; that is, it is just $y_2(x) = xy_1(x)$. [You can verify that this works by substituting $y_2(x)$ into formula (2).] But doesn't this make y_2 a multiple of y_1, hence dependent on y_1? No, not at all. The function $f(x) = x$ is not a constant. Thus y_2 is not a *constant* multiple of y_1.

Let us prove that y_1 and y_2 are independent. The technique we use is the most common method for proving functions independent. We begin by considering the dependency equation

$$c_1 y_1 + c_2 y_2 = 0$$

It is important to realize that $\mathbf{0}$ represents the zero function. This is the function f that satisfies $f(x) = 0$ for all x. Thus, from this equation, we may conclude that *for all x,*

$$c_1 e^x + c_2 x e^x = 0 \qquad (3)$$

Since this is true for all x, we may differentiate both sides of this equation, producing:

$$c_1 e^x + c_2 (e^x + x e^x) = 0$$
$$(c_1 + c_2) e^x + c_2 x e^x = 0 \qquad (4)$$

Since equations (3) and (4) are both true for all x, we may set $x = 0$ in each, producing the system

$$\begin{aligned} c_1 \quad\; &= 0 \\ c_1 + c_2 &= 0 \end{aligned}$$

This yields $c_1 = c_2 = 0$, proving independence.

Now that we know independence, it follows from Theorem 1 that the general solution to equation (2) is

$$y(x) = C_1 e^x + C_2 x e^x \tag{5}$$

◄

There is a version of Theorem 1 that applies to higher order differential equations:

Theorem 2.   *Consider the differential equation*

$$a_n y^{(n)} + a_{n-1} y^{(n-1)} + \ldots a_0 y = 0$$

where $y^{(k)}$ denotes the kth derivative of y and $a_n \neq 0$. Then the set of solutions to this equation is a subspace of $C^\infty(\mathbb{R})$ of dimension n.   ■

▶ **EXAMPLE 2:**   Find all solutions to the following equation:

$$y''' + 3y'' + 2y' = 0$$

Solution: Again we hope for a solution of the form $y(x) = e^{rx}$ Substitution into the differential equation yields

$$r^3 e^{rx} + 3r^2 e^{rx} + 2r e^{rx} = 0$$

This yields

$$\begin{aligned} r^3 + 3r^2 + 2r &= 0 \\ r(r+1)(r+2) &= 0 \end{aligned}$$

The roots are, $r = 0$, $r = -1$, and $r = -2$ corresponding to the solutions

$$y_1(x) = e^0 = 1, \quad y_2(x) = e^{-x}, \quad y_3(x) = e^{-2x}$$

According to Theorem 2, the solution space is a three-dimensional subspace. If these three functions are independent, they will form a basis. To prove the independence, we consider the dependency equation

$$c_1 y_1 + c_2 y_2 + c_3 y_3 = \mathbf{0}$$

Then, for all x we have:

$$c_1 1 + c_2 e^{-x} + c_3 e^{-2x} = 0 \tag{6}$$

We need three equations to determine the three unknown constants. We differentiate equation (6) twice, obtaining:

$$-c_2 e^{-x} - 2c_3 e^{-2x} = 0$$
$$c_2 e^{-x} + 4c_3 e^{-2x} = 0$$

Substituting $x = 0$ into these two equations as well as into formula (6) yields the system

$$c_1 + c_2 + c_3 = 0$$
$$-c_2 - 2c_3 = 0$$
$$c_2 + 4c_3 = 0$$

This is easily solved to yield $c_1 = c_2 = c_3 = 0$, proving independence.
It follows that the general solution to our equation is

$$y(x) = c_1 + c_2 e^{-x} + c_3 e^{-2x}$$

where the c_i are arbitrary constants. ◀

The differential equation in Theorem 2 is homogeneous in that the function on the right side of the equality is $\mathbf{0}$. If this function had been non-zero, then the equation would be non-homogeneous. We may express the general solution to a non-homogeneous differential equation in terms of the general solution to the corresponding homogeneous system in much the same manner as is done with systems of linear equations.

▶ EXAMPLE 3:   Find all solutions to

$$y'' + 2y' + y = 2\cos x$$

Solution: Notice that

$$(\sin x)'' + 2(\sin x)' + \sin x = 2\cos x \tag{7}$$

Thus, $y = \sin x$ is one solution to our differential equation.

On the basis of the previous examples, we might guess that multiples of $\sin x$ are also solutions. This is not true. For example

$$(2 \sin x)'' + 2(2 \sin x)' + 2 \sin x = 4 \cos x \neq 2 \cos x$$

showing that $y = 2 \sin x$ is *not* a solution to our equation. This demonstrates a fundamental difference between homogeneous and non-homogeneous linear differential equations: the general solution to a non-homogeneous differential equation will not be a subspace.

To find the general solution to our differential equation, let y be another solution so that

$$y'' + 2y' + y = 2 \cos x$$

We subtract formula (7) from this equation, producing

$$(y - \sin x)'' + 2(y - \sin x)' + (y - \sin x) = 0$$

Letting $w = y - \sin x$, we see that

$$w'' + 2w' + w = 0$$

This is exactly the equation we solved in Example 1. The general solution is given by formula (5). Since $y = \sin x + w$, it follows that

$$y(x) = \sin x + C_1 e^x + C_2 x e^x$$

Conversely, direct substitution shows that any function of this form solves our differential equation. Hence, this formula represents the general solution.   ◀

The reasoning from Example 3 proves the following theorem. The reader should compare this result with the translation theorem from Section 1.5.

Theorem 3.   Let y_p *be any solution to the differential equation*

$$a_n y^{(n)} + a_{n-1} y^{(n-1)} + \cdots + a_0 y = b$$

where a_n *and* b *are elements of* $C^\infty(\mathbb{R})$. *Then* y *satisfies this equation if and only if* $y = y_p + w$ *where* w *satisfies*

$$a_n w^{(n)} + a_{n-1} w^{(n-1)} + \cdots + a_0 w = 0$$

Proof: Since y_p is a solution

$$a_n(y_p)^{(n)} + a_{n-1}(y_p)^{(n-1)} + \cdots a_0 y_p = b$$

Let y be any other solution. Then

$$a_n y^{(n)} + a_{n-1} y^{(n-1)} + \cdots a_0 y = b$$

Subtracting the first equation from the second yields

$$a_n(y - y_p)^{(n)} + a_{n-1}(y - y_p)^{(n-1)} + \cdots a_0(y - y_p) = \mathbf{0}$$

This says exactly that $w = y - y_p$ is a solution to the homogeneous equation. Hence, $y = y_p + w$ as claimed.

Conversely, if w is any solution to the homogeneous equation then

$$a_n w^{(n)} + a_{n-1} w^{(n-1)} + \cdots a_0 w = \mathbf{0}$$
$$a_n(y_p)^{(n)} + a_{n-1}(y_p)^{(n-1)} + \cdots a_0 y_p = b$$

Adding these two equalities produces

$$a_n(w + y_p)^{(n)} + a_{n-1}(w + y_p)^{(n-1)} + \cdots a_0(w + y_p) = b$$

which shows that $y_p + w$ is a solution to the non-homogeneous equation. This finishes our proof. ∎

EXERCISES

(1) Show that the following sets of functions are linearly dependent in $C^\infty(\mathbb{R})$ by expressing one of them as a linear combination of the others. (You may need to refer to a calculus book for the definitions of the sinh and cosh functions, as well as some trig identities.)

    (a) $\{e^x, e^{-x}, \cosh x\}$ (b) $\{\cos(2x), \sin^2 x, \cos^2 x\}$

    (c) $\{\sinh x, e^x, \cosh x\}$ (d) $\{\cos(2x), 1, \sin^2 x\}$

    (e) $\{\ln(3x), \ln x, 1\}$ (f) $\{x(x - 2)^2, 1, x, x^2, x^3\}$

    (g) $\{1, (x - 1), (x - 1)^2, 2x^2 + 5x + 3\}$

(2) Prove vector space properties (c) (e) (h) and (i), Section 1.2, for $C^\infty(\mathbb{R})$.

(3) Consider the following differential equations. Prove that the set of solutions is a subspace of $C^\infty(\mathbb{R})$. Your proof should follow the same pattern as in the discussion

preceding Theorem 1. (Do not use Theorem 1.) In each case, what is the dimension of this subspace?

(a) $y'' + 3y' + 2y = 0$

(b) $y'' + 4y' + 13y = 0$

(c) $y''' + y'' - y' - y = 0$

(d) $y''' + 3y'' + 3y' + y = 0$

(e) $y''' = 0$

(4) In each of the equations in Exercise 3, find all solutions of the form $y(x) = e^{rx}$ where r is a real number. In part (a), prove that the functions found span the entire solution space by proving that they are linearly independent. Prove that the functions found in part (c) are also linearly independent. Do they span the solution space? Why? In parts (d) and (e), find the general solution. [*Hint*: Make an educated guess using Example 2 as a model. Then prove that your guess works.]

(5) Show that the functions $y_1(x) = e^{-2x} \cos 3x$ and $y_2(x) = e^{-2x} \sin 3x$ both satisfy the equation in part (b) of Exercise 3. Prove that these functions are independent. What is the general solution to the differential equation in part (b)?

(6) Consider the differential equation in part (c) of Exercise 3. Let $\mathcal{W}$ denote the set of all solutions y of this equation that satisfy $y(1) = 0$. Prove that this set is a subspace of $C^\infty(\mathbb{R})$. What is the dimension of this subspace?

(7) Prove that the set of all solutions to the following differential equation *do not* form a subspace of $C^\infty(\mathbb{R})$.

$$y'' + 3y' + 2y = x^2$$

(8) Find a y of the form $y = ax^2 + bx + c$ that solves the differential equation in Exercise 7. Use this, together with Theorem 3, to find the general solution to the differential equation in Exercise 7.

(9) Find all solutions to the equation

$$y'' + 3y' + 2y = \cos x$$

[*Hint*: Try $y = A \cos x + B \sin x$. See Theorem 3.]

(10) A function of the form $p(x) = a_n x^n + a_{n-1} x^{n-1} + \cdots + a_0$ is said to be a polynomial of degree less than or equal to n. (It has degree n if $a_n \neq 0$.) The space of all polynomials of degree less than or equal to n is denoted P_n. Prove that P_n is a subspace of $C^\infty(\mathbb{R})$.

(11) Is the set of all polynomial functions (regardless of degree) a subspace of $C^\infty(\mathbb{R})$? What about the set of all polynomial functions with integral coefficients?

(12) Consider the set of C^∞ functions $S = \{1, x, x^2, x^3\}$. Prove that S is independent. For this, you should set up the dependency equation and differentiate as in Example 1. Prove that this set spans P_3. What is the dimension of P_3?

(13) Show that all the functions in Exercise 12 satisfy the differential equation $y^{(4)} = \mathbf{0}$. On the basis of Theorem 2, what is the general solution to this equation?

(14) Prove that for all n, the set $\{1, x, x^2, \ldots, x^n\}$ is linearly independent. How does this prove that $C^\infty(\mathbb{R})$ is infinite-dimensional?

(15) What is the dimension of P_n? What is the dimension of the space of all polynomial functions?

(16) Let $\mathcal{W}$ denote the set of all functions $f \in C^\infty(\mathbb{R})$ such that $f(1) = 0$. Prove that $\mathcal{W}$ is a subspace.

(17) In Exercise 16, prove that $\mathcal{W}$ is infinite-dimensional. [*Hint*: Note that for all natural numbers n, the function $f_n(x) = (x - 1)^n$ belongs to $\mathcal{W}$. What more do you need to show to conclude that the dimension is infinite?]

(18) Let $\mathcal{W}$ denote the set of all functions $f \in C^\infty(\mathbb{R})$ such that $f(1) = f(2) = 0$. Prove that $\mathcal{W}$ is a subspace. What is its dimension? Prove your answer.

(19) Let $\mathcal{W}$ denote the set of all functions $f \in C^\infty(\mathbb{R})$ such that $f(1) = 2$.
   (a) Give an example of a function that is in $\mathcal{W}$.
   (b) Let f and g be in $\mathcal{W}$. For which scalars a and b is it true that $af + bg$ belongs to $\mathcal{W}$?
   (c) Is $\mathcal{W}$ a subspace?

ON LINE

Imagine that we have a box of mass 1g on a frictionless table and attached to a spring as shown in Figure 1. Initially, the spring is unstretched. We pull the box 3 cm to the right and let it go. We expect that it will oscillate left and right across the table. In this exercise set, we would like to quantify this.

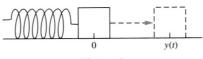

0   y(t)

Figure 1

The physical law which we use is

$$F = ma$$

where F is the force on the box, m is its mass (so, in our case, $m = 1$), and a is its acceleration. Explicitly, we let $y(t)$ denote the displacement of the box away from its rest position at time t. (Displacements to the right are considered to be positive.) Then, acceleration at time t is $y''(t)$.

The only force on the box which acts in the direction of the motion is that exerted by the spring. We shall assume that this force is proportional to the stretching. Specifically, we shall assume that this force is approximately 0.25 dynes per centimeter of stretching. This force will, of course, also be in the opposite direction from the stretching. Thus, $F = ma$ translates to

$$-0.25y = y'' \tag{8}$$

which is equivalent with

$$y'' + 0.25y = \mathbf{0}.$$

This is a differential equation similar to the ones studied in this section.

MATLAB has an excellent facility for computing numerical approximations to solutions of differential equations. MATLAB, however, can only deal with one derivative at a time. We can avoid this difficulty by introducing the velocity $v = y'$. Then $F = ma$ is also equivalent with the pair of equations

$$-0.25y = v'$$
$$v = y'$$

This is equivalent with the matrix equation

$$\begin{bmatrix} 0 & -.25 \\ 1 & 0 \end{bmatrix} \begin{bmatrix} v \\ y \end{bmatrix} = \begin{bmatrix} v' \\ y' \end{bmatrix}$$

To enter this equation into MATLAB, we first create a MATLAB function which represents the left side of this equation. This is done by pulling down the File menu in MATLAB and selecting New/M-File. A window will appear into which you should type the following sequence of commands. Once you are finished, again pull down the file menu and select "Save As...". When prompted, save the file as "yy.m". (Leave this window open as you will need to change the definition of yy several times during this exercise set. Every time you change this file, you will need to save the changes by selecting "Save" from the file menu.)

```
function g=yy(t,x)
A=[0,-.25;1,0];
g=A*x;
```

You have just created a function "yy" which, for t and x given, multiplies x by A and does not use t. You can test your function by entering "yy(17, [1;2])" . The result should equal $A[1, 2]^t$.

To solve the differential equation you enter

```
[t,y]=ode23('yy',0,40,[0;3]);
```

The "0,40" tells MATLAB to solve the equation for $0 \leq t \leq 40$. The term "[0;3]" tells MATLAB that the initial velocity is 0 and the initial stretching is 3. We can plot the result with the command

```
plot(t,y(:,2));
```

ON LINE EXERCISES

(1) After completing the steps described above, enter "[t,y]" into the command window. You should discover that [t,y] is a very large matrix with three columns. The first column represents times, the second the corresponding velocities and the third displacements. The third column of [t,y] is the second column of y. If you were to enter "plot(t,y(:,1));" what would your graph represent?

(2) The period of a periodic motion is the amount of time before the motion repeats itself. Use your graph to estimate the period of the motion. It might help to enter the "grid on" command into MATLAB. Would you consider this slow or fast oscillation? (The time is in seconds.)

(3) Find a value of a for which the functions $\cos at$ and $\sin at$ both solve equation (8). What is the general solution to equation (8)? Find a formula for the solution y which satisfies $y(0) = 3$, $y'(0) = 0$. Use your answer to find the exact value for the period of the motion you estimated in the previous exercise.

(4) The period of the oscillations is determined by the "stiffness" of the spring. (A stiff spring is one which takes a large force to stretch it.) How do you guess stiffness should relate the period of the motion: should stiff springs oscillate faster or slower? Test your guess by graphing the solution curve for a stiff spring and a non-stiff spring. Note: This will require that you change the coefficient .25 in equation (8) and in yy.m. How should you change it to model a stiffer spring? (Don't forget to save your changes to yy.m.)

(5) Prove mathematically that your guess in the last exercise is really correct. For this, you should compute the general solution to the equation $y'' = ky$ and then find a formula for the period.

(6) Imagine that our box is sitting near our stereo that is generating a tone which is causing the box to vibrate. We model this as applying an external force $F(t) = \epsilon \sin(\omega t)$ where ω is determined by the pitch of the tone and ϵ is determined by the volume of the tone. This changes formula (8) to

$$-0.25y + \epsilon \sin(\omega t) = y'' \qquad (9)$$

Assume initially that $\epsilon = .4$ and $\omega = .3$. Plot the solution to this equation over the interval $0 \le t \le 100$. Describe the motion of the box. What is its maximum displacement?

Note: You will need to change the line "g=A*x;" to "g=A*x+ [.4*sin(.3*t);0];" in the file yy.m and save your changes.

(7) Repeat the previous exercise with $\epsilon = .4$ and $\omega = .5$. Plot the solution over a large enough interval to allow you to guess the maximum value of the displacement.

(8) Surely, the behavior described by the graph in the previous exercise would not happen with a real box. What factors would limit the actual displacement of the box?

Remark: Exercises 6 and 7 demonstrate the concept of resonance: if we vibrate the spring with just the right tone, the displacements can become very large. The same principle applies to structures such as bridges. The most famous example of this occurred on November 7, 1940, when the Tacoma Narrows bridge at Puget Sound self-destructed in a high wind.

The effect of resonance on the Tacoma Narrows Bridge.

2.3 APPLICATIONS TO SYSTEMS

Let us take a moment to review an idea discussed in Section 1.3. Consider the following system of equations:

$$\begin{aligned} x + 2y + z &= 1 \quad I \\ 2x + 4y + z &= 3 \quad II \\ x + 2y + 2z &= 0 \quad III \end{aligned}$$

This system is dependent because $3I - II = III$. Thus, equation III contributes no information, and we may consider our system to consist of equations I and II. If we prefer, we can write the dependency equation as $II = 3I - III$. Thus, we can ignore II, considering only equations I and III. In either case, we obtain only two independent equations.

In Section 1.3, we stated a remarkable fact: if one begins with any dependent system of equations and eliminates the dependent equations, one at a time, the final number of equations so obtained is the same, regardless of which equations were eliminated and which were kept. In this section, we will present the proof using the concept of dimension. In the process, we will obtain a very efficient method for finding bases for vector spaces. Let us begin with the following fundamental definition:

Definition. *The row space of a matrix A is the span of the rows of A.* ■

Since spans are subspaces, the row space is a subspace. Specifically, if A is $m \times n$, then its row space will be a subspace of $\mathbb{R}^n$. If the matrix is the augmented matrix of some system of equations, then the row space represents all possible equations that are dependent on the original system. The row space contains all rows that could be produced by applying arbitrary sequences of elementary row operations to A.

▶ EXAMPLE 1: Show that the row vector $[1, 2, 3, 5]$ belongs to the row space of the matrix

$$A = \begin{bmatrix} 1 & 1 & 1 & 1 \\ 0 & 1 & 2 & 4 \\ 1 & 3 & 1 & 7 \end{bmatrix}$$

Solution: In this case, the problem is easily solved by inspection. Direct computation shows:

$$[1, 2, 3, 5] = [1, 1, 1, 1] + [0, 1, 2, 4] + 0[1, 3, 1, 7]$$ ◀

▶ EXAMPLE 2:   Determine whether the row vector [2, 3, 5, 10] belongs to the row space of the matrix A in Example 1.

Solution: This will be the case if and only if there are constants u, v, and w such that:

$$[2, 3, 5, 10] = u[1, 1, 1, 1] + v[0, 1, 2, 4] + w[1, 3, 1, 7]$$

This results in the system

$$
\begin{aligned}
u + 0v + \ w &= 2 \\
u + \ v + 3w &= 3 \\
u + 2v + \ w &= 5 \\
u + 4v + 7w &= 10
\end{aligned}
$$

The augmented matrix for this system is:

$$
\begin{bmatrix}
1 & 0 & 1 & 2 \\
1 & 1 & 3 & 3 \\
1 & 2 & 1 & 5 \\
1 & 4 & 7 & 10
\end{bmatrix}
$$

We row-reduce this matrix, producing

$$
\begin{bmatrix}
1 & 0 & 1 & 2 \\
0 & 1 & 2 & 1 \\
0 & 0 & -4 & 1 \\
0 & 0 & 0 & 1
\end{bmatrix}
$$

This is the matrix of an inconsistent system, since it contains the equation $0 = 1$. This says that the given row vector is not in the row space of A.   ◄

Let A be some matrix and let B be some other matrix that is row equivalent with A. (Thus B is obtained from A by a sequence of elementary row operations.) Then the systems of equations represented by A and B are equivalent. We certainly hope that the set of equations that are dependent on the system represented by A are the same as those dependent on the system represented by B. Thus, the following proposition should seem neither mysterious nor surprising:

Proposition 1.   *Let A and B be two row-equivalent matrices. Then A and B have the same row space.*

Proof: We begin by considering the effect of applying a single row operation to A. Thus, suppose that the matrix A_1 was produced by applying a single elementary row operation to A. We noted in Section 1.4 that every row of A_1 is a linear combination of rows of A. Thus, the rows of A_1 all belong to the row space of A. Since the row space of A is a subspace, it follows that all linear combinations of rows of A_1 also belong to the row space of A. Thus, the row space of A_1 is contained in that of A.

We also noted in Section 1.4 that every row of A is a linear combination of rows of A_1. Hence, the row space of A is also contained in that of A_1, proving that A and A_1 have the same row space. Thus, applying a single row operation to a matrix does not change its row space.

It follows that applying sequences of row operations to a given matrix will also not change its row space: if we apply a sequence of row operations to A producing matrices $A_1, A_2, \ldots, A_n$, then the row space of A equals that of A_1, which equals that of A_2, which equals... Thus, any matrix which is row equivalent with A will have the same row space as A, proving our proposition. ∎

Now, let us consider the process of eliminating dependent equations. In terms of row spaces, eliminating dependent equations corresponds to eliminating dependent rows of the augmented matrix. This will not, of course, change the span of the rows. Once we have eliminated all dependencies, we are left with a basis. But the number of elements in any basis is the dimension of the space. This is why we always wind up with the same number of independent equations, no matter which specific equations are eliminated. The final number of equations is just the dimension of the row space. Actually, this dimension is just the number of non-zero rows in the echelon form of the matrix. This follows from the following important theorem.

Non-zero Rows Theorem.   *The non-zero rows of any echelon form of a matrix A form a basis for the row space of A.*

Proof: We will first prove the theorem in the case of the row-reduced form of A. Thus, let the row-reduced form of A be R. From Proposition 1, the rows of R span the row space of A. Thus, we need only show that the non-zero rows of R are independent. It helps to visualize the typical row echelon matrix:

$$R = \begin{bmatrix} 1 & * & 0 & * & 0 & * \\ 0 & 0 & 1 & * & 0 & * \\ 0 & 0 & 0 & 0 & 1 & * \\ 0 & 0 & 0 & 0 & 0 & 0 \end{bmatrix}$$

The crucial observation is that in each non-zero row there is a one in a position where every other row has a zero. Thus, no row could be a linear combination of the other rows. This proves the theorem for R.

The general case follows from the observation that since any echelon form for A will have the same number of non-zero rows as the reduced form, the number of non-zero rows of the echelon form is the dimension of the row space. Moreover, the rows of the echelon form also span the row space. Hence, from Proposition 1 in Section 2.2, they are also independent, and thus form a basis for the row space of A. ■

In Section 1.3, we defined the rank of a system of equations to be the number of equations left after dependent equations have been eliminated. Here is a much better way of saying this:

Definition.   The rank of a matrix A is the dimension of the row space. It is computable as the number of non-zero rows in an echelon form of the matrix. ■

The non-zero rows theorem provides one of the most important ways of constructing bases.

▶ EXAMPLE 3:   Find a basis for the subspace $\mathcal{W}$ of $\mathbb{R}^5$ spanned by the following vectors. What is the dimension of this subspace?

$$\begin{bmatrix} 1 \\ -2 \\ 0 \\ 2 \\ 3 \end{bmatrix} \quad \begin{bmatrix} 1 \\ -6 \\ 1 \\ 4 \\ 7 \end{bmatrix} \quad \begin{bmatrix} 2 \\ 0 \\ -1 \\ 2 \\ 2 \end{bmatrix} \quad \begin{bmatrix} 1 \\ 2 \\ 1 \\ 2 \\ 1 \end{bmatrix} \quad \begin{bmatrix} 5 \\ -6 \\ 1 \\ 10 \\ 13 \end{bmatrix}$$

Solution: We interpret these vectors as the *rows* of the matrix A where

$$A = \begin{bmatrix} 1 & -2 & 0 & 2 & 3 \\ 1 & -6 & 1 & 4 & 7 \\ 2 & 0 & -1 & 2 & 2 \\ 1 & 2 & 1 & 2 & 1 \\ 5 & -6 & 1 & 10 & 13 \end{bmatrix}$$

We reduce A, obtaining

$$R = \begin{bmatrix} 1 & 0 & 0 & \frac{3}{2} & \frac{3}{2} \\ 0 & 1 & 0 & -\frac{1}{4} & -\frac{3}{4} \\ 0 & 0 & 1 & 1 & 1 \\ 0 & 0 & 0 & 0 & 0 \\ 0 & 0 & 0 & 0 & 0 \end{bmatrix}$$

The three non-zero rows of R will form a basis for the row space. To obtain a basis for $\mathcal{W}$, we reinterpret these vectors as columns, yielding the basis

$$X_1 = \begin{bmatrix} 1 \\ 0 \\ 0 \\ \frac{3}{2} \\ \frac{3}{2} \end{bmatrix} \quad X_2 = \begin{bmatrix} 0 \\ 1 \\ 0 \\ -\frac{1}{4} \\ -\frac{3}{4} \end{bmatrix} \quad X_3 = \begin{bmatrix} 0 \\ 0 \\ 1 \\ 1 \\ 1 \end{bmatrix}$$

The dimension of the subspace is 3. ◄

We can use Example 3 to explain an important theorem. From the reduced form of A, we see that the first three columns of A are pivot columns. Thus, Theorem 1 from Section 2.1 says that these columns are independent and the last two columns are linear combinations of them. This means that the space spanned by the *columns* of A is three-dimensional. This space was called the column space in Section 1.5. The row space of A is also three-dimensional, since there are only three non-zero rows in R.

It is true in general that the row space and the column space have the same dimension. To prove this, note that it follows from Theorem 1 in Section 2.1 that the pivot columns form a basis for the column space. Hence the dimension of the column space is the number of pivots. This, of course, is the same as the number of non-zero rows in the reduced form, which is the dimension of the *row space*. Thus, our comments can be summarized in the following result. This result is quite remarkable in that there is no a priori reason to expect any relation between the dimensions of the row and column spaces.

Theorem 1. *For any matrix A, the row space and the column space have the same dimension. This dimension is the rank of the matrix.* ■

In some linear algebra texts, the dimension of the row space for a matrix A is called the "row rank of A" and the dimension of the column space is called the "column rank of A." For this reason, Theorem 1 is often stated as follows: "The row rank of A equals the column rank of A."

There is also a relationship between the rank of an $m \times n$ matrix A and the "size" of the solution set to the system $AX = B$. According to the translation theorem, the general solution to $AX = B$ (assuming that the system is consistent) is obtained by translating the nullspace by any particular solution. Since the nullspace is a subspace (hence a vector space), it has a dimension. We consider the dimension of the nullspace as a measure of the size of the solution space. Finding the dimension of the nullspace is not hard.

▶ EXAMPLE 4:　Find the dimension of the nullspace of the matrix A from Example 3.

Solution: The nullspace is the solution set to $AX = 0$. The reduced form of the augmented matrix for this system is

$$
\begin{bmatrix}
1 & 0 & 0 & \frac{3}{2} & \frac{3}{2} & 0 \\
0 & 1 & 0 & -\frac{1}{4} & -\frac{3}{4} & 0 \\
0 & 0 & 1 & 1 & 1 & 0 \\
0 & 0 & 0 & 0 & 0 & 0 \\
0 & 0 & 0 & 0 & 0 & 0
\end{bmatrix}
$$

The general solution is

$$
\begin{bmatrix} x_1 \\ x_2 \\ x_3 \\ x_4 \\ x_5 \end{bmatrix} = x_4 \begin{bmatrix} -\frac{3}{2} \\ \frac{1}{4} \\ -1 \\ 1 \\ 0 \end{bmatrix} + x_5 \begin{bmatrix} -\frac{3}{2} \\ \frac{3}{4} \\ -1 \\ 0 \\ 1 \end{bmatrix} \tag{1}
$$

where x_4 and x_5 are arbitrary. The two column vectors on the right side of this equality span the nullspace. They are independent because each has a zero in a position where the other has a one. Thus, the dimension of the nullspace is 2.   ◄

The nullspace had dimension 2 because there are two free variables in the system $AX = 0$. In general, there will be one spanning vector for each free variable. We will prove shortly that the spanning vectors are independent. Hence, the dimension of the nullspace is just the number of free variables. This makes sense. We said that the dimension of a set should be the smallest number of variables necessary to describe the set. Every assignment of values to the free variables yields a solution to the system. Thus, the number of free variables *should* be the dimension of the nullspace.

It follows that among matrices of the same size, there is a complementary relationship between the rank of the matrix and the dimension of the nullspace. The matrix A above has rank 3 because there are three pivot variables. The nullspace has dimension 2 because there are two free variables. The sum of the rank and the dimension of the nullspace must equal the total number of variables since a variable is either free or pivot; hence $2 + 3 = 5$, which is the total number of variables. If A had had rank 2, then the nullspace would have had dimension 3 since the total number of variables would not have changed. In general we have the following:

Rank-Nullity Theorem.   *For any matrix A, the rank of A plus the dimension of the nullspace of A is the total number of columns of A.*   ■

If we recall that the dimension of the nullspace measures the size of the solution set, we see that the smaller the rank, the larger the solution set. This too makes sense.

The rank measures the number of independent equations. We certainly expect that fewer equations put fewer constraints on the solutions, allowing for the possibility of more solutions.

To finish our discussion, we still must prove the independence of the spanning vectors.

Proposition 2.   *The spanning vectors for any system $AX = 0$ are independent.*   ■

The general ideal of the proof is illustrated by Example 4, where we noted that the spanning vectors were independent because each had a one in a position where the other had a zero. The reason for this is not difficult to understand. From formula (1), the first spanning vector is the solution obtained by setting $x_4 = 1$ and $x_5 = 0$. It follows that this vector *must* have a one in the fourth position and a zero in the fifth position. Similarly, the second spanning vector is the solution obtained by setting $x_4 = 0$ and $x_5 = 1$, hence *must* have a zero in the fourth position and a one in the fifth position.

In the general case, our solution will be written

$$\begin{bmatrix} x_1 \\ x_2 \\ x_3 \\ \vdots \\ x_n \end{bmatrix} = x_{i_1} X_1 + x_{i_2} X_2 + \cdots + x_{i_k} X_k$$

where the x_{i_j} are the free variables and the X_i are the spanning vectors. Looking at the right side of this equality, we see that the vector X_1 will be obtained by setting $x_{i_1} = 1$ and all other $x_{i_j} = 0$. This means that X_1 will have a one in the i_1 position and a zero in all of the other i_j positions, just as in our example. Similarly, X_2 is produced by setting $x_{i_2} = 1$ and all the other free variables equal to zero. Hence, X_2 has a one in the i_2 position and zeros in all of the other "free" positions. In general, X_j will have similar behavior with respect to the i_j position, proving independence.

Summary: This section introduced a large number of important ideas. It is worthwhile to summarize the main points:

(1) The row space is the span of the rows of a matrix. A basis for the row space may be found by reducing the matrix to echelon form and using the non-zero rows. The dimension of the row space is the rank of the matrix.

(2) The dimension of the column space is also equal to the rank.

(3) The dimension of the nullspace is the number of free variables in the system $AX = 0$. The spanning vectors for the solution set to this system form a basis for the nullspace.

EXERCISES

(1) For the matrix A, decide whether the given row vectors X and Y belong to the row space.

$$A = \begin{bmatrix} 2 & 1 & 3 & 1 \\ 1 & 1 & 3 & 0 \\ 0 & 1 & 2 & 1 \\ 3 & 3 & 8 & 2 \end{bmatrix} \quad X = \begin{bmatrix} 4 & 1 & 2 & 5 \end{bmatrix} \quad Y = \begin{bmatrix} 1 & 2 & 3 & 4 \end{bmatrix}$$

(2) For the matrix A in Exercise 1,

    (a) Compute the *reduced* echelon form of A. Call it R.

    (b) Use your answer to find a basis for the row space of A.

    (c) Express each row of A as a linear combination of these basis elements. *Note*: This is not as tedious as it sounds: each basis element has a 1 in a position where the others have a 0. (This is why we had you compute the *reduced* form of A.)

    (d) Do the columns of R span the column space of A? Explain.

(3) Decide whether the vector $[7, 5, 4, 13]$ belongs to the row space of A^t where A is as in Exercise 1.

(4) For A as in Exercise 1, check (by row reduction) that A and A^t have the same rank.

(5) Is it possible to find a matrix A such that A and A^t have different ranks? If you feel that it is possible, find an example. If you feel that it is impossible, prove that you are right.

(6) Let $\mathcal{W}$ be the subspace of $\mathbb{R}^5$ spanned by the vectors in Example 3 of this section. Find a basis for $\mathcal{W}$ by using the method of Example 3 in Section 2.2.

(7) Use the non-zero rows theorem to find a basis for the subspace $\mathcal{W}$ of $\mathbb{R}^4$ spanned by the following vectors. What is the dimension of $\mathcal{W}$? Express each of these vectors as a linear combination of the basis elements. (This is easy if you reduce all the way to reduced form to find the basis.)

$$[2, 3, 1, 2]^t \quad [5, 2, 1, 2]^t \quad [1, -4, -1, -2]^t \quad [11, 0, 1, 2]^t$$

(8) Use the method of Section 2.2, Example 3, to find a basis for $\mathcal{W}$ from Exercise 7.

(9) Use row reduction to find a basis for the subspace of $M(3, 1)$ spanned by the matrices

$$\begin{bmatrix} 1 \\ 2 \\ 3 \end{bmatrix} \quad \begin{bmatrix} 2 \\ 1 \\ -1 \end{bmatrix} \quad \begin{bmatrix} 0 \\ 3 \\ 7 \end{bmatrix}$$

(10) Let

$$A = \begin{bmatrix} 1 & 1 & 3 & 2 \\ 2 & 2 & 6 & 4 \\ 10 & 2 & 14 & 20 \\ 2\sqrt{2} & -\sqrt{2} & 0 & 4\sqrt{2} \\ \pi & e & \pi + 2e & 2\pi \\ \sqrt{2} & \sqrt{3} & \sqrt{2} + 2\sqrt{3} & 2\sqrt{2} \\ \ln 5 & 6 & \ln 5 + 12 & 2\ln 5 \\ -7 & 4 & 1 & -14 \\ 17 & -24 & -31 & 34 \\ 2 & 2 & 6 & 4 \end{bmatrix}$$

(a) What is the rank of A? Explain.
(b) What is the dimension of the nullspace of A?
(c) Will the equation $AX = B$ be solvable for all B?
(d) Will the equation $AX = B$ have at most one solution? Explain.
(e) Find *two* different bases for the row space of A. Use some theorems from linear algebra to justify your answer.
(f) I claim that the nullspace of A is the solution set of the following system of equations. Explain.

$$x + y + 3z + 2w = 0$$
$$-7x + 4y + z - 14w = 0$$

(11) Consider the accompanying matrix A. Without doing any row reduction at all, answer the following. Be sure to justify your answers.

(a) What is the rank of A?
(b) Find two different bases for the column space of A.
(c) Show that the vectors $X_1 = [-\pi, 0, 0, 0, 0, 1]^t$ and $X_2 = [-2, 1, 0, 0, 0, 0]^t$ belong to the nullspace of A. Do they span the nullspace?
(d) Find a vector T such that $AT = [1, 1, 2, 3]^t$.
(e) Find an infinite number of vectors X such that $AX = [1, 1, 2, 3]^t$. Remember, no row reduction allowed!

$$A = \begin{bmatrix} 1 & 2 & e & \ln 4 & 7 & \pi \\ 1 & 2 & \pi & \sqrt{2} & 1 & \pi \\ 2 & 4 & e + \pi & \sqrt{2} + \ln 4 & 8 & 2\pi \\ 3 & 6 & 2e + \pi & 2\ln 4 + \sqrt{2} & 15 & 3\pi \end{bmatrix}$$

(12) Suppose that A is a 4×5 matrix with rank 2. Suppose that all the following vectors satisfy $AX = \mathbf{0}$. Do these vectors span the nullspace of A?

$$X_1 = [1, 2, 3, 2, 1]^t \quad X_2 = [1, -1, 0, 1, 2]^t \quad X_3 = [2, 1, 4, 4, 4]^t$$

(13) An $m \times n$ matrix A has a d-dimensional nullspace. What is the dimension of the nullspace of A^t?

(14) Suppose that A is $n \times n$ and has rank n. What is its row space? What is its nullspace? What is its column space?

(15) Let A be an $m \times n$ matrix. Prove that the rank of A is less than or equal to both m and n.

(16) Let A be an $m \times n$ matrix. Suppose that A has rank m. Prove that the rows of A form a basis of the row space.

(17) Let A be an $m \times n$ matrix. Suppose that A has rank m. Prove that $m \leq n$.

(18) The column space of a matrix is the space spanned by the columns of A. Does row reduction change column spaces? *i.e.* If B is row equivalent to A, does B have the same column space as A? Explain. [*Hint*: Try some examples.]

 ## ON LINE

(1) In MATLAB, construct a random 3×5 matrix M. (Use the command "M= rand(3,5)"). What do you expect the rank of M to be? Check your guess using MATLAB. Is it conceivable that the rank could have turned out otherwise? Why is this unlikely?

(2) Let M be as in the preceding exercise. Let s and t be random 1×3 matrices (constructed using the "rand" command). Set

```
>> M(4,:)=s(1)*M(1,:)+s(2)*M(2,:)+s(3)*M(3,:)
```

Similarly, let

```
>> M(5,:)=t(1)*M(1,:)+t(2)*M(2,:)+t(3)*M(3,:)
```

Note that M is now 5×5. What is the rank of M? Check this using "rref". What is the maximal number of linearly independent *columns* in M?

(3) For the preceding 5×5 matrix M, find a set of columns of M that forms a basis for the column space. Express the other columns of M as linear combinations of these columns. Use the technique of Example 3 in Section 2.1.

(4) Are the coefficients you computed in the previous exercise exact? To check this, issue the command "format long" and then form the linear combinations of the basis columns using the coefficients you computed. Do you reproduce the columns of *M*? You can return to the regular format with the command "format short".

(5) For M as above, find a basis for the column space of M by reducing the transpose of M (this is denoted "M'" in MATLAB) and then using the nonzero rows theorem. Try to express each of the basis columns you found in Exercise 3 as linear combinations of these columns. Again, you may need to include a tolerance in "rref".

(6) According to the comments in the text, there is an inverse relationship between the dimension of the nullspace of a matrix and its rank. Demonstrate this by creating (as in Exercise 1) four random 4×4 matrices with rank 1, 2, 3, and 4, respectively. For each of your matrices use the "null" command to find a basis for the nullspace. (See On Line, Section 1.5, Exercise 3.)

2 CHAPTER SUMMARY

The key concept from this chapter is *dimension* (Section 2.2). The dimension of a vector space W is the smallest number of elements that it takes to span W. Given a spanning set for a space, we can produce an independent spanning set by deleting dependent elements. The *test for independence* provides a systematic way of doing this (Example 3 in Section 2.2). An independent spanning set is called a *basis*. Bases are important because they span the space as efficiently as possible. Bases also allow us to find the dimension of a vector space: the number of elements in any basis is the dimension of the space (the *dimension theorem*, Section 2.2).

The dimension of a space is one of the most fundamental parameters in determining the properties of the space: in an *n*-dimensional space, there can exist no more than *n* independent elements (Theorem 1, Section 2.2); in an *n*-dimensional space, *n* independent elements will always span and *n* elements that span must be independent (Theorem 2, Section 2.2).

Dimension is also a powerful tool for studying matrices. The *row space* of a matrix is the span of its rows and the *column space* is the span of its columns. We may find bases for both of these spaces by row reducing the given matrix. The non-zero rows of the row-reduced form form a basis for the row space of the matrix (the *non-zero rows theorem* in Section 2.3), and the pivot columns of the original matrix form a basis for the column space (Theorem 1, Section 2.1). Since the number of such non-zero rows is the same as the number of pivot variables, we see that the dimension of the row space is the same as the dimension of the column space that, in turn, is the rank of the matrix (Theorem 1, Section 2.3).

Another important space that we associate with a matrix A is its *nullspace*, which is the set of vectors X such that $AX = \mathbf{0}$. The nullspace measures the size of the solution set to the equation $AX = B$ in that the general solution to this equation is a translate of the nullspace. The spanning vectors for the system $AX = \mathbf{0}$ form a basis for the nullspace, and the dimension of the nullspace is the number of free variables found in solving the system $AX = \mathbf{0}$ (Proposition 2, Section 2.3). The dimension of the nullspace is also $n - r$ where r is the rank of A and n is the number of columns of A (the *rank-nullity theorem*, Section 2.3). Hence, for matrices of the same size, the larger the rank, the smaller the dimension of the nullspace.

In Section 2.2.1, we applied the theory of dimension to the study of differential equations. The basic idea is that since the solution set to an nth order homogeneous differential equation is n-dimensional (Theorem 2, Section 2.2.1), we can find the general solution if we can find n linearly independent solutions to the equation. The general solution will then be the set of all linear combinations of these n solutions since n independent elements will span an n-dimensional space.

CHAPTER 3
LINEAR TRANSFORMATIONS

3.1 LINEAR TRANSFORMATIONS

Suppose that the triangle pictured in Figure 1 appears on a computer screen and we wish to rotate it about the origin by $\frac{\pi}{6}$ radian counterclockwise. How should we instruct the computer to plot the new image of the triangle?

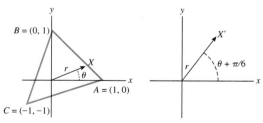

Figure 1

135

To answer this question, suppose that $X = [x, y]'$ is some point on the triangle and $X' = [x', y']'$ is X, rotated by $\frac{\pi}{6}$ radian. Let r and θ be the polar coordinates of X so that

$$x = r \cos \theta$$
$$y = r \sin \theta$$

Then, the polar coordinates of X' will be r and $\theta + \frac{\pi}{6}$. Hence

$$x' = r \cos(\theta + \frac{\pi}{6}) = r \cos \theta \cos \frac{\pi}{6} - r \sin \theta \sin \frac{\pi}{6} = x \frac{\sqrt{3}}{2} - y \frac{1}{2}$$

$$y' = r \sin(\theta + \frac{\pi}{6}) = r \cos \theta \sin \frac{\pi}{6} + r \sin \theta \cos \frac{\pi}{6} = x \frac{1}{2} + y \frac{\sqrt{3}}{2}$$

(We used the angle addition formulas for the sine and cosine functions.)

Now comes the main point. The preceding formulas may be written as a matrix equality as

$$\begin{bmatrix} x' \\ y' \end{bmatrix} = x \begin{bmatrix} \frac{\sqrt{3}}{2} \\ \frac{1}{2} \end{bmatrix} + y \begin{bmatrix} -\frac{1}{2} \\ \frac{\sqrt{3}}{2} \end{bmatrix} = \begin{bmatrix} \frac{\sqrt{3}}{2} & -\frac{1}{2} \\ \frac{1}{2} & \frac{\sqrt{3}}{2} \end{bmatrix} \begin{bmatrix} x \\ y \end{bmatrix}$$

Multiplication of points by this 2×2 matrix rotates them by $\frac{\pi}{6}$ radian. Thus, we solve our problem by multiplying each vertex of the triangle by this matrix. We obtain:

$$A' = \begin{bmatrix} \frac{\sqrt{3}}{2} & -\frac{1}{2} \\ \frac{1}{2} & \frac{\sqrt{3}}{2} \end{bmatrix} \begin{bmatrix} 1 \\ 0 \end{bmatrix} = \begin{bmatrix} \frac{\sqrt{3}}{2} \\ \frac{1}{2} \end{bmatrix} \approx \begin{bmatrix} 0.87 \\ 0.5 \end{bmatrix}$$

$$B' = \begin{bmatrix} \frac{\sqrt{3}}{2} & -\frac{1}{2} \\ \frac{1}{2} & \frac{\sqrt{3}}{2} \end{bmatrix} \begin{bmatrix} 0 \\ 1 \end{bmatrix} = \begin{bmatrix} -\frac{1}{2} \\ \frac{\sqrt{3}}{2} \end{bmatrix} \approx \begin{bmatrix} -0.5 \\ 0.87 \end{bmatrix}$$

$$C' = \begin{bmatrix} \frac{\sqrt{3}}{2} & -\frac{1}{2} \\ \frac{1}{2} & \frac{\sqrt{3}}{2} \end{bmatrix} \begin{bmatrix} -1 \\ -1 \end{bmatrix} = \begin{bmatrix} -\frac{\sqrt{3}}{2} + \frac{1}{2} \\ -\frac{1}{2} - \frac{\sqrt{3}}{2} \end{bmatrix} \approx \begin{bmatrix} -0.37 \\ -1.37 \end{bmatrix}$$

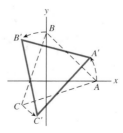

Figure 2

We then instruct the computer to plot the triangle with these vertices. Figure 2 shows our new image.

In general, multiplication by the matrix

$$R_\psi = \begin{bmatrix} \cos \psi & -\sin \psi \\ \sin \psi & \cos \psi \end{bmatrix}$$

rotates points by ψ radian counterclockwise about the origin.

Rotation is an example of a transformation. In general, a **transformation** is a well-defined process for taking elements from one set (the **domain**) and using them to produce elements of another set (the **target space**). "Well-defined" here means that each element of the domain is transformed onto only one element of the target space. For rotation, the domain and the target space are $\mathbb{R}^2$.

Linear transformations are used extensively in computer-aided design.

Generally, transformations will have names, (e.g., T). If the transformation is called T, then $T(X)$ denotes the element produced from X by the transformation. Thus, if T denotes rotation by $\frac{\pi}{6}$ radian, then, for all X in $\mathbb{R}^2$,

$$T(X) = R_{\frac{\pi}{6}} X$$

Since rotation is describable in terms of matrix multiplication, we say that it is a matrix transformation. In general, a transformation T with domain $\mathbb{R}^n$ and target space $\mathbb{R}^m$ is a **matrix transformation** if there is an $m \times n$ matrix A such that $T(X) = AX$ for all X in $\mathbb{R}^n$.

Notice that to rotate our triangle, it was not necessary to rotate each point on the triangle—only the vertices need to be rotated. This works because rotation transforms line segments onto line segments. The same is true for any matrix transformation. To understand this, let X and Y be points in $\mathbb{R}^n$. As t varies over all real numbers,

$$tX + (1 - t)Y = Y + t(X - Y)$$

describes a line that passes through both X and Y. (We hit Y at $t = 0$ and X at $t = 1$.) The segment between X and Y (Figure 3) is obtained by $0 \le t \le 1$.

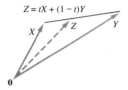

The segment between X and Y

Figure 3

If A is some $m \times n$ matrix, then, from the linearity properties of matrix multiplication,

$$A(tX + (1 - t)Y) = tAX + (1 - t)AY$$

As t varies between 0 and 1, the term on the right exactly describes the line segment between AX and AY. Thus multiplication by A transforms the segment from X to Y onto the segment from AX to AY. We state this as a proposition:

Proposition 1.   *Matrix transformations transform line segments onto line segments.*   ▦

Suppose that S is a subset of the domain of some transformation T. Then $T(S)$ will denote the set obtained by transforming each point of S by T. This set is referred to as the **image** of S under T. Thus, our rotated triangle in Figure 2 is the image of the triangle under multiplication by $R_{\frac{\pi}{6}}$. Proposition 1 is very useful in computing images.

► EXAMPLE 1:   Let

$$M = \begin{bmatrix} 1 & 0 \\ 1 & 1 \end{bmatrix}$$

Let S be the square in $\mathbb{R}^2$ with vertices $A = [0, 0]'$, $B = [1, 0]'$, $C = [1, 1]'$, and $D = [0, 1]'$. What is the image of S under multiplication by M?

Solution: It is easily computed that multiplication by M transforms the four vertices of S onto the points $A' = [0, 0]'$, $B' = [1, 1]'$, $C' = [1, 2]'$, and $D' = [0, 1]'$, respectively. These points form the vertices of a parallelogram as shown in Figure 4. Multiplication by M will transform the square onto the parallelogram, because segments transform onto segments. We picture this transformation as "tilting" the square. Such a transformation is called a "shear."

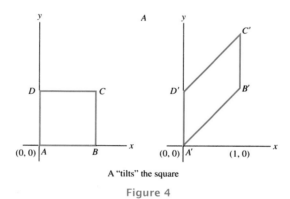

A "tilts" the square

Figure 4

◄

Example 1 demonstrates the importance of the linearity properties of matrix multiplication. It is both interesting and important that among transformations from $\mathbb{R}^n$ into $\mathbb{R}^m$, matrix transformations are the *only* transformations that exhibit these properties. First, however, let us make a general definition:

Definition.   *Let V and W be vector spaces. Let $T : V \rightarrow W$ be a transformation. We say that T is a linear transformation if it satisfies two properties:*

(1) *For all X and Y in V, $T(X + Y) = T(X) + T(Y)$. (additivity)*

(2) *For all X in V and all scalars c, $T(cX) = cT(X)$. (scalar)*   ■

Matrix transformations are linear transformations since the properties just defined are simply restatements of the linearity properties for matrix multiplication. The following theorem states that matrix transformations are the *only* transformations from $\mathbb{R}^n$ into $\mathbb{R}^m$ with these properties. For an example of a non-linear transformation, see Exercise 18.

Matrix Representation Theorem.   Let $T : \mathbb{R}^n \to \mathbb{R}^m$ be any linear transformation. Then there is a unique matrix A such that $T(X) = AX$ for all $X \in \mathbb{R}^n$.

Proof: If $X = [x_1, x_2, \ldots, x_n]^t$, then we may write

$$X = x_1 I_1 + x_2 I_2 + \ldots + x_n I_n$$

where I_j is the jth standard basis element for $\mathbb{R}^n$.

Applying the additivity and scalar properties of T, we see:

$$T(X) = x_1 T(I_1) + x_2 T(I_2) + \ldots + x_n T(I_n)$$

The elements $T(I_j)$ are column vectors. From the definition of matrix multiplication, the term on the right is exactly AX where $A = [T(I_1), T(I_2), \ldots, T(I_n)]$. This proves the existence of A.

To prove uniqueness, suppose that B is some other matrix that defines the same transformation as A. Then, for all j

$$BI_j = AI_j$$

However, for any matrix C, CI_j is just the jth column of C. Thus, each column of A equals each column of B, proving that $A = B$.   ■

▶ EXAMPLE 2:   Let $T : \mathbb{R}^2 \to \mathbb{R}^3$ be defined by $T([x, y]^t) = [2x + 3y, x - y, x]^t$. Use the definition of linearity to show that T is linear, and find a matrix A that represents T.

Solution: To prove linearity, we must prove the additivity and scalar properties. For additivity let $X = [x_1, y_1]^t$ and $Y = [x_2, y_2]^t$ be two points in $\mathbb{R}^2$. Then

$$T(X) = [2x_1 + 3y_1, x_1 - y_1, x_1]^t$$
$$T(Y) = [2x_2 + 3y_2, x_2 - y_2, x_2]^t$$

Hence

$$T(X) + T(Y) = [2x_1 + 3y_1, x_1 - y_1, x_1]^t + [2x_2 + 3y_2, x_2 - y_2, x_2]^t$$

$$= [2(x_1 + x_2) + 3(y_1 + y_2), (x_1 + x_2) - (y_1 + y_2), (x_1 + x_2)]^t$$

On the other hand,

$$T(X + Y) = T([x_1 + x_2, y_1 + y_2]^t)$$
$$= [2(x_1 + x_2) + 3(y_1 + y_2), (x_1 + x_2) - (y_1 + y_2), (x_1 + x_2)]^t$$

These two expression are clearly equal.

For the scalar property, let $X = [x, y]^t$. We compute

$$T(c[x, y]^t) = T([cx, cy]^t) = [2cx + 3cy, cx - cy, cx]^t$$
$$= c[2x + 3y, x - y, x]^t = cT(X)$$

To find the matrix that describes T, we write

$$T([x, y]^t) = \begin{bmatrix} 2x + 3y \\ x - y \\ x \end{bmatrix} = x \begin{bmatrix} 2 \\ 1 \\ 1 \end{bmatrix} + y \begin{bmatrix} 3 \\ -1 \\ 0 \end{bmatrix} = \begin{bmatrix} 2 & 3 \\ 1 & -1 \\ 1 & 0 \end{bmatrix} \begin{bmatrix} x \\ y \end{bmatrix}$$

The 3×2 matrix on the right is A.   ◄

The linearity properties are often useful in piecing together incomplete information about linear transformations.

▶ EXAMPLE 3:   Let A be a 2×3 matrix such that multiplication by A transforms $X_1 = [4, 7, 3]^t$, onto $[1, 3]^t$, $X_2 = [1, 1, 0]^t$ onto $[1, 4]^t$, and $X_3 = [1, 0, 0]^t$ onto $[1, 1]^t$. Determine what multiplication by A transforms $[1, 2, 1]^t$ onto.

Solution: The vectors X_i are easily seen to form an independent set and hence a basis for $\mathbb{R}^3$. We may therefore write $[1, 2, 1]^t$ as a linear combination of them. Specifically, the vector equation

$$[1, 2, 1]^t = x[4, 7, 3]^t + y[1, 1, 0]^t + z[1, 0, 0]^t$$

yields a system of three equations in x, y, and z which we solve, finding that $x = \frac{1}{3}$, $y = -\frac{1}{3}$, and $z = 0$. Hence

$$[1, 2, 1]^t = \frac{1}{3}[4, 7, 3]^t - \frac{1}{3}[1, 1, 0]^t$$

Multiplying both sides of this equality by A and using the linearity properties, we see that

$$A[1, 2, 1]^t = A(\frac{1}{3}[4, 7, 3]^t - \frac{1}{3}[1, 1, 0]^t) = \frac{1}{3}A[4, 7, 3]^t - \frac{1}{3}A[1, 1, 0]^t$$

$$= \frac{1}{3}[1, 3]' - \frac{1}{3}[1, 4]' = [0, -\frac{1}{3}]'.$$

◀

What allowed us to answer the question asked in Example 3 was that the X_i formed a basis for $\mathbb{R}^3$. It is a general principle that *a linear transformation is determined by how it transforms a basis.* Specifically, suppose that V is a vector space and $\{X_1, X_2, \ldots, X_n\}$ is a basis for V. Suppose also that $T : V \to W$ is a linear transformation of V into another vector space W. Then, we can compute $T(X)$ for any X in V, provided only that we are given $T(X_i)$ for each basis element X_i. This is because any X in V may be written as a linear combination of the X_i:

$$X = x_1 X_1 + x_2 X_2 + \cdots + x_n X_n$$

Hence, $T(X)$ equals

$$T(x_1 X_1 + x_2 X_2 + \cdots + x_n X_n) = x_1 T(X_1) + x_2 T(X_2) + \cdots + x_n T(X_n)$$

EXERCISES

(1) Indicate on a graph the image of the square in Example 1 above under multiplication by each of the following matrices.

(a) $A = \begin{bmatrix} -2 & 0 \\ 0 & -3 \end{bmatrix}$

(b) $A = \begin{bmatrix} 1 & 3 \\ 0 & 1 \end{bmatrix}$

(c) $A = \begin{bmatrix} 1 & 0 \\ 0 & 0 \end{bmatrix}$

(d) $A = \begin{bmatrix} 1 & 0 \\ 1 & 0 \end{bmatrix}$

(2) Repeat Exercise 1 for the parallelogram with vertices $[1, 1]'$, $[1, 2]'$, $[2, 2]'$, and $[2, 3]'$. Sketch both the parallelogram and its image.

(3) Let

$$A = \begin{bmatrix} 2 & 0 \\ 0 & 3 \end{bmatrix}$$

Show that the image of the circle $x^2 + y^2 = 1$ under multiplication by A is the ellipse pictured in Figure 5. [*Hint:* It is easily computed that $A([x, y]') = [2x, 3y]'$. Let $u = 2x$ and $v = 3y$. If x and y lie on the circle, what equation must u and v satisfy?]

(4) Indicate on a graph the image of all of $\mathbb{R}^2$ under multiplication by the matrix from Exercise 1d above. How, geometrically, would you describe the image of the general point $[x, y]'$?

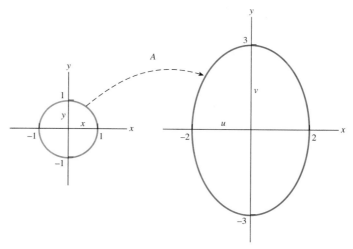

Figure 5

On the same graph, draw the circle $x^2 + y^2 = 1$ and indicate what you would expect its image to be. A transformation is said to be "two-to-one" if every point in the image is the image of exactly two points in the domain. Is this transformation two-to-one on the circle? Explain.

(5) Find a matrix A such that multiplication by A transforms the square in Figure 4 onto the parallelogram indicated in Figure 6.

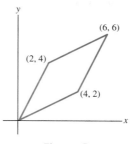

Figure 6

(6) Find a matrix B such that multiplication by B transforms the parallelogram in Figure 6 onto the square in Figure 4.

(7) Show (using a diagram) that the transformation that reflects points in $\mathbb{R}^2$ in the x axis is linear. Find a matrix that describes this transformation.

(8) Show (using a diagram) that the transformation that reflects points in $\mathbb{R}^2$ in the y axis is linear. Find a matrix that describes this transformation.

(9) Show (using a diagram) that the transformation that reflects points in $\mathbb{R}^2$ about the line $y = x$ is linear. Find a matrix that describes this transformation.

(10) Describe geometrically the effect of the transformation of $\mathbb{R}^3$ into $\mathbb{R}^3$ defined by multiplication by the following matrix.

$$R_z(\theta) = \begin{bmatrix} \cos\theta & -\sin\theta & 0 \\ \sin\theta & \cos\theta & 0 \\ 0 & 0 & 1 \end{bmatrix}$$

[*Hint*: How does this transformation transform points of the form $[x, y, 0]^t$? $[0, 0, z]^t$?]

(11) Describe geometrically the effect of the transformation of $\mathbb{R}^3$ into $\mathbb{R}^3$ defined by multiplication by the following matrix.

$$R_x(\theta) = \begin{bmatrix} 1 & 0 & 0 \\ 0 & \cos\theta & -\sin\theta \\ 0 & \sin\theta & \cos\theta \end{bmatrix}$$

[*Hint*: How does this transformation transform points of the form $[0, y, z]^t$? $[x, 0, 0]^t$?]

(12) Describe geometrically the effect of the transformation of $\mathbb{R}^3$ into $\mathbb{R}^3$ defined by multiplication by the following matrix.

$$R_y(\theta) = \begin{bmatrix} \cos\theta & 0 & -\sin\theta \\ 0 & 1 & 0 \\ \sin\theta & 0 & \cos\theta \end{bmatrix}$$

[*Hint*: How does this transformation transform points of the form $[x, 0, z]^t$? $[0, y, 0]^t$?]

(13) Consider the points

$$X_1 = [1, 1]^t \ \ X_2 = [2, 2]^t$$
$$Y_1 = [4, 5]^t \ \ Y_2 = [5, 6]^t$$

Is it possible to find a 2×2 matrix A for which multiplication by A transforms X_1 into Y_1 and X_2 into Y_2? [*Hint*: How are X_1 and X_2 related?]

(14) Consider the points

$$X_1 = [1, 1]^t \quad X_2 = [2, 3]^t \quad X_3 = [3, 4]^t$$
$$Y_1 = [1, 0]^t \quad Y_2 = [0, 1]^t \quad Y_3 = [2, 2]^t$$

Is it possible to find a 2×2 matrix A for which multiplication by A transforms $X_1 \rightarrow Y_1$, $X_2 \rightarrow Y_2$, and $X_3 \rightarrow Y_3$?

(15) Let A be a 2×3 matrix for which multiplication by A transforms $[2, 1, 1]^t$ onto $[1, 1]^t$ and $[1, 1, 1]^t$ onto $[1, -2]^t$.

(a) Determine, if possible, what multiplication by A transforms $[0, 1, 1]^t$ onto. [*Hint*: How does $[0, 1, 1]^t$ relate to $[1, 1, 1]^t$ and $[2, 1, 1]^t$?]

(b) Suppose that it is also true that multiplication by A transforms $[0, 0, 1]^t$ onto $[3, -5]^t$. Determine what A transforms $[2, 2, 3]^t$ onto.

(c) Find A. [*Hint*: The columns of A are AI_1, AI_2 and AI_3.]

(16) Suppose that A is a 3×2 matrix such that multiplication by A transforms $[2, -1]^t$ onto Y_1 and $[2, 3]^t$ onto Y_2. What does multiplication by A transform $[3, -5]$ onto? Can we, in principle, determine what multiplication by A transforms any given vector onto? Explain.

(17) Although it seems silly, it is possible to do elementary row operations on $n \times 1$ matrices. Every such operation defines a transformation of $\mathbb{R}^n$ into $\mathbb{R}^n$. For example, if we define a transformation of $\mathbb{R}^3$ into $\mathbb{R}^3$ by "add twice row 1 to row 3," this transformation transforms

$$
\begin{bmatrix} x_1 \\ x_2 \\ x_3 \end{bmatrix} \text{ to } \begin{bmatrix} x_1 \\ x_2 \\ x_3 + 2x_1 \end{bmatrix} = \begin{bmatrix} 1 & 0 & 0 \\ 0 & 1 & 0 \\ 2 & 0 & 1 \end{bmatrix} \begin{bmatrix} x_1 \\ x_2 \\ x_3 \end{bmatrix}
$$

Since this transformation is described by a matrix, we see that our elementary row operation defines a *linear* transformation. A transformation defined by a single elementary row operation is called an **elementary transformation** and the matrix that describes such a transformation is called an **elementary matrix**.

Find matrices that describe the following elementary row operations on $\mathbb{R}^n$ for the given value of n.

(a) Add twice row 3 to row 2 in $\mathbb{R}^4$.

(b) Multiply row 2 by 17 in $\mathbb{R}^3$.

(c) Interchange rows 1 and 2 in $\mathbb{R}^4$.

(18) Let $T : \mathbb{R}^2 \rightarrow \mathbb{R}^2$ be defined by

$$
T([x, y]^t) = [x, y(1 + x^2)]^t
$$

(a) Show that T is not a linear transformation by finding two specific points (reader's choice) X and Y such that $T(X + Y) \neq T(X) + T(Y)$.

(b) On a graph, draw the lines $x = n$ and $y = m$ for n and m having the values 0, 1, 2. (There will be six lines.) On a separate graph, draw the image of these lines under T. The purpose here is to understand how T transforms squares. You will notice that the squares become increasingly "warped" as you move away from the origin.

(19) Prove that each of the transformations in (a)–(e) is linear, and find a matrix A such that $T(X) = AX$.

(a) $T : \mathbb{R}^3 \to \mathbb{R}^2$, $T([x, y, z]^t) = [2x + 3y - 7z, 0]^t$
(b) $T : \mathbb{R}^2 \to \mathbb{R}^3$, $T([x, y]^t) = [x + y, x - y, x + 2y]^t$
(c) $T : \mathbb{R}^5 \to \mathbb{R}^3$, $T([x_1, x_2, x_3, x_4, x_5]^t) = [x_1, x_2, x_3]^t$

(20) What is the image of all of $\mathbb{R}^3$ under the transformation T in Exercise 19a? The equation $2x + 3y - 7z = 0$ describes a plane in $\mathbb{R}^3$. What is the image of this plane under T? How, geometrically, would you describe the set of $[x, y, z]^t$ such that $T([x, y, z]^t) = [-3, 0]^t$?

(21) What matrix describes rotation of $\mathbb{R}^2$ clockwise by θ radian?

(22) In the text, it was shown that a linear transformation from $\mathbb{R}^n$ to $\mathbb{R}^m$ transforms line segments onto line segments. Prove that such a transformation will in fact transform midpoints onto midpoints. [The midpoint of the segment from X to Y is $(X + Y)/2$.]

(23) Integration may be thought of as a transformation that transforms functions into numbers. Specifically, if f is a continuous function on the closed interval $[-1, 1]$, then we define

$$T(f) = \int_{-1}^{1} f(x)dx$$

Thus, for example,

$$T(x^3) = \int_{-1}^{1} x^3 dx = \frac{x^4}{4} \bigg|_{-1}^{1} = \frac{1}{4}$$

The space $\mathcal{V}$ of all continuous functions on $[-1, 1]$ is a vector space; if we add two continuous functions, we get a continuous function. If we multiply a continuous function by a scalar, the result is still continuous. Certainly, addition of functions satisfies all the algebraic properties (a)–(j) from Section 1.2. Thus, T is a transformation from the vector space $\mathcal{V}$ into the vector space $\mathbb{R}$.

(a) Let $f(x) = e^x$ and $g(x) = x^2$ for $-1 < x \leq 1$. Show by direct computation that

$$T(2f + 3g) = 2T(f) + 3T(g)$$

(b) Is T linear? Why?
(c) Let S be the transformation defined by

$$S(f) = \int_{-1}^{1} (f(x))^2 dx$$

Show by explicit computation that $S(2e^x) \neq 2S(e^x)$. Is S linear?

(d) Let U be defined by

$$U(f) = \int_{-1}^{1} xf(x)dx$$

Is U linear?

(24) Differentiation may be thought of as a transformation that transforms functions to functions. Specifically, if f is an infinitely differentiable function on $\mathbb{R}$, then we define

$$D(f) = \frac{df}{dx}$$

Thus, for example,

$$D(x^3) = 3x^2 \text{ and } D(e^{2x}) = 2e^{2x}$$

The space $\mathcal{W}$ of all infinitely differentiable function on $\mathbb{R}$ is a vector space; if we add two such functions, we get an infinitely differentiable function on $\mathbb{R}$. If we multiply such a function by a scalar, the result is still infinitely differentiable. Again, addition of functions satisfies all the algebraic properties (a)–(j) from Section 1.2. Thus, D is a transformation from the vector space $\mathcal{W}$ into itself.

(a) Let $f(x) = e^x$ and $g(x) = x^2$. Show by direct computation that

$$D(2f + 3g) = 2D(f) + 3D(g)$$

(b) Is D linear? Why?

(c) Let S be the transformation defined by

$$S(f) = (f)^2$$

Show by explicit computation that $S(e^x + x^2) \neq S(e^x) + S(x^2)$. Is S linear?

(d) Let U be defined by

$$U(f) = D(xf)$$

Is U linear?

(25) Let $\mathcal{V}$ and $\mathcal{W}$ be vector spaces. Let $T : \mathcal{V} \to \mathcal{W}$ be a linear transformation. Let X_1, X_2, and X_3 be three dependent elements in $\mathcal{V}$. Prove that the elements $T(X_1)$, $T(X_2)$, and $T(X_3)$ are also dependent.

(26) Let $\mathcal{V}$ and $\mathcal{W}$ be vector spaces. Let $T : \mathcal{V} \to \mathcal{W}$ be a linear transformation. Prove that $T(\mathbf{0}) = \mathbf{0}$. [*Note*: You should not assume that T is a matrix transformation. Instead, think about the property that in any vector space $0X = \mathbf{0}$.]

(27) Let $T : \mathcal{V} \to \mathcal{W}$ be a linear transformation between two vector spaces. We define the nullspace of T to be the set of X such that $T(X) = \mathbf{0}$. Show that the nullspace of T is a subspace.

ON LINE

Here we present a very simple-minded program for drawing stick figures in MATLAB. To enter this into MATLAB, select the "File" box and pull down the menu, selecting "New M-File". A new window will open. Type the commands below into this window. When finished, select "File" and then "Save As". Save it as "stick.m". Then select "Close" from the File menu. [*Note:* You might want to save your file to a disc as this program will be used again in Section 3.2.]

To run "stick," simply type "stick" in the MATLAB command window. You should see a blank graph with axes $0 \le x \le 1$ and $0 \le y \le 1$. You place points on the graph by pointing to where you want each one and clicking with the (left) mouse button. Each subsequent point will be connected to the previous one by a line. At the last point of your figure, use the right mouse button instead of the left. This will cause the program to end. (On a Macintosh, you would hold down the "shift" button while clicking the mouse.)

The program saves each of your points, in the order that you placed them, as the columns of a large matrix called FIG.

```
CLA; hold on; axis([0 1 0 1]); Grid on;
FIG=[0;0]; i=1; z=1;
while z==1,
    [x,y,z]=ginput(1);
    FIG(:,i)=[x,y] ';
    plot(FIG(1,:),FIG(2,:));
    i=i+1;
end
```

(1) Use "stick" to draw the first letter of your initials. It is necessary to go over some lines twice to produce some letters (such as B). After exiting stick, save your work in an appropriately named matrix. For example, my initials are RCP. I use stick to draw an R. After exiting stick, I save my work in a matrix called *R* by entering "R=FIG" at the MATLAB prompt.

(2) To see the fruits of your labor from Exercise 1, you can use the MATLAB "plot" command. For example, to see my (presumably beautiful) R, I write

```
cla;
plot(R(1,:),R(2,:))
```

This tells MATLAB to first clear the figure and then plot the first row of *R* as *x*-values and the second as *y*-values. Plot your first initial. [*Note:* If you have exited the Figure window, you will need to first enter the line "hold on;axis equal" in order to get the right proportions on your initial.]

(3) The transformation defined by the matrix in Example 1 above is a "shear along the y axis." To see what a shear would do to my R, I enter the matrix M from Example 1 into MATLAB and set "SR=M*R". This has the effect of multiplying each column of R by M and storing in a matrix SR. I then plot SR just as in Exercise 2. Plot the image of your first initial under the shear. [*Note:* You will need to enlarge the viewing window to see your output. The command "axis([0,2,0,2])" should make it large enough.]

(4) Use the rotation matrix to plot the image of your first initial under rotation by 20 degrees. [*Notes*: MATLAB works in radians, so you will need to convert from degrees to radians. You might want to read the MATLAB help entry for "cos." See the note in Exercise 3 about enlarging the viewing window.]

(5) Use "stick" to create appropriately named matrices to represent your other initials. For example, I would draw a C and save it in a matrix called C and draw a P and save it in matrix P. Then get MATLAB to plot each of the letters you created on the same graph: quite a mess.

Try to find a way of shifting the letters over so that your initials come out right. [*Note*: If M is a matrix in MATLAB, then the command "M=M+1" will add 1 onto every entry of M. (You will need to change the size of the viewing window. See the note in Exercise 3.)]

(6) Let S be the transformation of $\mathbb{R}^2$ into itself defined by stipulating that $S(X)$ is the result of shifting X one unit to the right. Show graphically that S is not linear. Specifically, use the first letter of your initial to show that $S(2X) \neq 2S(X)$.

(7) For items (a)–(d), find a matrix M for which multiplication by M would:
   (a) Flip your initials upside down
   (b) Flip your initials left-to-right
   (c) Rotate your initials by 20 degrees
   (d) Shear your initials along the x axis

For each of your matrices, plot the effect of the matrix on your *first* initial.

(8) Plot the effects on your *first* initial of the following transformations:
   (a) A shear along the x axis followed by a rotation by 20 degrees.
   (b) A rotation of 20 degrees followed by a shear along the x axis.
   (c) A shear along the x axis followed by a shear along the y axis.
   (d) A shear along the y axis followed by a shear along the x axis.

3.2 MATRIX MULTIPLICATION (COMPOSITION)

In applications, one sometimes needs to construct transformations that transform one specific set onto another.

▶ EXAMPLE 1: Find a matrix C such that multiplication by C transforms the circle $x^2 + y^2 = 1$ onto the ellipse indicated in Figure 1.

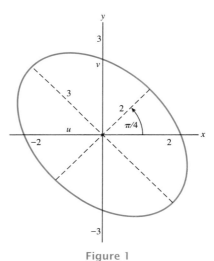

Figure 1

Solution: In Exercise 3, Section 3.1, we found that multiplication by the following matrix A "stretches" the circle $x^2 + y^2 = 1$ onto the ellipse on the right in Figure 2.

$$A = \begin{bmatrix} 2 & 0 \\ 0 & 3 \end{bmatrix}$$

The ellipse in Figure 1 is this ellipse rotated by $\frac{\pi}{4}$ radian counterclockwise. This rotation is described by multiplication by

$$R = \begin{bmatrix} \cos\frac{\pi}{4} & -\sin\frac{\pi}{4} \\ \sin\frac{\pi}{4} & \cos\frac{\pi}{4} \end{bmatrix} = \begin{bmatrix} \frac{\sqrt{2}}{2} & -\frac{\sqrt{2}}{2} \\ \frac{\sqrt{2}}{2} & \frac{\sqrt{2}}{2} \end{bmatrix}$$

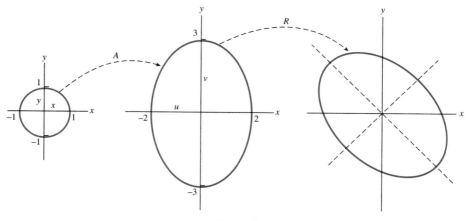

Figure 2

Thus, the transformation that transforms X into

$$Y = R(AX)$$

will transform the circle onto the rotated ellipse.

To express this as a matrix transformation, let the columns of A be A_1 and A_2 and let $X = [x_1, x_2]^t$. Then,

$$Y = R(AX) = R(x_1 A_1 + x_2 A_2) = x_1 R A_1 + x_2 R A_2$$

This formula describes multiplication by the matrix C whose columns are RA_1 and RA_2. Explicitly, then, we have

$$C = \left[\begin{bmatrix} \frac{\sqrt{2}}{2} & -\frac{\sqrt{2}}{2} \\ \frac{\sqrt{2}}{2} & \frac{\sqrt{2}}{2} \end{bmatrix} \begin{bmatrix} 2 \\ 0 \end{bmatrix} \quad \begin{bmatrix} \frac{\sqrt{2}}{2} & -\frac{\sqrt{2}}{2} \\ \frac{\sqrt{2}}{2} & \frac{\sqrt{2}}{2} \end{bmatrix} \begin{bmatrix} 0 \\ 3 \end{bmatrix} \right] = \begin{bmatrix} \sqrt{2} & -\frac{3\sqrt{2}}{2} \\ \sqrt{2} & \frac{3\sqrt{2}}{2} \end{bmatrix}$$

In words, we would say that C is "R times the columns of A." ◄

In Example 1, the desired transformation was constructed by successively applying two transformations: first we stretched the circle; then we rotated the result. The process of successively applying transformations is called **composition**, and the resulting transformation is called the **composite transformation.**

More precisely, suppose that S and T are transformations. Formally, the **composite** of S with T (which is denoted $S \circ T$) is defined by the formula

$$S \circ T(x) = S(T(x))$$

Notice, however, that this is meaningful only if $T(x)$ belongs to the domain of S. Thus, *we will define $S \circ T$ only if the target space of T is the domain of S.* The resulting transformation will have the same domain as that of T and the same target space as that of S.

If S and T are matrix transformations, the stipulation that the target space of T should be the domain of S puts restrictions on the sizes of the matrices that define S and T. If T is defined by an $m \times n$ matrix, then its target space is $\mathbb{R}^m$. This forces S to be defined by an $n \times p$ matrix. Hence, the composite transformation will transform $\mathbb{R}^n$ into $\mathbb{R}^p$ (see Figure 3).

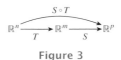

Figure 3

The reflection of an object can be thought of as the image of the object under a transformation. The reflection of a reflection is the image of the object under a composite transformation. Each subsequent reflection is obtained by applying an additional transformation.

It is crucially important that the composite of two matrix transformations (if defined) can be expressed using a single matrix. To see this, let A and B be matrices of size $p \times m$ and $m \times n$, respectively. If we successively multiply $X = [x_1, x_2, \ldots, x_n]^t$ by B and then by A, we obtain (using the linearity properties from Section 1.5)

$$A(BX) = A(x_1 B_1 + x_2 B_2 + \cdots + x_n B_n) = x_1 A B_1 + x_2 A B_2 + \cdots + x_n A B_n$$

where B_i are the columns of B. The terms AB_i on the right are columns and this expression is exactly the product of X with the matrix

$$AB = [AB_1, AB_2, \ldots, AB_n] \tag{M}$$

We refer to this matrix as the product of A with B. In words, AB is "A times the columns of B."

What we have said may be summarized in the formula

$$A(BX) = (AB)X \tag{1}$$

which is valid for all matrices A and B of size $p \times m$ and $m \times n$, respectively, and all $X \in \mathbb{R}^n$.

Notice that in (M), it is possible to multiply the columns of B by A since, by assumption, A has size $p \times m$ and the columns of B have length m. The final result will have size $p \times n$. Symbolically, then, the sizes change according to the rule:

$$(p \times m) \cdot (m \times n) = p \times n$$

Matrix multiplication is extremely important. Fortunately, it is not difficult.

▶ **EXAMPLE 2:**   Compute AB where A and B are as follows.

(a)
$$A = \begin{bmatrix} 1 & 2 \\ 2 & 3 \end{bmatrix} \quad B = \begin{bmatrix} 1 & 2 & 3 \\ 4 & 5 & 6 \end{bmatrix}$$

(b)
$$A = \begin{bmatrix} 1 & 2 & 1 \\ 2 & -3 & 3 \\ 6 & -5 & 4 \end{bmatrix} \quad B = \begin{bmatrix} 1 & 0 \\ 3 & -1 \\ 7 & 1 \end{bmatrix}$$

Solution: From formula (M), the product in part (a) is

$$AB = \begin{bmatrix} A\begin{bmatrix} 1 \\ 4 \end{bmatrix} & A\begin{bmatrix} 2 \\ 5 \end{bmatrix} & A\begin{bmatrix} 3 \\ 6 \end{bmatrix} \end{bmatrix} = \begin{bmatrix} 9 & 12 & 15 \\ 14 & 19 & 24 \end{bmatrix}$$

Similarly, for part (b),

$$AB = \begin{bmatrix} A\begin{bmatrix} 1 \\ 3 \\ 7 \end{bmatrix} & A\begin{bmatrix} 0 \\ -1 \\ 1 \end{bmatrix} \end{bmatrix} = \begin{bmatrix} 14 & -1 \\ 14 & 6 \\ 19 & 9 \end{bmatrix}$$

◄

We have described AB as "A times the columns of B." There is another description of matrix multiplication that can be demonstrated using the matrix product in part (a). Let us compute the product of the first row of the left matrix with the right matrix. We get

$$\begin{bmatrix} 1 & 2 \end{bmatrix}\begin{bmatrix} 1 & 2 & 3 \\ 4 & 5 & 6 \end{bmatrix} = \begin{bmatrix} \begin{bmatrix} 1 & 2 \end{bmatrix}\begin{bmatrix} 1 \\ 4 \end{bmatrix} & \begin{bmatrix} 1 & 2 \end{bmatrix}\begin{bmatrix} 2 \\ 5 \end{bmatrix} & \begin{bmatrix} 1 & 2 \end{bmatrix}\begin{bmatrix} 3 \\ 6 \end{bmatrix} \end{bmatrix}$$
$$= \begin{bmatrix} 9 & 12 & 15 \end{bmatrix}$$

which is the first row of the answer.

Similarly, if we multiply the second row of the left matrix by the right matrix, we obtain the second row of the product:

$$\begin{bmatrix} 2 & 3 \end{bmatrix}\begin{bmatrix} 1 & 2 & 3 \\ 4 & 5 & 6 \end{bmatrix} = \begin{bmatrix} 14 & 19 & 24 \end{bmatrix}$$

This example demonstrates the general principle that *the rows of AB are the rows of A times B*. In symbols, if A_i are the rows of A, then

$$\begin{bmatrix} A_1 \\ A_2 \\ \vdots \\ A_m \end{bmatrix} B = \begin{bmatrix} A_1 B \\ A_2 B \\ \vdots \\ A_m B \end{bmatrix}$$

Proving this property is just a matter of thinking about how one multiplies matrices. The jth column of AB is AB_j where B_j is the jth column of B. We compute this by multiplying each row of A by B_j. Hence, the (i, j) entry of AB is

$$A_i B_j \tag{2}$$

where A_i is the ith row of A. Formula (2) says that each row of AB is obtained by multiplying the corresponding row of A by each column of B. Thus, the rows of AB are the rows of A times B, as claimed.

Remark: Formula (2) is useful in its own right. It is also useful to note that formula (2) says that $AB = C$ where

$$c_{ij} = a_{i1}b_{1j} + a_{i2}b_{2j} + \cdots + a_{in}b_{nj} \tag{3}$$

We can also do multistep transformations. We might, for example, begin with the circle in Example 1, stretch it into an ellipse (as in Example 1), rotate counterclockwise by $\frac{\pi}{4}$ radian (as in Example 1) and, finally apply the shear from Example 1 in Section 3.1. This shear is described by the matrix

$$M = \begin{bmatrix} 1 & 0 \\ 1 & 1 \end{bmatrix}$$

In total, then, we would be transforming vectors X in $\mathbb{R}^2$ to

$$Y = M(R(AX)) \tag{4}$$

Composites of any number of matrix transformations may always be described using a single matrix. For example, in the three-step transformation just given, the composite of the first two transformations is describable with the single matrix RA, which was computed in Example 1. Thus, the transformation that transforms X to Y is describable by the product

$$M(RA) = \begin{bmatrix} 1 & 0 \\ 1 & 1 \end{bmatrix} \begin{bmatrix} \sqrt{2} & -\frac{3\sqrt{2}}{2} \\ \sqrt{2} & \frac{3\sqrt{2}}{2} \end{bmatrix} = \begin{bmatrix} \sqrt{2} & -\frac{3\sqrt{2}}{2} \\ 2\sqrt{2} & 0 \end{bmatrix}$$

Actually, there is another way of producing this matrix. We can think of Y in formula (4) as having been obtained by transforming AX. From this point of view,

$$Y = MR(AX) = ((MR)A)X$$

Thus, the transformation that transforms X into Y is describable by the matrix $(MR)A$. We compute this as

$$(MR)A = \left(\begin{bmatrix} 1 & 0 \\ 1 & 1 \end{bmatrix} \begin{bmatrix} \frac{\sqrt{2}}{2} & -\frac{\sqrt{2}}{2} \\ \frac{\sqrt{2}}{2} & \frac{\sqrt{2}}{2} \end{bmatrix} \right) \begin{bmatrix} 2 & 0 \\ 0 & 3 \end{bmatrix}$$

$$= \begin{bmatrix} \frac{\sqrt{2}}{2} & -\frac{\sqrt{2}}{2} \\ \sqrt{2} & 0 \end{bmatrix} \begin{bmatrix} 2 & 0 \\ 0 & 3 \end{bmatrix} = \begin{bmatrix} \sqrt{2} & -\frac{3\sqrt{2}}{2} \\ 2\sqrt{2} & 0 \end{bmatrix}$$

We get the same answer as before because the matrix that describes a given linear transformation is unique.

The equality $M(RA) = (MR)A$ demonstrated in Example 2 is a consequence of a general property of transformations. Suppose that T, U, and V are transformations such that the target space of V is the domain of U and the target space of U is the domain of T. Then, we may define a transformation $T \circ U \circ V$ by

$$T \circ U \circ V(x) = T(U(V(x)))$$

This is the same as

$$T(U \circ V(x)) = T \circ (U \circ V)(x)$$

It is also the same as

$$T \circ U(V(x)) = (T \circ U) \circ V(x)$$

Two transformations R and S are said to be equal if they have the same domain and target space and $R(x) = S(x)$ for all x in the domain. It follows that

$$T \circ U \circ V = T \circ (U \circ V) = (T \circ U) \circ V$$

This formula is called the associative law for transformations. Applied to matrix transformations, it proves the following very important theorem.

Associative Law. Let A, B, and C be matrices such that the product $A(BC)$ is defined. Then $A(BC) = (AB)C$. ∎

Matrix multiplication also has some important "distributive laws" which we state next, in Proposition 1. If C and D are column vectors, then the left distributive and scalar laws are just restatements of the linearity properties from Section 1.5. Proving them in the general case is simply a matter of combining the known results and definition (M) (see Exercise 8 on page 158). The proof of the right distributive law is indicated in Exercise 9. It is necessary to distinguish between the left and right distributive laws because, as the reader will see in the exercises, *matrix multiplication is not commutative.*

Proposition 1. Let A and B be $m \times n$ matrices and let C and D be $n \times q$ matrices. Let a be a scalar. Then

$$A(C + D) = AC + AD \quad \text{(left distributive law)}$$
$$(A + B)C = AC + BC \quad \text{(right distributive law)}$$
$$A(aC) = aAC \quad \text{(scalar law)}$$

Before closing, we need to mention one final result concerning matrix multiplication which will be used later. It is, of course, possible to multiply elements of $M(1, n)$ with elements of $M(n, 1)$. Thus, for example

$$[x_1, x_2, x_3] \begin{bmatrix} y_1 \\ y_2 \\ y_3 \end{bmatrix} = x_1 y_1 + x_2 y_2 + x_3 y_3$$

The final result is a 1×1 matrix, which we interpret as just a number. Note that this is the same as

$$[y_1, y_2, y_3] \begin{bmatrix} x_1 \\ x_2 \\ x_3 \end{bmatrix} = y_1 x_1 + y_2 x_2 + y_3 x_3$$

The property demonstrated here is not exactly commutativity. Instead, what we are saying is that if A is a row and B is a column of the same length as A, then

$$AB = B^t A^t$$

More generally, suppose that A and B are arbitrary matrices for which the product AB is defined. From equation (3), the (i, j) entry of AB is

$$A_i B_j = B_j^t A_i^t$$

where A_i is the ith row of A and B_j is the jth column of B. The term on the right is the (j, i) entry of $B^t A^t$. These comments prove an extremely important formula:

$$(AB)^t = B^t A^t \tag{5}$$

EXERCISES

A number of the exercises below refer to the "unit square." This is the square S defined in Example 1, Section 3.1.

(1) For the matrices A, B, and C, demonstrate the associative law $A(BC) = (AB)C$ by directly computing the given products in the given orders.

$$A = \begin{bmatrix} 1 & 2 \\ 2 & 2 \end{bmatrix} \quad B = \begin{bmatrix} 2 & 3 \\ 0 & -1 \end{bmatrix} \quad C = \begin{bmatrix} 1 & 2 & 0 \\ 2 & 3 & -1 \end{bmatrix}$$

(2) In Exercise 1, is $(CA)B$ defined? $B(AC)$? $A(CB)$?

(3) For the matrices A and B in Exercise 1, show by direct calculation that $(AB)^t = B^t A^t \neq A^t B^t$.

(4) Repeat Exercise 3 for the matrices B and C from Exercise 1.

(5) What is the analogue of formula (5) for a product of three matrices? Test your formula for the matrices A, B, and C in Exercise 1. Prove your formula. [*Hint:* You may use formula (5) in your proof.]

(6) What is the analogue of formula (5) for a product of n matrices? Prove your formula.

(7) Demonstrate both the left and right distributive laws from Proposition 1 using specific, non-zero, 2×2 matrices A and B and 2×1 matrices C and D.

(8) Prove the left distributive law from Proposition 1. [*Hint:* Use formula (M) from this section along with the observation that if $C = [C_1, C_2, \ldots, C_q]$ and $D = [D_1, D_2, \ldots, D_q]$, then $C + D = [C_1 + D_1, C_2 + D_2, \ldots, C_q + D_q]$.]

(9) Prove the right distributive law from Proposition 1 in two steps:

   (a) Prove the right distributive law in the case that A and B are $m \times n$ matrices and C is an $n \times 1$ matrix. [*Hint:* Use formula (M) from Section 1.5 along with the observation that if $A = [A_1, A_2, \ldots, A_n]$ and $B = [B_1, B_2, \ldots, B_n]$, then $A + B = [A_1 + B_1, A_2 + B_2, \ldots, A_n + B_n]$.]
   (b) Use part (a) to prove the general case. For this part you will use formula (M) from this section.

(10) Define a transformation $T : \mathbb{R}^2 \to \mathbb{R}^2$ by the following rule: $T(X)$ is the result of first rotating X counterclockwise by $\frac{\pi}{6}$ radian and then multiplying by

$$A = \begin{bmatrix} 2 & 0 \\ 0 & 3 \end{bmatrix}$$

   (a) What is the image of the circle $x^2 + y^2 = 1$ under T?
   (b) What is the image of the unit square under T?
   (c) Find a matrix B such that $T(X) = BX$ for all $X \in \mathbb{R}^2$.

(11) Define a transformation $T : \mathbb{R}^2 \to \mathbb{R}^2$ by the following rule: $T(X)$ is the result of first rotating X counterclockwise by $\frac{\pi}{4}$ radian and then multiplying by

$$A = \begin{bmatrix} 1 & 1 \\ 0 & 1 \end{bmatrix}$$

   (a) What is the image of the unit square under T?
   (b) Find a matrix B such that $T(X) = BX$ for all $X \in \mathbb{R}^2$.

(12) Suppose that in Exercise 11 we multiply first and then rotate. Are your answers to parts (a) and (b) different? How?

(13) Define a transformation $U : \mathbb{R}^2 \to \mathbb{R}^2$ by $U(X) = A(AX)$ where A is the matrix from Exercise 10.

   (a) What is the image of the circle $x^2 + y^2 = 1$ under U?
   (b) Find a matrix C such that $U(X) = CX$ for all $X \in \mathbb{R}^2$.

(14) Define a transformation $T : \mathbb{R}^2 \to \mathbb{R}^2$ by $T(X) = A(AX)$ where A is the matrix from Exercise 11. What is the image of the unit square under T? Find a matrix that represents this transformation. What if we multiply X by A n times? Find a matrix that represents this transformation.

(15) Let S be the unit square. I rotate S by $\frac{\pi}{4}$ radian counterclockwise and then reflect the result in the x axis. You, first reflect S in the x axis and then rotate.

    (a) Indicate on separate graphs my result and your result.
    (b) Compute matrices M and Y that describe my transformation and yours.

(16) For the given matrix A, find a 3×2, non-zero matrix B such that $AB = 0$. Prove that any such matrix B must have rank 1. [*Hint:* The columns of B belong to the nullspace of A.]

$$A = \begin{bmatrix} 1 & 2 & 1 \\ 1 & 1 & 1 \end{bmatrix}$$

(17) Find a pair of 2×2 matrices A and B of your own choice such that $AB \neq BA$. [*Hint:* This isn't hard. It is a theorem that with probability 1, any two randomly selected matrices will not commute.]

(18) Find all 2×2 matrices B such that $AB = BA$ where

$$A = \begin{bmatrix} 1 & 2 \\ 0 & 3 \end{bmatrix}$$

(19) Find a pair of 2×2 matrices A and B of your own choice with $A \neq B$ such that $AB = BA$. (Do not use the matrix from Exercise 18.)

(20) Find a pair of 2×2 matrices A and B such that $(A + B)(A + B) \neq A^2 + 2AB + B^2$. Under what conditions does this equality hold?

(21) Find a pair of 2×2 matrices A and B such that $(A + B)(A - B) \neq A^2 - B^2$. Under what conditions does this equality hold?

(22) Find a 2×2 non-zero matrix A such that $A^2 = 0$. [*Hint:* Try making most of the entries equal to zero.]

(23) Find a 3×3 matrix A such that $A^3 = 0$ but $A^2 \neq 0$. [*Hint:* Try making most of the entries equal to zero.]

(24) Find an $n \times n$ matrix A such that $A^n = 0$ but $A^{n-1} \neq 0$.

(25) An $n \times n$ matrix is said to be diagonal if its only non-zero entries lie on the diagonal. Thus, an $n \times n$ diagonal matrix would have the form

$$A = \begin{bmatrix} a_{11} & 0 & 0 & \cdots & 0 \\ 0 & a_{22} & 0 & \cdots & 0 \\ 0 & 0 & a_{33} & \cdots & 0 \\ \vdots & \vdots & \vdots & \ddots & \vdots \\ 0 & 0 & 0 & \cdots & a_{nn} \end{bmatrix}$$

Find eight different 3×3 diagonal matrices A such that $A^2 = I$ where I is the 3×3 identity matrix.

(26) We saw in Section 3.1 (Exercise 25) that the set $\mathcal{W}$ of infinitely differentiable functions on $\mathbb{R}$ is a vector space and that differentiation may be thought of as a transformation that transforms $\mathcal{W}$ into itself. Specifically, if f is a differentiable function on $\mathbb{R}$, then we define

$$D(f) = \frac{df}{dx}$$

Similarly, we define a transformation S from $\mathcal{W}$ into $\mathcal{W}$ by

$$S(f) = xf$$

(By "x" we mean the function $y = x$.) Thus,

$$S(x^3) = x^4 \text{ and } S(e^x) = xe^x$$

(a) Compute $S \circ D(f)$ and $D \circ S(f)$ for (i) $f(x) = \sin x$, (ii) $f(x) = e^{3x}$, and (iii) $f(x) = \ln x$. Does $D \circ S = S \circ D$? [Two transformations U and V are defined to be equal if they (i) have the same domain, (ii) have the same target space, and (iii) satisfy $U(X) = V(X)$ for all X in the domain.]

(b) Let I be the transformation of $\mathcal{W}$ into $\mathcal{W}$ defined by $I(f) = f$. Prove that

$$D \circ S = S \circ D + I$$

(27) Suppose that $\mathcal{U}$, $\mathcal{V}$, and $\mathcal{W}$ are vector spaces and $T : \mathcal{U} \to \mathcal{V}$ and $S : \mathcal{V} \to \mathcal{W}$ are linear transformations. Prove that $S \circ T$ is a linear transformation.

ON LINE

(1) Use the program "stick" from On Line, Section 3.1, to draw a stick silhouette of a car somewhat as shown below and store in a matrix called C. Then transform C into $\mathbb{R}^3$ by setting "D=[C(1,:); 0*C(1,:);C(2,:)]". Plot your car in three dimensions with "plot3(D(1,:),D(2,:),D(3,:))". You should first close the figure window and then enter "hold on" and "axis([0 1 0 1 0 1])".

Figure 4

(2) It's not much fun to picture flat cars. To add some dimension to your car, set "E=D" and then "E(2,:)=E(2,:)+.25" and "F=[D,E]". Plot F. It still is not much like a car.

(3) To add some substance to the car, enter the following commands:

```
>>d=size(C,2);
>>for i=1:d,
   F(:,2*d+2*i-1)=D(:,i);
   F(:,2*d+2*i)=E(:,i);
>>end
```

Then plot F. You should get a reasonably good stick picture of a car.

(4) You are about ready to start "playing with" your car. First, however, you should move it so that the origin is at the center of the car. This will insure that as you rotate it, it will not move out of the viewing window. For this, you simply need to subtract a suitable constant from each row of F. (You should determine what "suitable" means.) You will need to clear the screen (with "cla") and center the axes (try "axis([-.6 .6 -.6 .6 -.6 .6])").

Rotate F by 30 degrees about the x axis and then rotate this image by 20 degrees about the z axis. Plot both images. Be careful to save your original F! (The three-dimensional rotation matrices are in Exercises 10, 11, and 12 in Section 3.1.)

(5) Find a single matrix that transforms your car into the final image from the last exercise. Plot the image of F under this transformation. (Be careful to save F.)

(6) How do you suppose the image would appear if you were to transform F by a rank 2 matrix transformation? Create a random rank 2 matrix and test your guess. Clear the figure window first. (See Exercise 2 in On Line, Section 2.3, for information on creating random matrices with specific ranks.) Plot F on a separate graph and compare this plot with your plot of the image. Get both plots printed. Attempt to label on the plot of F several points where the transformation is many-to-one. (This will only be approximate.)

(7) How do you suppose the image would appear if you were to transform F by a rank 1 matrix transformation? Create a random rank 1 matrix and test your guess.

3.3   IMAGE OF A TRANSFORMATION

Rank is just as important in studying transformations as it is in studying systems. Let us consider an example of a transformation defined by a rank 1 matrix.

▶ **EXAMPLE 1:**   Compute the image of the square S in Figure 1 under multiplication by

$$A = \begin{bmatrix} 2 & 3 \\ 2 & 3 \end{bmatrix}$$

Solution: It is easily computed that multiplication by A transforms the four vertices **0**, A, B, and C of S onto the points **0**, $A' = [2, 2]^t$, $B' = [5, 5]^t$, and $C' = [3, 3]^t$, respectively. All these points lie on the line $y = x$. The image of the square is the segment of this line between the points $[0, 0]^t$ and $[5, 5]$.

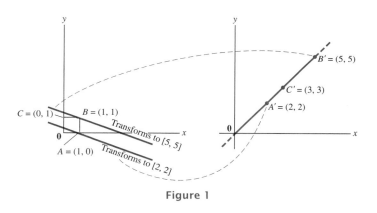

Figure 1

The transformation in Example 1 "smashed" the square into a line. This transformation in fact smashes all of $\mathbb{R}^2$ onto this line! To see why, note that

$$\begin{bmatrix} 2 & 3 \\ 2 & 3 \end{bmatrix} \begin{bmatrix} x \\ y \end{bmatrix} = \begin{bmatrix} 2 \\ 2 \end{bmatrix} x + \begin{bmatrix} 3 \\ 3 \end{bmatrix} y = (2x + 3y) \begin{bmatrix} 1 \\ 1 \end{bmatrix} \tag{1}$$

Thus, the image of all of $\mathbb{R}^2$ is the line spanned by $[1, 1]^t$. In general, the image of the whole domain under a transformation is called **the image** of the transformation. Hence, the image of this transformation is the line $y = x$.

The term on the right in formula (1) is just the general linear combination of the columns of A. Thus, the line $y = x$ is the column space of A. The column space is a

The shadow cast by a three-dimensional object on a sunny day can be thought of the image of the object under a linear transformation. Since the image is two-dimensional, the corresponding matrix must be rank two.

line because A has rank 1. (Recall that the dimension of the column space is the rank of the matrix.)

In the general case, if A is an $m \times n$ matrix with columns A_i, then multiplication by A transforms the vector $X = [x_1, x_2, \ldots, x_n]$ into

$$AX = x_1 A_1 + x_2 A_2 + \cdots + x_n A_n$$

Thus, *multiplication by an $m \times n$ matrix will transform all of $\mathbb{R}^n$ onto its column space.* In particular, the image of the transformation defined by A is an r-dimensional subspace of $\mathbb{R}^m$ where r is the rank of A.

One consequence of this is that *a matrix transformation will not increase dimension.* For example, a 3×2 matrix defines a transformation from $\mathbb{R}^2$ into $\mathbb{R}^3$. However, since there are only two columns, the image can be at most two-dimensional. In general, the image of a transformation defined by an $m \times n$ matrix can be at most n-dimensional.

Notice that in Example 1, we transformed a two-dimensional space ($\mathbb{R}^2$) onto a line. Thus, we lost a dimension in the transformation process. To understand this better, note that the vertex $[1, 0]^t$ of S is transformed onto the point $[2, 2]^t$. Let us ask, "What are all points of $\mathbb{R}^2$ that get transformed onto $[2, 2]^t$?" We answer this by solving the system

$$\begin{bmatrix} 2 & 3 \\ 2 & 3 \end{bmatrix} \begin{bmatrix} x \\ y \end{bmatrix} = \begin{bmatrix} 2 \\ 2 \end{bmatrix} \tag{2}$$

We find that y is arbitrary and $x = (2 - 3y)/2$. Hence, letting $y = s$, we see that all points of the form

$$[x, y]' = [1, 0]' + s[-\frac{3}{2}, 1]'$$

transform onto $[2, 2]'$. This equation describes the line through $[1, 0]'$ in the direction of the vector $[-\frac{3}{2}, 1]'$. (See Figure 1.) All points on this line are transformed into the single point $[2, 2]'$ under multiplication by A.

The "structure" of this line can be understood in terms of the translation theorem from Section 1.5. The vector $[1, 0]'$ is a particular solution to equation (2) and (from the Translation Theorem) "$s[-\frac{3}{2}, 1]'$" is the general solution to the corresponding homogeneous system.

If we are asked for all points that are transformed onto $[5, 5]'$, we can answer without further work. We saw earlier that the vertex $[1, 1]'$ is one point that is transformed onto $[5, 5]'$. This forms the particular solution. From the Translation Theorem, the general solution is $[1, 1]' + s[-\frac{3}{2}, 1]'$, which is the line through $[1, 1]'$, parallel to the preceding line. In general, multiplication by A will transform all of the points on any line parallel to the nullspace of A into a single point.

In general, the set of vectors an $m \times n$ matrix A transforms onto a given vector B in $\mathbb{R}^m$ may be found by solving the equation $AX = B$. The translation theorem says that the general solution to this equation may be expressed as $T + \mathcal{W}$, where T is any particular solution and $\mathcal{W}$ is the nullspace of A. Thus, just as in Example 1, multiplication by A transforms each translate of the nullspace onto a single point.

It is also true that in the general case, the dimension of the nullspace measures the number of dimensions that are lost in the transformation process. In fact, from the rank-nullity theorem in Section 2.3, we know that if A has rank r, the dimension d of the nullspace is $d = n - r$. This is equivalent with $r = n - d$. In words,

Theorem 1. *The dimension of the image of the matrix transformation defined by an $m \times n$ matrix A is n minus the dimension of the nullspace of A.* ∎

Since $\mathbb{R}^n$ is n-dimensional, this says exactly that we lose d dimensions in the transformation process.

These ideas can be used to give us a different perspective on the question of existence and uniqueness of solutions to systems of equations. Let us revisit Example 1 from Section 1.5.

▶ EXAMPLE 2: Are there vectors $B = [b_1, b_2, b_3]'$ for which the system below is not solvable? Is the solution (when it exists) unique? Do not do any row reduction to find the answer.

$$\begin{aligned} x \quad\;\; + 2z &= b_1 \\ x + y + 3z &= b_2 \\ y + \;\; z &= b_3 \end{aligned}$$

Solution: Let A be the coefficient matrix for this system. Thus

$$A = \begin{bmatrix} 1 & 0 & 2 \\ 1 & 1 & 3 \\ 0 & 1 & 1 \end{bmatrix}$$

Notice that the columns of A are dependent. In fact $A_3 = 2A_1 + A_2$. Since A_1 and A_2 are independent, we see that A has rank 2.

The system will be solvable for a given vector B if and only if there is an X such that $AX = B$. This is the same as saying that multiplication by A transforms X onto B which is, in turn, the same as saying that B is in the image of the transformation defined by A.

Since A has rank 2, the image is two-dimensional, hence is a plane. In fact, since the image is the column space, this plane is spanned by the first two columns of A. It is graphed in Figure 2. The system will be solvable if and only if B lies in this plane. Hence, there are many B for which the system is not solvable.

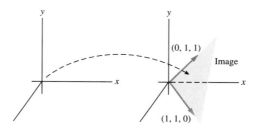

A smashes all of R^3 onto the plane

Figure 2

Since multiplication by A transforms all of $\mathbb{R}^3$ onto a plane, we have lost one dimension in the transformation process. It follows that the nullspace of A is a line. Since the general solution to $AX = B$ (when a solution exists) is a translate of the nullspace, we see that the solution is definitely not unique.   ◄

We noted that the transformation in Example 1 transforms every point on the line defined by $[1, 0]^t + s[-\frac{3}{2}, 1]^t$ onto the point $[2, 2]^t$. Thus, this transformation is capable of transforming many different points onto one point. Because of this, this transformation is said to be "many-to-one". In general, a transformation T is many-to-one if there is at least one point in the image of T that is the image of at least two different points in the domain of T.

The opposite of many-to-one is "one-to-one". A transformation is one-to-one if each point in the image is the image of only one point in the domain. Counterclockwise rotation by a fixed angle (say $\frac{\pi}{4}$ radian) is an example of a one-to-one transformation. There is only one point that rotates onto a given point.

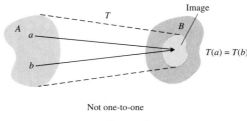

Not one-to-one

Figure 3

In the general case, multiplication by an $n \times n$ matrix A transforms each translate of its nullspace onto a single point. It follows that if the nullspace of A is not zero, then the corresponding transformation is multi-to-one. Conversely, if the nullspace of A is zero, then for each B in the image of the corresponding transformation, there is only one vector X such that $AX = B$. (Recall that the general solution is the particular solution plus the nullspace.) Thus, we arrive at the following theorem.

Theorem 2.   The transformation defined by a matrix A is one-to-one if and only if the nullspace of A is $\{\mathbf{0}\}$. ∎

▶ **EXAMPLE 3:**   Let A be as shown. Is the equation $AX = B$ solvable for every choice of $B \in \mathbb{R}^3$? Is the solution (when it exists) unique?

$$A = \begin{bmatrix} 1 & 2 & 3 \\ 1 & 0 & 1 \\ 4 & 5 & 0 \end{bmatrix}$$

Solution: The columns of A are easily seen to be independent. Any three independent vectors in $\mathbb{R}^3$ will span all of $\mathbb{R}^3$. Thus, the column space is all of $\mathbb{R}^3$ and the equation $AX = B$ is solvable for all B in $\mathbb{R}^3$.

Concerning the uniqueness, note that since the image is three-dimensional, and the domain of the transformation defined by A is $\mathbb{R}^3$, we didn't lose any dimensions in passing from the domain to the image. It follows that the nullspace must be $\{\mathbf{0}\}$; hence the transformation is one-to-one. This means exactly that for each B, there is at most one X such that $AX = B$. Thus, when the solution exists, it is unique. ◀

The transformation in Example 3 has the property that *its image equals the whole target space*. In other words, it transforms the domain *onto* the target space, rather than just into the target space. Transformations with this property are called onto. This transformation, in fact, is both one-to-one and onto. In general, a transformation that is both one-to-one and onto is called invertible. A matrix is called invertible if it defines an invertible transformation.

Invertible matrices are by far the most important. This is because they have the property that the system $AX = B$ is solvable for all B (this is the "onto" property) and the solution is unique (due to the "one-to-one" property).

Only square matrices can be invertible. To see this, suppose that A is $m \times n$. If $n > m$, then multiplication by A smashes $\mathbb{R}^n$ onto a smaller dimensional space, which contradicts one-to-one. If $n < m$, the transformation cannot be onto because linear transformations cannot increase dimension. Thus, $n = m$ is the only possibility.

Of course, not all square matrices are invertible. The matrix in Example 2 is not invertible because the corresponding transformation is not onto—the image is two-dimensional. This transformation is also not one-to-one since the nullspace is one-dimensional. It is clear that a matrix transformation from $\mathbb{R}^n$ into $\mathbb{R}^n$ that is not onto cannot be one-to-one either. If the dimension of the image is r where $r < n$, then the dimension of the nullspace will be $n - r \neq 0$. For similar reasons, if the transformation is not one-to-one, it cannot be onto. These comments yield the following theorem.

Inverse Theorem.   Let A be an $n \times n$ matrix. Then the following statements are equivalent to each other[1]:

(a) *The transformation defined by A is one-to-one.*

(b) *The transformation defined by A is onto.*

(c) *A has rank n.*

(d) *A is invertible.*   ■

We shall study invertible matrices in depth in Section 3.4.

We can use transformations to understand how the rank of a product of matrices relates to the rank of each individual matrix. Suppose that A and B are two matrices such that the product AB is defined. Assume that B is $n \times p$. For all X in $\mathbb{R}^p$,

$$(AB)X = A(BX)$$

Thus, the image of multiplication by AB is the set of all vectors of the form

$$A(BX)$$

where $X \in \mathbb{R}^p$. This set is contained in the set of all vectors of the form AY where Y is in $\mathbb{R}^n$. This latter set is the image of multiplication by A (Figure 4). Since the dimension of the image is the rank of the corresponding transformation, we see that

$$\text{rank}(AB) \leq \text{rank}(A)$$

---

[1] In mathematics, saying that two statements are equivalent means that each implies the other.

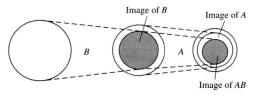

Figure 4

It is also true that

$$\text{rank}(AB) \leq \text{rank}(B)$$

There are many ways of proving this. One is to note that

$$\text{rank}(AB) = \text{rank}(AB)^t = \text{rank}(B^t A^t) \leq \text{rank}(B^t) = \text{rank}(B)$$

The exercises go into several different proofs. Of particular note is the direct proof from the definition of matrix multiplication (Exercise 14).

These facts are surprisingly important—important enough to be called a theorem.

Rank of Products Theorem. *The rank of the product of two matrices is less than or equal to the rank of each of the matrices.* ∎

EXERCISES

(1) For each matrix (a)–(d) find the rank, a basis for the image of the transformation defined by the matrix, and a basis for the nullspace.

(a) $\begin{bmatrix} 1 & 2 & 0 & -2 \\ 2 & 3 & 2 & 3 \\ 2 & 10 & 4 & 10 \end{bmatrix}$
(b) $\begin{bmatrix} -1 & 4 & -2 \\ 4 & 4 & 2 \\ 3 & 0 & -3 \end{bmatrix}$

(c) $\begin{bmatrix} 2 & 1 & 1 \\ 1 & 3 & 2 \\ 5 & 0 & 1 \end{bmatrix}$
(d) $\begin{bmatrix} 1 & 2 & 2 \\ 2 & 4 & 4 \\ 3 & 6 & 6 \\ -2 & -4 & -4 \end{bmatrix}$

(2) It is claimed that the vectors $[1, 0, 0]^t$, $[0, 1, 0]^t$, and $[0, 0, 1]^t$ form a basis for the image in part (b) of Exercise 1. Explain.

(3) Let A be the matrix in Example 2 on page 164. Do the pivot columns of the reduced form of A belong to the image of the transformation defined by A? Prove your answer.

(4) Let A be the matrix in Exercise 11, Section 2.3. Without doing any row reduction, show that the vectors $[1, 1, 2, 3]^t$ and $[7, 1, 8, 15]^t$ form a basis for the image of the corresponding transformation. Is the equation $AX = B$ solvable for all B in $\mathbb{R}^4$? Is the solution (when it exists) unique?

(5) Give an example of a 3×4 matrix A such that corresponding transformation transforms $\mathbb{R}^4$ onto the line spanned by $[1, 2, 3]^t$ in $\mathbb{R}^3$. Choose your matrix so that all its entries are different. [*Hint:* The image is the column space.] Use an argument similar to that made in Example 1 to prove that your answer is correct. What is the dimension of the nullspace of A?

(6) Find a 3×5 matrix A such that the corresponding transformation transforms $\mathbb{R}^5$ onto the space spanned by the vectors $[1, 0, 1]^t$ and $[3, -1, 0]^t$. Build your matrix A in such a way that *none of its entries is equal to zero.* What is the dimension of the nullspace of A?

(7) Give an example of a 2×3 matrix A such that the image of the corresponding transformation is the line $y = 2x$. What is the dimension of the nullspace of A?

(8) Give an example of a 3×4 matrix A such that the image of the corresponding transformation is the plane $2x + y - z = 0$ in $\mathbb{R}^3$. What is the dimension of the nullspace of A?

(9) Is it possible to find a 3×4 matrix A such that the image of the corresponding transformation is the plane $2x + y - z = 1$ in $\mathbb{R}^3$? Explain.

(10) Each row of the table indicated in Figure 5 on page 170 summarizes data collected for some matrix A and the corresponding transformation. The "always solvable" column refers to whether the system $AX = B$ is solvable for all B and the "unique solution" column refers to whether or not there will be at most one solution. Your assignment is to fill in all the missing data from the table. Where the data is inconsistent, write "impossible."

(11) Which of the rows in Exercise 10 describe invertible matrices?

(12) Is it possible to find a 3×3 invertible matrix A such that A^t is not invertible? If so, give an example. If not, why not?

(13) Let A be an $m \times n$ matrix. Suppose also that the equation $AX = B$ has *at most* one solution for each $m \times 1$ matrix B. What is the dimension of the nullspace of A?

Size	Always Solvable	Unique Solution	Dimension of Image	Dimension of Nullspace	Rank
4 x 3	yes				
3 x 4	yes				
5 x 5				1	
5 x 5			5		
3 x 2		yes			
4 x 4	yes				
5 x 4					3
5 x 4					5

Figure 5

What is the dimension of the image of the transformation defined by A? What is the rank of A? What can you say about the relative sizes of n and m?

(14) Let A and B be the indicated 3×2 and 2×3 matrices and let $C = AB$:

$$AB = \begin{bmatrix} 1 & 1 \\ 2 & 1 \\ 0 & -3 \end{bmatrix} \begin{bmatrix} 1 & 0 & 1 \\ -2 & 1 & 0 \end{bmatrix} = \begin{bmatrix} -1 & 1 & 1 \\ 0 & 1 & 2 \\ 6 & -3 & 0 \end{bmatrix} = C$$

(a) Express one column of B as a linear combination of the other two columns of B. Show that the columns of C satisfy the same dependency relation. Explain this fact in terms of the definition of matrix multiplication [(M) in Section 3.2].

(b) Use an argument similar to that in part (a) to prove that if A and B are any matrices such that the product AB is defined, then rank $AB \leq$ rank B. (Think in terms of how many independent columns AB can have compared to how many B can have.)

(c) Again, let A, B, and C be as given. Express one row of A as a linear combination of the other two. Show that the rows of C satisfy the same dependency relation. Explain this fact in terms of the definition of C as AB.

(d) Use an argument similar to that in part (b) to prove that if A and B are any matrices such that the product AB is defined, then rank $AB \leq$ rank A. (Think in terms of how many independent rows AB can have as compared to how many A can have.)

(15) Let A and B be matrices such that AB is defined. Show that if $BX = \mathbf{0}$, then X belongs to the nullspace of AB. What does this say about the nullspace of B in relation to that of AB? What, in turn, does this say about the rank of AB in relation to that of B?

(16) Let A and B be matrices such that AB is defined. Show that if A is invertible, then AB and B have the same nullspace. Use this to prove that rank $(AB) =$ rank (B).

(17) Prove that if B is invertible and AB is defined, then rank $(AB) =$ rank (A). [*Hint:* You can use Exercise 16, along with an argument involving transposes.]

(18) Let $\mathcal{W}$ be a subspace of $\mathbb{R}^n$ and let A be an $m \times n$ matrix. Prove that $A\mathcal{W}$ is a subspace (of $\mathbb{R}^m$).

(19) In the text, we proved that a matrix transformation from $\mathbb{R}^n$ into $\mathbb{R}^m$ cannot increase dimension in the sense that the image will be a subspace of dimension at most n. There is a stronger sense in which the transformation cannot increase dimension. Let A be an $m \times n$ matrix and let $\mathcal{W}$ be a k-dimensional subspace of $\mathbb{R}^n$. Show that $A\mathcal{W}$ is a subspace of dimension at most k. For your proof, let $X_1, X_2, \ldots, X_k$ be a basis for $\mathcal{W}$ and let $Y_i = AX_i$. Show that the Y_i span $A\mathcal{W}$. How does this solve the problem?

(20) Suppose that in Exercise 19, A is invertible. Show that $A\mathcal{W}$ is k-dimensional.

(21) Let A and B be matrices such that the product AB is defined. Explain why the image of the transformation defined by AB could be described as "A times the image of B." Use this fact and the result from Exercise 19 to prove that rank $(AB) \leq$ rank (B). What does the Exercise 20 tell you about rank (AB) when A is invertible?

On Line

(1) Create a "random" 2×3 matrix M with rank 1. (See Exercise 2 in On Line Section 2.3 for information on creating random matrices with specific ranks.) What should the dimension of the image of the transformation defined by M be? Verify this by creating 100 random points in $\mathbb{R}^3$, multiplying them by M, and getting MATLAB to plot the points. To create your 100 points, you can enter "P=rand(3,100)". Asking MATLAB to compute "S=M*P" will have the effect of multiplying each column of P by M. You can plot the result with "plot(S(1,:),S(2,:),'.')".

Question: Why are you getting only a line segment for the image rather than a whole line? How can you get more of the line?

(2) In Exercise 1, the image should be the span of any non-zero column of M. Demonstrate this by choosing a column of M and plotting 100 random points in its span on the same graph as before. (Use "hold on". Try to plot a large part of the span, not just a small segment.)

(3) Use your information from the preceding exercises to find (a) a specific vector $B \in \mathbb{R}^2$ such that the equation $MX = B$ is not solvable and (b) a specific vector $C \in \mathbb{R}^2$ such that the equation $MX = C$ is solvable. Indicate these vectors on your graph of the image. Verify your answers by computing "rref([M,B])" and "rref([M,C])".

(4) Plot 100 random elements of the nullspace of M. Note that you can find a basis for the nullspace using "null(M)". (See On Line, Section 1.5, Exercise 3.) Explain how your plot relates to the rank-nullity theorem.

(5) Create a "random" 3×3 matrix M with rank 2. Repeat the last three exercises (suitably modified to take account of the different dimensions) for this M. (See the last exercise in On Line, Section 1.2 for help in three-dimensional plotting.)

 INVERSES

In Section 3.3, we said that an $n \times n$ matrix A is invertible if it defines a one-to-one and onto transformation. Explicitly, this means that for all Y in $\mathbb{R}^n$, there is one, and only one, X such that $AX = Y$. In principle, then, if A is invertible, there is a transformation that computes X in terms of Y. This transformation is called the inverse transformation. Finding inverse transformations (when they exist) is one very important technique for solving systems.

▶ EXAMPLE 1:   Find a formula for the inverse transformation for the following matrix:

$$A = \begin{bmatrix} 1 & 2 & 1 \\ 1 & 3 & 4 \\ 2 & 4 & 1 \end{bmatrix} \tag{1}$$

Solution: The inverse transformation computes X in terms of Y in the system $AX = Y$. If $Y = [y_1, y_2, y_3]'$, then this system has augmented matrix:

$$\begin{bmatrix} 1 & 2 & 1 & y_1 \\ 1 & 3 & 4 & y_2 \\ 2 & 4 & 1 & y_3 \end{bmatrix}$$

We row-reduce, obtaining (after some work):

$$\begin{bmatrix} 1 & 0 & 0 & 13y_1 - 2y_2 - 5y_3 \\ 0 & 1 & 0 & -7y_1 + y_2 + 3y_3 \\ 0 & 0 & 1 & 2y_1 - y_3 \end{bmatrix}$$

This yields the answer

$$\begin{bmatrix} x_1 \\ x_2 \\ x_3 \end{bmatrix} = \begin{bmatrix} 13y_1 - 2y_2 - 5y_3 \\ -7y_1 + y_2 + 3y_3 \\ 2y_1 - y_3 \end{bmatrix} = y_1 \begin{bmatrix} 13 \\ -7 \\ 2 \end{bmatrix} + y_2 \begin{bmatrix} -2 \\ 1 \\ 0 \end{bmatrix} + y_3 \begin{bmatrix} -5 \\ 3 \\ -1 \end{bmatrix}$$

This may be written in the form

$$\begin{bmatrix} x_1 \\ x_2 \\ x_3 \end{bmatrix} = \begin{bmatrix} 13 & -2 & -5 \\ -7 & 1 & 3 \\ 2 & 0 & -1 \end{bmatrix} \begin{bmatrix} y_1 \\ y_2 \\ y_3 \end{bmatrix}$$

Hence the transformation that expresses X in terms of Y is defined by multiplication by the matrix

$$B = \begin{bmatrix} 13 & -2 & -5 \\ -7 & 1 & 3 \\ 2 & 0 & -1 \end{bmatrix}$$

◀

In general, when a matrix transformation has an inverse, the inverse transformation is a matrix transformation. (This will follow from the general inversion algorithm described below.) The matrix that defines the inverse transformation is referred to as the inverse matrix. If the original transformation is defined by a matrix A, then the inverse matrix is denoted A^{-1}. Thus, if A is as in Example 1, then A^{-1} is the matrix B just computed.

If we can find A^{-1}, then solving any system of equations that has A as its coefficient matrix is simple. For example, if A is as in Example 1, then the unique solution to $AX = [1, 2, 3]^t$ is

$$X = \begin{bmatrix} 13 & -2 & -5 \\ -7 & 1 & 3 \\ 2 & 0 & -1 \end{bmatrix} \begin{bmatrix} 1 \\ 2 \\ 3 \end{bmatrix} = \begin{bmatrix} -6 \\ 4 \\ -1 \end{bmatrix} \tag{2}$$

The reader may check that this solution really works.

We noted in Chapter 1, that in solving a system, it is not necessary to keep repeating the names of the variables. We can represent the whole solution process in terms of the augmented matrix. The same is true for computing inverse matrices. The only difference is that now we need to keep track of the coefficients of both the x_i and the y_i. To explain this, consider the first few steps in solving Example 1. The system we are solving is

$$
\begin{aligned}
x_1 + 2x_2 + x_3 &= y_1 + 0y_2 + 0y_3 \\
x_1 + 3x_2 + 4x_3 &= 0y_1 + y_2 + 0y_3 \\
2x_1 + 4x_2 + x_3 &= 0y_1 + 0y_2 + y_3
\end{aligned}
$$

We represent this system by a "double" matrix:

$$
\begin{bmatrix}
1 & 2 & 1 & | & 1 & 0 & 0 \\
1 & 3 & 4 & | & 0 & 1 & 0 \\
2 & 4 & 1 & | & 0 & 0 & 1
\end{bmatrix}
$$

We subtract the first equation in our system from the second, producing

$$
\begin{aligned}
x_1 + 2x_2 + x_3 &= y_1 + 0y_2 + 0y_3 \\
x_2 + 3x_3 &= -y_1 + y_2 + 0y_3 \\
2x_1 + 4x_2 + x_3 &= 0y_1 + 0y_2 + y_3
\end{aligned}
$$

The double matrix that represents this system is

$$
\begin{bmatrix}
1 & 2 & 1 & | & 1 & 0 & 0 \\
0 & 1 & 3 & | & -1 & 1 & 0 \\
2 & 4 & 1 & | & 0 & 0 & 1
\end{bmatrix}
$$

Notice that the effect was to subtract the first row from the second in the whole double matrix.

Next we might subtract twice the first equation from the third. In terms of the double matrix, this would be described by subtracting twice the whole first row from the third, producing

$$
\begin{bmatrix}
1 & 2 & 1 & | & 1 & 0 & 0 \\
0 & 1 & 3 & | & -1 & 1 & 0 \\
0 & 0 & -1 & | & -2 & 0 & 1
\end{bmatrix}
$$

We keep reducing our double matrix until the 3×3 matrix on its left side is reduced to the identity matrix:

$$
\begin{bmatrix}
1 & 0 & 0 & | & 13 & -2 & -5 \\
0 & 1 & 0 & | & -7 & 1 & 3 \\
0 & 0 & 1 & | & 2 & 0 & -1
\end{bmatrix}
$$

This tells us that

$$x_1 = 13y_1 - 2y_2 - 5y_3$$
$$x_2 = -7y_1 + y_2 + 3y_3$$
$$x_3 = 2y_1 \quad\;\; - \;\; y_3$$

which, of course, is the same answer as before. Notice that the 3×3 matrix on the right of the final double matrix is just A^{-1}.

In general, our process for inverting some $n \times n$ matrix A will be to form the double matrix $[A \mid I]$ where I is the $n \times n$ identity matrix. If we can row-reduce this double matrix until A is reduced to the identity matrix, then the matrix on the right side will be A^{-1}. This process cannot fail unless during the reduction A, some row reduces to **0**, in which case A does not have rank n, hence is not invertible. Here is another example of this process.

▶ **EXAMPLE 2:** Find A^{-1} where A is as follows.

$$A = \begin{bmatrix} 1 & 2 & 1 \\ 1 & 2 & 2 \\ 3 & 3 & 1 \end{bmatrix}$$

Solution:

$$\left[\begin{array}{ccc|ccc} 1 & 2 & 1 & 1 & 0 & 0 \\ 1 & 2 & 2 & 0 & 1 & 0 \\ 3 & 3 & 1 & 0 & 0 & 1 \end{array}\right]$$

$$\left[\begin{array}{ccc|ccc} 1 & 2 & 1 & 1 & 0 & 0 \\ 0 & 0 & 1 & -1 & 1 & 0 \\ 0 & -3 & -2 & -3 & 0 & 1 \end{array}\right]$$

$$\left[\begin{array}{ccc|ccc} 1 & 2 & 1 & 1 & 0 & 0 \\ 0 & -3 & -2 & -3 & 0 & 1 \\ 0 & 0 & 1 & -1 & 1 & 0 \end{array}\right]$$

$$\left[\begin{array}{ccc|ccc} 1 & 2 & 0 & 2 & -1 & 0 \\ 0 & -3 & 0 & -5 & 2 & 1 \\ 0 & 0 & 1 & -1 & 1 & 0 \end{array}\right]$$

$$\left[\begin{array}{ccc|ccc} 1 & 2 & 0 & 2 & -1 & 0 \\ 0 & 1 & 0 & \frac{5}{3} & -\frac{2}{3} & -\frac{1}{3} \\ 0 & 0 & 1 & -1 & 1 & 0 \end{array}\right]$$

$$\left[\begin{array}{ccc|ccc} 1 & 0 & 0 & -\frac{4}{3} & \frac{1}{3} & \frac{2}{3} \\ 0 & 1 & 0 & \frac{5}{3} & -\frac{2}{3} & -\frac{1}{3} \\ 0 & 0 & 1 & -1 & 1 & 0 \end{array}\right]$$

Thus, the inverse is

$$A^{-1} = \begin{bmatrix} -\frac{4}{3} & \frac{1}{3} & \frac{2}{3} \\ \frac{5}{3} & -\frac{2}{3} & -\frac{1}{3} \\ -1 & 1 & 0 \end{bmatrix}$$

◀

There is an important relation between an $n \times n$ matrix A and its inverse B. Saying that multiplication by B solves the equation $AX = Y$ means that for all X and Y in $\mathbb{R}^n$,

$$AX = Y \text{ if and only if } BY = X$$

We can put these equations together in two different ways. We may substitute for X in the first equation, getting

$$A(BY) = Y$$

or we may substitute for Y in the second getting

$$B(AX) = X$$

Thus, our equations say that for all X and all Y in $\mathbb{R}^n$

$$(AB)Y = Y \text{ and } (BA)X = X$$

The only $n \times n$ matrix C with the property that $CX = X$ for all X is the identity matrix. (The matrix that describes a transformation is unique!) Hence, $AB = BA = I$. We state this as a theorem:

Theorem 1.   *Let A be an invertible matrix. Then $AA^{-1} = A^{-1}A = I$.*   ■

The foregoing result offers a different way of thinking about inverses. Suppose we wish to solve the equation $2x = y$. We simply multiply both sides by $\frac{1}{2}$, producing $x = \frac{1}{2}y$.

The use of inverses to solve systems may be thought of in precisely the same manner. Suppose we wish to solve the equation

$$AX = Y$$

where A is invertible. We multiply both sides *on the left* by A^{-1}, producing

$$A^{-1}Y = A^{-1}(AX) = (A^{-1}A)X = IX = X$$

Thus, $X = A^{-1}Y$, which is the same answer as before.

Notice that in the preceding calculation, we used only $A^{-1}A = I$, not $AA^{-1} = I$. In general, a matrix B is said to be a left inverse for A if $BA = I$. Our calculation suggests that perhaps all we really need to solve systems are left inverses. It turns out, however, if an $n \times n$ matrix has a left inverse, the matrix is invertible and the left inverse is, in fact, just the usual inverse. To see this, recall that the rank of BA is less than or equal to the ranks of both A and B. Since the $n \times n$ identity matrix has rank n, it follows that if $BA = I$, then both B and A must have rank n, proving that each is invertible. The fact that $B = A^{-1}$ follows by multiplying both sides of $BA = I$ on the right by A^{-1}. These arguments prove part (b) of Theorem 2; part (a) is similar.

Theorem 2.   *Let A be a square matrix. If either of the following statements hold, then A is invertible and $B = A^{-1}$.*

(a) *There is a matrix B such that $AB = I$.*

(b) *There is a matrix B such that $BA = I$.*

EXERCISES

(1) Use the method of Example 1 to find the inverse of

$$\begin{bmatrix} 1 & 1 & 1 \\ 1 & 2 & 2 \\ 2 & 3 & 4 \end{bmatrix}$$

(2) Invert the following matrices (if possible).

(a) $$\begin{bmatrix} 1 & 0 & 3 \\ 4 & 4 & 2 \\ 2 & 5 & -4 \end{bmatrix}$$

(b) $$\begin{bmatrix} 0 & 1 & 1 \\ 1 & 0 & 1 \\ 1 & 1 & 0 \end{bmatrix}$$

(c) $$\begin{bmatrix} 1 & 2 & 3 \\ 2 & 4 & 6 \\ 3 & 6 & 9 \end{bmatrix}$$

(d) $$\begin{bmatrix} 2 & 0 & -1 \\ 1 & 1 & 0 \\ 1 & 1 & 2 \end{bmatrix}$$

(e)

$$\begin{bmatrix} 3 & 10 & 3 & 8 \\ 3 & -2 & 8 & 7 \\ 2 & 1 & 4 & -5 \\ 5 & 11 & 7 & 3 \end{bmatrix}$$

(f)

$$\begin{bmatrix} 2 & 1 & 0 & 3 \\ 0 & 3 & 1 & 1 \\ 1 & -1 & 1 & 2 \\ 2 & 7 & 0 & 4 \end{bmatrix}$$

(g)

$$\begin{bmatrix} 1 & 1 & 1 & 1 \\ 0 & 2 & 2 & 2 \\ 0 & 0 & 3 & 3 \\ 0 & 0 & 0 & 4 \end{bmatrix}$$

(h)

$$\begin{bmatrix} a & 0 & 0 & 0 \\ 0 & b & 0 & 0 \\ 0 & 0 & c & 0 \\ 0 & 0 & 0 & d \end{bmatrix}$$

where none of a, b, c, and d are zero.

(3) For each invertible, 3×3 matrix A from Exercise 2, use the inverse to solve the equation $AX = Y$, where $Y = [1, 2, 3]'$. For each invertible, 4×4 matrix, solve the equation $AX = Y$, where $Y = [1, 2, 3, 4]'$.

(4) In Exercise 2, for each noninvertible matrix, express one row as a linear combination of the others. Then do the same for the columns. [For part (e), the technique of Example 3 in Section 2.1 might be useful.]

(5) For each invertible matrix A in Exercise 2, check directly that $AA^{-1} = A^{-1}A = I$.

(6) Solve the system in Exercise 5a, Section 1.3, using inverses. Do you think this was less work than simply reducing the augmented matrix? Explain.

(7) Attempt to solve the system in Exercise 5b, Section 1.3, using inverses. Why does your attempt fail?

(8) Give an example of a noninvertible 5×5 matrix in which none of the entries is equal to zero. How can you be sure that this matrix is not invertible?

(9) Compute the inverse of the matrix A:

$$A = \begin{bmatrix} 1 & a & b \\ 0 & 1 & c \\ 0 & 0 & 1 \end{bmatrix}$$

Remark: An upper (or lower) triangular matrix is called unipotent if it has only ones on the diagonal. This exercise proves that the inverse of a 3×3, upper triangular, unipotent matrix is unipotent.

(10) Explain why the inverse of an $n \times n$, upper triangular, unipotent matrix is unipotent.

(11) Assume that $ad - bc \neq 0$. Find the inverse of the matrix

$$A = \begin{bmatrix} a & b \\ c & d \end{bmatrix}$$

(12) A certain matrix A has inverse

$$B = \begin{bmatrix} 2 & 1 & 3 \\ 1 & 1 & 1 \\ 4 & 2 & 1 \end{bmatrix}$$

Find a matrix X such that $XA = C$ where

$$C = \begin{bmatrix} 1 & 2 & 3 \\ 1 & 0 & 1 \\ 0 & 0 & 0 \end{bmatrix}$$

(13) In Exercise 12, find a matrix Y such that $AY = C$.

(14) Suppose that A is invertible. Prove that A^{-1} is invertible and $(A^{-1})^{-1} = A$.

(15) Let A and B be invertible $n \times n$ matrices. Prove that AB is invertible and $(AB)^{-1} = B^{-1}A^{-1}$. For your proof, recall that $(AB)^{-1}$ is the matrix that expresses Y in terms of X in the equation $ABX = Y$. If you know A^{-1}, what would be your first step in solving this equation?

(16) Using properties of inverses and matrix multiplication, prove that if A and B are invertible $n \times n$ matrices, then

$$(B^{-1}A^{-1})(AB) = I$$

How does this prove that $(AB)^{-1} = B^{-1}A^{-1}$?

(17) Using properties of inverses and matrix multiplication, prove that if A and B are invertible $n \times n$ matrices, then

$$(AB)(B^{-1}A^{-1}) = I$$

How does this prove that $(AB)^{-1} = B^{-1}A^{-1}$?

(18) Prove that if A is invertible, then so are A^2, A^3, and A^4. What are the inverses of these matrices? (Assume that you know A^{-1}.)

(19) Let A be invertible. Prove that $(A^t)^{-1} = (A^{-1})^t$. [*Hint:* Simplify $(A^{-1})^t A^t$.]

(20) Suppose that A and B are invertible $n \times n$ matrices. Prove that $(AB)^2 = A^2 B^2$ if and only if $AB = BA$.

(21) Let Q and D be $n \times n$ matrices with Q invertible.
   (a) Prove that $(Q^{-1}DQ)^2 = Q^{-1}D^2 Q$.

   *Warning: Matrix multiplication is not necessarily commutative! It is **not** usually true that $Q^{-1}DQ = D$. It is also not in general true that for matrices, $(AB)^2 = A^2 B^2$. (See Exercise 20.)*
   (b) What is the corresponding formula for $(Q^{-1}DQ)^3$? $(Q^{-1}DQ)^n$? Prove your answer.

(22) Compute a formula for A^n where

$$A = \begin{bmatrix} 3 & 1 \\ 1 & 3 \end{bmatrix}$$

Suggestion: Show that $A = QDQ^{-1}$ where

$$Q = \begin{bmatrix} 1 & 1 \\ -1 & 1 \end{bmatrix} \text{ and } D = \begin{bmatrix} 2 & 0 \\ 0 & 4 \end{bmatrix}$$

Then use the result from Exercise 21.

(23) We know that only square matrices can be invertible. We also know that if a square matrix has a right inverse, the right inverse is also a left inverse. It is possible, however, for a nonsquare matrix to have either a right inverse or a left inverse (but not both). Parts (a)–(d) explore these possibilities.
   (a) For the given matrix A find a 3×2 matrix B such that $AB = I$ where I is the 2×2 identity matrix. [*Hint:* If B_1 and B_2 are the columns of B, then $AB_j = I_j$.]

$$A = \begin{bmatrix} 1 & 2 & 1 \\ 1 & 1 & 1 \end{bmatrix}$$

   (b) Suppose that A is any 2×3 matrix with rank 2. Prove that A has a right inverse. [*Hint:* The matrix transformation defined by A is onto.]
   (c) Show conversely that if A is a 2×3 matrix that has a right inverse, then A has rank 2.

(d) Under what circumstances will an $m \times n$ matrix have a right inverse? State your condition in terms of rank and prove your answer.

(24) For A' as shown, find a matrix B' with $B'A' = I$ where I is the 2×2 identity matrix. (The answer to Exercise 23a might help.) Under what circumstances will an $m \times n$ matrix have a left inverse?

$$A' = \begin{bmatrix} 1 & 1 \\ 2 & 1 \\ 1 & 1 \end{bmatrix}$$

(25) Without using the result from Exercise 24, prove that if A is an $m \times n$ matrix that has a left inverse, the matrix transformation defined by A is one-to-one.

(26) Let C be the 3×3 matrix shown. Let A and B be, respectively, the indicated 3×2 and 2×3 matrices such that $AB = C$. In this exercise we explore a few of the numerous ways of seeing that C cannot be invertible.

$$AB = \begin{bmatrix} 1 & 1 \\ 2 & 1 \\ 0 & -3 \end{bmatrix} \begin{bmatrix} 1 & 0 & 1 \\ -2 & 1 & 0 \end{bmatrix} = \begin{bmatrix} -1 & 1 & 1 \\ 0 & 1 & 2 \\ 6 & -3 & 0 \end{bmatrix} = C$$

(a) Since the columns of B are three vectors in $\mathbb{R}^2$, they cannot be independent. Express one of them as a linear combination of the other two. Show that the columns of C satisfy the same dependency relation. Explain this fact in terms of the definition of C as AB. (Recall that the columns of AB are A times the columns of B.) How does this show that C is not invertible?
(b) What general theorem guarantees that the rows of A are dependent? Find an explicit dependency relation. Show that the rows of C satisfy the same dependency relation. Explain this fact in terms of the definition of C as AB. How does this prove that C is not invertible?
(c) Can the dimension of the nullspace of B be zero? Suppose that $BX = 0$. Prove that $CX = 0$. What can you say about the nullspace of C? How does this prove that C is not invertible?
(d) Can the image of the transformation corresponding to A be three-dimensional? What does this tell you about the image of the transformation corresponding to C? How does this prove that C is not invertible?

(27) Let A be an $m \times n$ matrix and let B be $n \times m$ where $n < m$. Give four different proofs that AB is not invertible. Your proofs should parallel the four parts of the previous exercise.

(28) Use the rank of products theorem from Section 3.3 to prove the non-invertibility of the matrix AB from Exercise 26.

(29) Suppose that A is an $n \times n$ matrix such that $A^2 + 3A + I = 0$. Show that A is invertible and $A^{-1} = -A - 3I$. [*Hint:* Theorem 2 might help.]

(30) Suppose that A is an $n \times n$ matrix such that $A^3 + 3A^2 + 2A + 5I = 0$. Show that A is invertible.

(31) A matrix N is said to be nilpotent if there is a k such that $N^k = 0$. The smallest such k is called the **degree of nilpotency** of N.

(a) Prove that the following matrix is nilpotent of degree at most 3.

$$N = \begin{bmatrix} 0 & a & b \\ 0 & 0 & c \\ 0 & 0 & 0 \end{bmatrix}$$

(b) Suppose that N is nilpotent of degree 3. Prove that $(I - N)^{-1} = I + N + N^2$. Use this to explicitly compute the inverse of the matrix $I - N$ where N is as in part (a).

(c) Suppose that N is nilpotent of degree 4. What would be the corresponding formula for $(I - N)^{-1}$? Prove your answer.

ON LINE

MATLAB provides an excellent facility for solving systems of the form $AX = B$ where A is invertible. Once A and B have been entered, X is computed by the MATLAB command "X = A\B". The reader should think of this as shorthand for $A^{-1}B$, although MATLAB actually uses row reduction rather than inverses in solving this equation.

(1) Let A be the matrix from Exercise 2a above. Use the MATLAB command "X = A\B" to solve the system $AX = B$ where $B = [2.1, 3.2, -4.4]^t$.

(2) In most applications of linear algebra, we get our numerical data from measurements that are susceptible to error. Suppose that the vector B in the preceding problem was obtained by measuring a vector B' whose actual value is $B' = [2.1, 3.21, -4.4]^t$. Compute the solution to the equation $AX' = B'$. Which component of the X you just computed has the largest error? (We measure error as the absolute value of the difference between the computed value and the actual value.) Explain why this is to be expected in terms of the magnitude of the components of A^{-1}. [In MATLAB, A^{-1} is computed with the command "inv(A)".] Which component of B would you change to produce the greatest change in X? Why? Back up your answer with a numerical example. How much error could you tolerate in the measured values of the components of B if each entry of X is to have an error of at most ± 0.001?

(3) Let

$$
A = \begin{bmatrix} 1 & \frac{1}{2} & \frac{1}{3} \\ \frac{1}{2} & \frac{1}{3} & \frac{1}{4} \\ \frac{1}{3} & \frac{1}{4} & \frac{1}{5} \end{bmatrix}
$$

and let $B = [83, 46, 32]^t$. Use "X = A\B" to find a solution to $AX = B$. As before, suppose that the vector B was obtained by measuring a vector B' whose actual value is $B' = [82.9, 46.07, 31.3]^t$. Compute the solution to the equation $AX' = B'$. What is the percentage of error in the least accurate entry of X? What is there about the components of A^{-1} that accounts for the large error? How much error could you tolerate in the measured values of the components of B if each entry of X is to have an error of at most ± 0.001?

(4) The above exercise demonstrates that the process of solving a system of equations can magnify errors in disastrous ways. One quantitative measure of the inaccuracy of a calculation is *the ratio of the percentage of error of the final answer to the percentage of error of the input data.* But what do we mean by the "percentage of error in a vector" (such as X in Exercise 1) in which every entry might have an error of a different magnitude?

For vectors in $\mathbb{R}^3$, this question has a geometric answer. We think of X and X' as representing points in $\mathbb{R}^3$. The distance d between these points is one measure of the error. If $X = [x, y, z]^t$ and $X' = [x', y', z']^t$, then

$$
d = \sqrt{(x - x')^2 + (y - y')^2 + (z - z')^2}
$$

In MATLAB, this can be computed as "norm(X-X')". If X' is the computed answer and X is the actual answer, we define the percentage of error to be

```
P=100*norm(X-X')/norm(X)
```

(a) Let B, B', X, and X' be as in Exercise 2. Compute (i) the percentage of error in B, (ii) the percentage of error in X, and (iii) the ratio of the percentage of error in X to that in B. This is the inaccuracy of the calculation of X from B.

Question: Assuming that accuracy is desired, do we want this number to be large or small? Explain.

(b) Compute the inaccuracy of the computation of X from B in Exercise 3.

(c) For each $n \times n$, invertible matrix A there is a number "cond(A)" (the "conditioning number of A") such that the inaccuracy in solving the system $AX = B$ is at most cond(A), regardless of B and regardless of the amount of error in B. This means that if, say, cond(A)=20 and the error in B is 0.001 %, then the computed value of X will have at most $20 \times 0.001 = 0.02\%$ error. In general, cond(A) ≥ 1. (This says that we cannot expect the answer to be more accurate than the input data.) MATLAB recognizes the command "cond(A)". Compute the conditioning numbers for the matrices A in Exercises 2 and 3 above.

Remark: Matrices with large conditioning numbers are called "ill-conditioned." If the coefficient matrix of a system is ill-conditioned, any slight error in the input data can make the solution very inaccurate, and we must be extremely suspicious of answers obtained by solving the system. Notice that these inaccuracies are not related to round-off error. Ill-conditioning is inherent to the matrix and not to the method of solution.

3.4.1   ECONOMIC MODELS

Linear algebra is used extensively in the study of economics. Here, we describe an economic model (the Leontief open model), which was developed by Wassily Leontief, who won the Nobel Prize in economics in 1973. We begin with an example.

▶ EXAMPLE 1:   In many areas of eastern Europe, the main source of energy is coal. Coal is used both to heat homes and to produce electricity. One of the most basic jobs of any manager is to determine the levels of production necessary to meet demand.

In this example, we will imagine a small village somewhere in eastern Europe in which there is a coal mine and an electric plant. All the residents get their coal from the mine and their electricity from the electric company. Let us assume that the currency used by the people is the slug (S).

It is known that to produce 1 S of coal, the mine must purchase 0.20 S worth of electricity. To produce 1 S of electricity, the power plant must purchase 0.60 S of coal and 0.07 S of electricity (from itself). Assume that the the village demands 2,500 S worth of electricity and 10,000 S worth of coal. What level of production is necessary to meet this demand?

Solution: Let c and e denote the respective values of the coal and electricity produced to meet the demand of the village. The consumers demand 10,000 S worth of coal and the electric company demands 0.6e S worth of coal. Thus, the total demand for coal is

$$10,000 + 0.6e.$$

Similarly, the demand for electricity is

$$2,500 + 0.2c + 0.07e$$

Setting production equal to demand, we obtain the following system

$$\begin{aligned} c &= 10,000 + 0.6e \\ e &= 2,500 + 0.2c + 0.07e \end{aligned} \qquad (1)$$

which is equivalent with

$$c - 0.6e = 10,000$$
$$-0.2c + 0.93e = 2,500$$

We solve this system by inverting the coefficient matrix. We find that

$$\begin{bmatrix} c \\ e \end{bmatrix} = \begin{bmatrix} 1 & -0.6 \\ -0.2 & 0.93 \end{bmatrix}^{-1} \begin{bmatrix} 10,000 \\ 2,500 \end{bmatrix} \approx \begin{bmatrix} 1.1481 & 0.7407 \\ 0.2469 & 10.2346 \end{bmatrix} \begin{bmatrix} 10,000 \\ 2,500 \end{bmatrix} \approx \begin{bmatrix} 13,300 \\ 5,500 \end{bmatrix}$$

Thus, we must produce 13,300 S worth of coal and 5,500 S worth of electricity.   ◄

Notice that in the example, the computed values of both c and e turned out to be positive. This is due to the (somewhat surprising) property that the inverse of the coefficient matrix in the above system has only positive entries. Had this not been the case, there would be levels of demand resulting in a negative value for either c or e. If, say, c were negative, the coal company would be *importing* coal, which we certainly do not expect. Thus, a negative value of either c or e would suggest a serious flaw in our model. Hence, the positivity of the coefficients of the inverse matrix is one reality check on our model.

In the general case, we will have n industries numbered one through n. The production of our economy is described by an $n \times 1$ column vector X where x_i is the value produced by industry i. In Example 1, this was $[c, e]^t$.

Each industry must purchase raw materials from all the local industries (including itself). Let c_{ij} represent the cost (in dollars) of the goods or services which industry j must purchase from industry i in order to produce one dollar's worth of output. If the output vector is X, industry i will sell $c_{ij}x_j$ units to industry j. In total, then, industry i sells a total of

$$c_{i1}x_1 + c_{i2}x_2 + \cdots + c_{in}x_n$$

to itself and other industries. This is the industrial demand for industry i. Notice that this is just the ith row of the matrix product CX where $C = [c_{ij}]$. C is called the **consumption** matrix because CX gives the units of output consumed by industry. The vector CX is called the **intermediate demand vector**. In Example 1,

$$C = \begin{bmatrix} 0 & 0.6 \\ 0.2 & 0.07 \end{bmatrix}$$

Now, suppose that the consumers demand d_i units from industry i. Then the total demand for the ith industry is

$$d_i + c_{i1}x_1 + c_{i2}x_2 + \cdots + c_{in}x_n$$

We write this in matrix format as

$$D + CX.$$

The vector D is referred to as the **final demand** vector. In the above example, $D = [10,000, 2,500]'$.

As in the example, we set total demand equal to output, obtaining the equation

$$X = D + CX.$$

Usually, D and C will be given and our goal is to find X. The above equation is equivalent with $(I - C)X = D$, which we solve as $X = (I - C)^{-1}D$.

▶ EXAMPLE 2:   Suppose that we have an economy consisting of three sectors: Agriculture (Ag), Manufacturing (Mn), and Service (Sv). To produce a dollar's worth of output, each sector purchases units according to the following table.

	Industry		
Purchases from	Ag	Mn	Sv
Agriculture	0.2	0.1	0.3
Manufacturing	0.1	0.3	0.2
Service	0.2	0.2	0.2

Answer the following questions:

(a) If the production vector is $[Ag, Mn, Sv]' = [10, 20, 30]'$, what is the industrial demand for manufacturing?

(b) What is the consumption matrix for this problem?

(c) What level of production would be necessary to sustain a demand of $D = [70, 80, 90]'$?

Solution: From the table, we see that at the levels of production given in (a), manufacturing will sell a total of

$$(0.1 \cdot 10) + (0.3 \cdot 20) + (0.2 \cdot 30) = 13$$

units to industry. This is the demand for manufacturing.

For question (b), recall that the product of the consumption matrix with the output vector should equal the amount sold to industry. From part (a), it is clear that the rows of our table should be the rows of C. Thus, our consumption matrix is

$$C = \begin{bmatrix} 0.2 & 0.1 & 0.3 \\ 0.1 & 0.3 & 0.2 \\ 0.2 & 0.2 & 0.2 \end{bmatrix}$$

For part (c), we write $X = (I - C)^{-1}D$ as above. From MATLAB, we see

$$(I - C)^{-1} = \begin{bmatrix} 0.8 & -0.1 & -0.3 \\ -0.1 & 0.7 & -0.2 \\ -0.2 & -0.2 & 0.8 \end{bmatrix}^{-1} = \begin{bmatrix} 1.4607 & 0.3933 & 0.6461 \\ 0.3371 & 1.6292 & 0.5337 \\ 0.4494 & 0.5056 & 1.5449 \end{bmatrix}$$

It is of course striking that all the entries of $(I - C)^{-1}$ are positive. Multiplying by D and rounding yields the production vector $X = [192, 202, 211]^t$.

Recall that $D = [70, 80, 90]^t$. The need to produce at such high levels to meet a relatively small demand means that our industries are inefficient. Notice, for example, that the sum of the first column of C is 0.5. This means that 50% of our agricultural output is consumed by other industries! This would not be typical of a real economy.   ◀

The general method of finding the production vector works only if $I - C$ is invertible. It turns out, however, that the following theorem is true. It should be noted that the assumption on the column sums is very natural. Consider, for example, the economy in Example 2. To produce a dollar's worth of goods, Agriculture must purchase goods worth a total of $0.2 + 0.1 + 0.2 = 0.5$ dollars from itself and other industries. This is just the sum of the first column of C. Similarly, the second column sum is the amount Manufacturing must purchase to produce one unit of output, and the third column sum is the amount that Service must purchase. If any of these had been greater than 1, that sector would have been operating at a loss. If the sum were equal to 1, then that sector would have been breaking even. Thus, the condition in this theorem is described by saying that all our industries are profitable.

Theorem 1.   *Suppose that C is a matrix with non-negative entries such that the sum of each of its columns is strictly less than 1. Then $(I - C)^{-1}$ exists and has non-negative entries.*   ■

The proof of this theorem is based upon the well-known fact that for x a real number, $|x| < 1$,

$$(1 - x)^{-1} = 1 + x + x^2 + \cdots + x^n + \cdots$$

This formula is simple to prove, provided that we grant the convergence of the series. Specifically, note that

$$\begin{aligned}(1 - x)(1 + x + x^2 + \cdots + x^n + \cdots) &= 1(1 + x + x^2 + \cdots + x^n + \cdots) \\ &\quad - x(1 + x + x^2 + \cdots + x^n + \cdots) \\ &= 1 + x + x^2 + \cdots + x^n + \cdots \\ &\quad - (x + x^2 + \cdots + x^n + \cdots) = 1\end{aligned}$$

proving the formula.

Remarkably, exactly the same formula will work if we replace x with an $n \times n$ matrix C, as long as the series converges. In fact, the foregoing argument shows that if C is an $n \times n$ matrix for which the series

$$I + C + C^2 + \cdots + C^n + \cdots$$

converges to a matrix B, then $(I - C)B = I$. Hence $I - C$ is invertible and B is its inverse.

If we can prove that the series converges for matrices of the form stated in Theorem 1, then we have proven Theorem 1, since if C is positive, so is any power of C, and thus, so is B.

The idea behind the convergence proof is most easily demonstrated with a specific matrix. Let

$$C = \begin{bmatrix} 0 & 0.6 \\ 0.2 & 0.07 \end{bmatrix}$$

(This is the consumption matrix from Example 1.)

Let $S = [1, 1]$. Note that for any 2×2 matrix B,

$$SB = [1, 1] \begin{bmatrix} b_{11} & b_{12} \\ b_{21} & b_{22} \end{bmatrix} = [b_{11} + b_{21}, b_{12} + b_{22}]$$

Thus, the components of SB are the column sums of B. In particular, for C as above,

$$SC = [0.2, 0.67] \leq 0.67[1, 1] = 0.67S$$

where we define "$\leq$" for matrices by saying that each entry of the first matrix is less than or equal to the corresponding entry for the second.

Multiplying both sides of this inequality by C on the right, we get

$$SC^2 \leq 0.67SC \leq (0.67)^2 S$$

(You should think about why multiplying both sides of a matrix inequality by a matrix with non-negative coefficients is allowed.)

Continuing, we see that

$$SC^n \leq (0.67)^n S = [(0.67)^n, (0.67)^n]$$

This means that each of the column sums in C^n is $\leq (0.67)^n$, hence each entry of C^n is $\leq (0.67)^n$. The convergence follows from the comparison test, since $\sum_0^\infty (0.67)^n$ converges.

The argument for general C is similar. In the $n \times n$ case, one uses $S = [1, 1, \ldots, 1]$. Then, for any $n \times n$ matrix B, SB is, once again, the vector of column sums of B. In particular, $SC \leq mS$ where m is the largest column sum in C. The convergence of the series then follows exactly as before, since, by assumption, m is less than 1.

EXERCISES

(1) Suppose that we have an economy governed by three sectors: Lumber (Lu), Paper (Pa), and Labor (Lb). Make up a table using your own numbers similar to the one from Example 2 to describe the units each industry would need to purchase from the other industries to produce one unit of output. Try to keep your numbers reasonable. For example, how would the amount the lumber company spends on paper to produce one dollar's worth of lumber compare with the amount the paper company would spend on lumber? Would the labor sector typically spend more on paper or lumber? Note also the comments preceding Theorem 1.

(2) For the economy you invented in Exercise 1, what quantity of lumber does industry demand if the production vector is $[Lu, Pa, Lb] = [30, 20, 10]^t$?

(3) For the economy you invented in Exercise 1, give the consumption matrix. At what level does the economy need to produce if the demand is $[Lu, Pa, Lb]^t = [70, 80, 100]^t$? [*Note*: This is best done with software that can invert matrices. If none is available, express your answer in terms of the inverse of a specific matrix but do not compute the inverse.]

(4) Suppose that in Example 1, besides coal and electricity, there is a train company whose sole business is transporting coal from the mine to the electric company, at the electric company's expense. Assume also that the trains run on coal, hence use large amounts of coal and only a little electricity. Invent a consumption matrix that might correspond to this information. You may assume that the coal company does not use the train. Note also the comments preceding Theorem 1.

Find the output level for the three companies necessary to satisfy the demand described in Example 1. [*Note*: This is best done with software that can invert matrices. If none is available, express your answer in terms of the inverse of a specific matrix but do not compute the inverse.]

(5) Suppose that in Exercise 4, the trains run on electricity. Give a potential consumption matrix to describe this situation. What if the electric trains operate at the coal mine's expense? What if the mine and the electric company share the cost of the trains?

3.5   THE LU FACTORIZATION

You are assigned to run some manufacturing process that involves solving a system of n equations in n unknowns to determine how to set your machines. Let us suppose that the system is given in the form

$$AX = Y$$

where A is an $n \times n$ matrix. Let us also assume that the value of A is reset each hour while the value of Y changes each minute.

One approach to finding X would be to simply row-reduce the system $AX = Y$ once a minute. (Let's hope you have a computer!) This, however, is very inefficient. The steps in the row reduction depend only on the coefficient matrix A. Since A changes only every hour, the calculations done to compute X will be almost the same within a given hour. This suggests that we should compute A^{-1} each time A changes and multiply Y by A^{-1} each time Y changes.

It turns out, however, that even this more sophisticated approach is not the best solution to our problem. To understand why, suppose that A turned out to be upper triangular. Say

$$A = \begin{bmatrix} 1 & 2 & 3 & 4 \\ 0 & 5 & 6 & 7 \\ 0 & 0 & 8 & 9 \\ 0 & 0 & 0 & 10 \end{bmatrix}$$

If you were asked to find a solution to the system $AX = Y$ where $Y = [1, 1, 1, 1]^t$ by hand, you probably would not even consider computing A^{-1}. Since A is triangular, the augmented matrix is already in echelon form:

$$\begin{bmatrix} 1 & 2 & 3 & 4 & 1 \\ 0 & 5 & 6 & 7 & 1 \\ 0 & 0 & 8 & 9 & 1 \\ 0 & 0 & 0 & 10 & 1 \end{bmatrix}$$

We solve this by setting $x_4 = 1/10$, $x_3 = (1 - 9x_4)/8 = 0.0125$, etc. This process, called back substitution, involves essentially no more work than multiplying Y by A^{-1}. Thus, the step of computing A^{-1} is wasted time.

In general, computing A^{-1} involves first reducing A to echelon form and then reducing the resulting upper triangular matrix to the identity. Once the matrix is reduced to echelon form, however, the system is solved. Reducing the echelon form to the identity should be, in principle, wasted time. What we really want to do, then, is reduce A to upper triangular form each time A changes. In doing so, we keep close track of the steps we used so that it won't be necessary to redo them each time Y changes.

To explain this, suppose that

$$A = \begin{bmatrix} 1 & 2 & 1 \\ 1 & 3 & 4 \\ 2 & 7 & 8 \end{bmatrix}$$

The general system we wish to solve is, then,

$$x_1 + 2x_2 + x_3 = y_1$$
$$x_1 + 3x_2 + 4x_3 = y_2$$
$$2x_1 + 7x_2 + 8x_3 = y_3$$

We shall reduce this system, keeping track of both the coefficients of the x_i and y_i. As in Section 3.4, we describe this system with the double matrix

$$\begin{bmatrix} 1 & 2 & 1 & | & 1 & 0 & 0 \\ 1 & 3 & 4 & | & 0 & 1 & 0 \\ 2 & 7 & 8 & | & 0 & 0 & 1 \end{bmatrix} \tag{1}$$

where the 3×3 matrix on the left is the matrix of coefficients for the x_i and the one on the right is the matrix of coefficients for the y_i. We reduce this double matrix until the 3×3 matrix on the left is in echelon form. For example, we might first subtract multiples of the first row to reduce the (2, 1) and (3, 1) entries of A to zero:

$$\begin{bmatrix} 1 & 2 & 1 & | & 1 & 0 & 0 \\ 0 & 1 & 3 & | & -1 & 1 & 0 \\ 0 & 3 & 6 & | & -2 & 0 & 1 \end{bmatrix} \tag{2}$$

Next, we subtract 3 times the second row from the third to reduce the (3, 2) entry of the left-hand matrix to zero:

$$\begin{bmatrix} 1 & 2 & 1 & | & 1 & 0 & 0 \\ 0 & 1 & 3 & | & -1 & 1 & 0 \\ 0 & 0 & -3 & | & 1 & -3 & 1 \end{bmatrix} \tag{3}$$

Let U be the 3×3 matrix on the left of this double matrix and let B be the one on the right. As explained in Section 3.4, the system $AX = Y$ is equivalent with the system $UX = BY$. Thus, for example, to solve

$$x_1 + 2x_2 + x_3 = 1$$
$$x_1 + 3x_2 + 4x_3 = 2$$
$$2x_1 + 7x_2 + 8x_3 = 3$$

we compute $B[1, 2, 3]^t = [1, 1, -2]^t$ and solve

$$\begin{aligned} x_1 + 2x_2 + x_3 &= 1 \\ x_2 + 3x_3 &= 1 \\ -3x_3 &= -2 \end{aligned}$$

finding $x_3 = \frac{2}{3}$, $x_2 = -1$ and $x_1 = \frac{7}{3}$.

The idea just demonstrated, in principle, solves the problem with which we began this section. Every time the coefficient matrix A changes, we form the double matrix $[A|I]$ and reduce until we have obtained a matrix of the form $[U|B]$ where U is in echelon form. Then, X is found by solving the system $UX = BY$.

There is, however, an additional improvement we can make in this procedure. It turns out that in the general case

(a) The matrix B is invertible.

(b) There is a remarkably simple way of computing $L = B^{-1}$ as we reduce A that works for "many" matrices A. This process involves considerably fewer computations than computing B.

(c) We can solve the system $AX = Y$ using L.

To explain these comments, notice that B is produced by applying elementary row operations to the identity matrix I. In fact, we produce B by applying to I, the same row operations we use to reduce A to U. Elementary row operations are reversible. We could, if we wish, transform U back into A by applying the inverse row operations to U in reverse order. If we apply the same inverse operations (in the same order) to B, we transform B back into I. Thus B is invertible and we can compute B^{-1} by applying these inverse operations to the double matrix $[B|I]$.

Specifically, in our example, $[B|I]$ is

$$\begin{bmatrix} 1 & 0 & 0 & | & 1 & 0 & 0 \\ -1 & 1 & 0 & | & 0 & 1 & 0 \\ 1 & -3 & 1 & | & 0 & 0 & 1 \end{bmatrix} \tag{4}$$

The *last* entry to be eliminated in reducing A was the $(3, 2)$ entry. Thus, our *first* step will be to eliminate this entry from the left matrix by adding three times the second row to the third:

$$\begin{bmatrix} 1 & 0 & 0 & | & 1 & 0 & 0 \\ -1 & 1 & 0 & | & 0 & 1 & 0 \\ -2 & 0 & 1 & | & 0 & 3 & 1 \end{bmatrix} \tag{5}$$

Notice that the 3 in the right-hand matrix is just the scalar by which we multiplied row 2 before subtracting it from row 3 in the reduction of A. This number is called the

multiplier. Note also that it is in the $(3, 2)$ position, which is also the position of the variable that was reduced to zero in passing from matrix (2) to matrix (3).

Next we eliminate $(2, 1)$ and $(2, 2)$ entries of the left matrix since these were the next to last to be eliminated from A. Specifically, we add row 1 onto row 2 and add 2 times row 1 onto row 3, producing:

$$\left[\begin{array}{ccc|ccc} 1 & 0 & 0 & 1 & 0 & 0 \\ 0 & 1 & 0 & 1 & 1 & 0 \\ 0 & 0 & 1 & 2 & 3 & 1 \end{array}\right] \tag{6}$$

The matrix on the right is L. Notice that the 3 in the right-hand matrix does not change in passing from matrix (5) to (6). Notice also that the multipliers 1 and 2 appear in the new right-hand matrix. *Thus, the entries of L are just the multipliers obtained in the reduction of A.* Also, *the location of the multiplier in L corresponds to the entry that was being reduced to zero.*

These patterns exist because in reducing A, the only type of row operations we used involved subtracting multiples of rows from lower rows. We never interchanged two rows, and we never factored a constant out of any row. For example, the 3 in the right-hand side of matrix (5) did not change in passing to matrix (6) because we were adding multiples of the first row onto lower rows and the first row of the right side of (5) is zero in the last two positions.

In general, suppose that A is an $n \times n$ matrix and we reduce the double matrix $[A\,|\,I]$ to $[U\,|\,B]$ where U is in echelon form. Suppose also that in the reduction process, we never interchanged rows and never factored constants out of any rows. Then B^{-1} will be the matrix L described as follows: L is lower triangular with ones on the main diagonal, and the non-zero entries of L are the multipliers. The location of each multiplier is determined by the entry of A that was being reduced to zero. It follows that *we can compute L as we reduce A simply by recording each multiplier in the appropriate position.*

Remark: See Exercises 2–4 in the On Line section for a discussion of what happens when row interchanges are required.

▶ **EXAMPLE 1:**   Let A be as shown. Compute the matrices L and U described in the text.

$$A = \left[\begin{array}{ccc} 1 & 2 & 1 \\ 2 & 1 & 1 \\ 1 & 3 & 2 \end{array}\right]$$

Solution: As commented earlier, we want to reduce A without interchanging rows and without factoring scalars out of rows. We begin by subtracting 2 times row 1 from row 2 in order to eliminate the (2, 1) entry of A. Hence, we place a 2 in the (2, 1) position of L. We get:

$$A_1 = \begin{bmatrix} 1 & 2 & 1 \\ 0 & -3 & -1 \\ 1 & 3 & 2 \end{bmatrix} \quad L = \begin{bmatrix} 1 & 0 & 0 \\ 2 & 1 & 0 \\ & & 1 \end{bmatrix}$$

Our next step is to subtract row 1 from row 3 in A_1, putting a 1 in the (3, 1) position of L:

$$A_2 = \begin{bmatrix} 1 & 2 & 1 \\ 0 & -3 & -1 \\ 0 & 1 & 1 \end{bmatrix} \quad L = \begin{bmatrix} 1 & 0 & 0 \\ 2 & 1 & 0 \\ 1 & & 1 \end{bmatrix}$$

Now we are tempted to exchange rows 2 and 3. Again, however, row exchanges should be avoided. We also must not yield to the temptation to divide the second row by -3.

Instead, we add $\frac{1}{3}$ times the second row to the third row. The corresponding multiplier will be $-\frac{1}{3}$. (It is negative because we are *adding*.) This produces

$$A_3 = \begin{bmatrix} 1 & 2 & 1 \\ 0 & -3 & -1 \\ 0 & 0 & \frac{2}{3} \end{bmatrix} \quad L = \begin{bmatrix} 1 & 0 & 0 \\ 2 & 1 & 0 \\ 1 & -\frac{1}{3} & 1 \end{bmatrix}$$

Since A_3 is now in echelon form, we stop. We set $A_3 = U$.

It is, of course, very nice that L is so easy to compute. What is even nicer is that when we know L, our system is, for all intents and purposes, solved. To explain this, recall that the system $AX = Y$ is equivalent with

$$UX = BY$$

where U and B are described in the material preceding (a)–(c). Since $L = B^{-1}$, this equation can be written

$$L(UX) = Y$$

We solve this in two steps. First we find W such that $LW = Y$. Then we solve the system $UX = W$. Both these systems can be solved *without any row reduction*, since L is lower triangular and U is upper triangular. (This is why these matrices are called "L" and "U".) Specifically, since $Y = [1, 2, 3]'$, the system $LW = Y$ is the same as

$$\begin{bmatrix} 1 & 0 & 0 \\ 2 & 1 & 0 \\ 1 & -\frac{1}{3} & 1 \end{bmatrix} \begin{bmatrix} w_1 \\ w_2 \\ w_3 \end{bmatrix} = \begin{bmatrix} 1 \\ 2 \\ 3 \end{bmatrix}$$

The first equation says $w_1 = 1$. The second says $w_2 = 2 - 2w_1 = 0$ and $w_3 = 3 - w_1 + w_2/3 = 2$, yielding $W = [1, 0, 2]^t$.

Next, we solve $UX = W$. This is equivalent with

$$\begin{bmatrix} 1 & 2 & 1 \\ 0 & -3 & -1 \\ 0 & 0 & \frac{2}{3} \end{bmatrix} \begin{bmatrix} x_1 \\ x_2 \\ x_3 \end{bmatrix} = \begin{bmatrix} 1 \\ 0 \\ 2 \end{bmatrix}$$

Again, this is a "solved system." We find $x_3 = 3$, which yields $x_2 = -1$ and, finally, $x_1 = 0$.   ◀

We were able to solve for W without row reduction because L was lower triangular. This property is a consequence of our avoidance of row exchanges in the reduction of A.

As noted earlier, for all Y in $\mathbb{R}^n$, the system $AX = Y$ is equivalent with $(LU)X = Y$. Specifically, if X is any vector in $\mathbb{R}^n$ and $Y = AX$, then it is also true that $(LU)X = Y$. This means that $A = LU$, since these two matrices define the same linear transformation. These comments lead to the following theorem.

LU-Decomposition Theorem.   *Let A be an $m \times n$ matrix that may be reduced to an echelon matrix U without interchanging any rows. Then there is a lower triangular, invertible, matrix L with ones on its main diagonal, such that*

$$A = LU$$ ■

Remark: At this point, the reader might be questioning exactly how easy it is to solve equations using the **LU** factorization. Certainly, in most of the problems that follow, directly solving the system is less work. In fact, not being allowed to switch equations or divide out scalars can make hand calculations tedious. The reader should bear in mind that the **LU** factorization is designed for the convenience of computers, not humans. "Ugly" numbers are of no concern. The only reason we have exercises in computing the **LU** factorization is to demonstrate to the reader what the computer is doing. Bear in mind, as well, that using the **LU** factorization is relevant only if in the system $AX = Y$, we expect to use many different Y's with the same A. If A changes every time Y changes, we should simply reduce each system.

EXERCISES

(1) For the following matrices L, U, and Y, solve the system $AX = Y$ for A where $A = LU$. [*Note*: It is not necessary to compute A.]

(a)

$$L = \begin{bmatrix} 1 & 0 & 0 & 0 \\ 1 & 1 & 0 & 0 \\ 2 & 0 & 1 & 0 \\ 1 & 1 & 1 & 1 \end{bmatrix} \quad U = \begin{bmatrix} 3 & 2 & 0 & 4 \\ 0 & 2 & 1 & 0 \\ 0 & 0 & 2 & 1 \\ 0 & 0 & 0 & 1 \end{bmatrix} \quad Y = \begin{bmatrix} 6 \\ 5 \\ 4 \\ 3 \end{bmatrix}$$

(b)

$$L = \begin{bmatrix} 1 & 0 & 0 \\ 2 & 1 & 0 \\ 2 & 2 & 1 \end{bmatrix} \quad U = \begin{bmatrix} 2 & 1 & 0 & 1 \\ 0 & 1 & 1 & 2 \\ 0 & 0 & 1 & 1 \end{bmatrix} \quad Y = \begin{bmatrix} 3 \\ 2 \\ 1 \end{bmatrix}$$

(2) Compute the **LU** factorization for the following matrices:

$$A = \begin{bmatrix} 1 & 4 & -1 \\ 2 & 10 & 2 \\ 1 & 5 & 2 \end{bmatrix} \quad B = \begin{bmatrix} 1 & 2 & 0 & 1 \\ 2 & 3 & 4 & 1 \\ 1 & 0 & 0 & 4 \\ 3 & 7 & 3 & 4 \end{bmatrix}$$

(3) Attempt to compute the **LU** factorization for the matrix A:

$$A = \begin{bmatrix} 1 & 1 & 3 \\ 2 & 2 & 5 \\ 2 & 1 & 4 \end{bmatrix}$$

Is there a way of exchanging the rows of A so that it does work?

(4) We are attempting to compute the **LU** factorization of a certain 4×4 matrix A. After three reductions, we arrive at a matrix of the form below. In this case, the **LU** process breaks shown. What modification could we have made to our original matrix to avoid this problem? How can you be sure that this modification does avoid the problem?

This exercise demonstrates a general theorem: if A is invertible, then it is always possible to arrange the rows in such a way that the **LU** factorization can be accomplished.

$$A = \begin{bmatrix} * & * & * & * \\ 0 & 0 & * & * \\ 0 & 1 & * & * \\ 0 & * & * & * \end{bmatrix}$$

(5) A lower (or upper) triangular matrix is called unipotent if it has only ones on the diagonal. In Example 1 in the text, find a diagonal matrix D, a lower triangular, unipotent matrix L, and an upper triangular unipotent matrix U such that $A = LDU$. This is called the **LDU** factorization of A. If it exists, it is unique.

(6) Compute the inverse of matrix A. Note that this proves that the inverse of a 3×3, lower triangular, unipotent matrix is unipotent.

$$A = \begin{bmatrix} 1 & 0 & 0 \\ a & 1 & 0 \\ b & c & 1 \end{bmatrix}$$

(7) Explain why the inverse of an $n \times n$ unipotent matrix is unipotent.

(8) Let $[U|B]$ be the second double matrix in Example 2 in Section 3.4. Compute BA. You should get U. Why? (See the explanation preceding the statement of the **LU** decomposition theorem in the text.)

(9) What would be the result if you were to repeat Exercise 8 for the fifth double matrix? Why? Can you state a theorem about the double matrices one gets in computing inverses?

ON LINE

(1) Would the designers of MATLAB have omitted the **LU** decomposition? Certainly not! Enter the matrix A from Example 1 in the text and give the command "[L,U]=lu(A)". MATLAB should respond almost immediately with the **LU** factorization.

(2) We commented earlier that the **LU** factorization exists only if the matrix can be reduced without row interchanges. What would the "lu" command in MATLAB do if it came across a matrix that required row interchanges? To investigate this, enter a 3×3 upper triangular matrix into MATLAB. (Recall that a matrix is upper triangular if all the entries below the main diagonal are zero.) In order to avoid trivialities, make all of the entries on or above the main diagonal non-zero. Call your matrix A. Let

$$E = \begin{bmatrix} 0 & 1 & 0 \\ 1 & 0 & 0 \\ 0 & 0 & 1 \end{bmatrix}$$

In MATLAB, compute "B=E∗A". How are B and A related? The matrix E is called a **permutation matrix**. Can you explain why?

(3) The matrix B computed in Exercise 2 cannot be row-reduced without interchanging rows. Ask MATLAB to compute "[L,U]=B". Note that the L you get is just the permutation matrix E. Use this to explain why $LU = B$.

(4) Enter a 3×3 lower triangular, unipotent matrix into MATLAB. (Recall that a triangular matrix is unipotent if all the entries on the main diagonal are one.) To avoid trivialities, make all the entries below the main non-zero. Call your matrix V. Let "C=V*B" where B is as in Exercise 2 and then ask MATLAB to compute "[L,U]=lu(C)". Show that EL is lower triangular, where E is as above.

> *Remark:* In general, in the MATLAB command "[L,U]=lu(A)", the L will at least have the property that it becomes unipotent after interchanging a few rows. The U will always be upper triangular. It is a theorem that such a decomposition always exists.

(5) Enter a new upper triangular matrix A and again let $B = EA$ where E is as in Exercise 2. Then, using the same V as in Exercise 5, compute "[L,U]=lu(V*B)". You should get the same L as before. Why? [*Hint*: Compute the **LU** decomposition for VE.]

3 ◢ CHAPTER SUMMARY

In Section 3.1, we noted that multiplication by an $m \times n$ matrix A defines a *transformation* of $\mathbb{R}^n$ into $\mathbb{R}^m$, which is called a *matrix transformation*. Matrix transformations are *linear transformations* due to the linearity properties of matrix multiplication. The *matrix representation theorem* says that every linear transformation from $\mathbb{R}^n$ into $\mathbb{R}^m$ is a matrix transformation.

In Section 3.3, we saw that a matrix transformation cannot increase dimension: an $m \times n$ matrix A will transform $\mathbb{R}^n$ onto a subspace of $\mathbb{R}^m$ (the *image*), which has dimension at most n. In fact, the image is the column space of the matrix and its dimension is the rank r of the matrix. In this case, the dimension of the nullspace is $n - r$ (Theorem 1 in Section 3.3). Note that $n - r$ measures the number of dimensions lost in the transformation process in that we have transformed $\mathbb{R}^n$ onto an r-dimensional subspace. Thus, the dimension of the nullspace measures the number of dimensions lost in the transformation process.

Actually, much more is true. The nullspace itself tells us how the dimensions get lost in the sense that the transformation transforms each translate of the nullspace onto a single vector. Conversely, the set of all vectors that transform onto a given vector is a translate of the nullspace. A consequence of this is that a matrix transformation is *one-to-one* if and only if the nullspace is zero. (One-to-one means that each vector in the image is the image of only one point in the domain.)

In the study of linear algebra, a special role is played by those linear transformations that are both one-to-one and *onto*. (Onto means the image is the whole target space.) In Section 3.3, we called such transformations *invertible*. Matrices that define invertible transformations are called invertible. The significance of this concept is that if A is an invertible $n \times m$ matrix, then the equation $AX = Y$ has one and only one solution X for every Y in $\mathbb{R}^n$. In fact, in Section 3.4 we saw that if A is invertible, then there is a unique matrix (the *inverse matrix*) A^{-1} with the property that $AX = Y$ if and only if $X = A^{-1}Y$. This matrix is also describable as the unique matrix, such that $AA^{-1} = A^{-1}A = I$. A technique for computing matrix inverses was given in Section 3.4.

Not all $n \times n$ matrices A have an inverse: an $n \times n$ matrix A has an inverse if and only if its rank is n. The *inverse theorem* in Section 3.3 tells us that if an $n \times n$ matrix defines a one-to-one transformation, then the transformation it defines is also onto; hence invertible. Similarly, the matrix is invertible if it defines an onto transformation.

Inverses are not necessarily the most efficient way of solving systems. In Section 3.5, we studied the **LU**-factorization, which is a very efficient method of solving the system $AX = Y$ based upon writing A as a product $A = LU$ where L is lower triangular and U is upper triangular with ones on its diagonal. In this form, the **LU** factorization only exists if A can be reduced to echelon form without switching rows and without multiplying any row by a scalar. However, there are more general versions of this factorization that apply to more general matrices. (These were discussed in the On Line section for Section 3.5.)

In Section 3.2, we studied composition of transformations, which is the process of following one transformation with another. Composition of matrix transformations is described by *matrix multiplication*. Matrix multiplication has some very important algebraic properties, including the *left and right distributive laws* and the *associative law*. There are also several important descriptions of matrix multiplication, such as "the columns of AB are A times the columns of B" and "the rows of AB are the rows of A times B." One important result is the *product of ranks theorem* from Section 3.3 that says that the rank of AB is less than or equal to both the rank of A and the rank of B.

CHAPTER 4

ORTHOGONALITY

4.1 COORDINATES

When you first studied graphing in school, you began by introducing Cartesian coordinates into the plane. First, you chose a point **0** in the plane. You then put a horizontal line and a vertical line through **0**. You then chose a unit of measure on each line and proceeded to plot points as indicated in Figure 1 (left). In mathematics, one generally uses the same unit of measure on both axes, although in science it is often necessary to use different units.

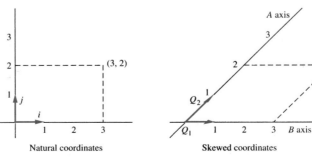

Figure 1

We would like to ask, "Was it really necessary that one axis be horizontal and the other vertical?" In fact, do the axes even need to be perpendicular? The answer, clearly, is no. The purpose of introducing coordinates is to describe points by pairs of numbers. Any method of corresponding points in the plane with pairs of numbers in a one-to-one manner is, in some sense, a "coordinatization" of the plane. This can be done quite effectively using nonperpendicular axes, as the right-hand side of Figure 1 demonstrates.

In this section, we assume that one specific choice of coordinates, using perpendicular axes with the same unit of measure, has been fixed. We shall call the coordinates with respect to this choice of axes **natural coordinates**. Practically speaking, this assumption has the effect of allowing us to think of the plane as equal to $\mathbb{R}^2$. We next suppose that new coordinate axes are chosen, having the same origin. Each point, then, has two sets of coordinates: its natural coordinates and its new coordinates. The most basic questions to be answered are: Given the natural coordinates, how do we compute the new coordinates? and Given the new coordinates, how do we compute the natural coordinates? To consider these questions, we must first consider how we shall represent such coordinates.

In the plane, it is common to use the vectors $\mathbf{i} = [1, 0]^t$ and $\mathbf{j} = [0, 1]^t$ to describe the natural coordinates. A point with natural coordinates $X = [x, y]^t$ may be written

$$X = x\mathbf{i} + y\mathbf{j}$$

The new coordinates may be similarly described using pairs of independent vectors. To explain this, suppose that we are given two axes A and B in the plane as shown in Figure 1. Let Q_1 be the vector of unit length (measured in the units of measure on the A axis) pointing in the positive direction the A axis. Let Q_2 be similarly defined with respect to the B axis. As Figure 1 indicates, a point X has coordinate vector $[x', y']^t$ in the new coordinate system if and only if

$$X = x'Q_1 + y'Q_2 \tag{1}$$

Conversely, any pair of independent vectors Q_1 and Q_2 in $\mathbb{R}^2$ can be used to define coordinates for $\mathbb{R}^2$. Specifically, Q_1 and Q_2 form a basis for $\mathbb{R}^2$; hence any vector X in $\mathbb{R}^2$ may be expressed as in formula (1). We shall show in the discussion following formula (3) that the numbers x' and y' are uniquely determined by X. The vector $X' = [x', y']^t$ is called the **coordinate vector** for X with respect to the vectors Q_1 and Q_2. Notice that in formula (1), the coefficient of the first basis vector (Q_1) defines the first entry of X' and the coefficient of the second basis vector (Q_2) defines the second entry. It follows that the coordinates depend both on the the basis vectors and the order in which they are listed. In general, an **ordered basis** for a vector space is a basis in which the vectors are listed in a particular order.

▶ **EXAMPLE 1:** In $\mathbb{R}^2$, let coordinates be defined by the ordered basis formed by $Q_1 = [1, 0]^t$ and $Q_2 = [1, 1]^t$. What point X has coordinate vector $[-3, 2]^t$?

Solution: From Formula (1) we can write

$$X = -3Q_1 + 2Q_2 = -3\begin{bmatrix}1\\0\end{bmatrix} + 2\begin{bmatrix}1\\1\end{bmatrix} = \begin{bmatrix}-1\\2\end{bmatrix}$$ ◀

Unfortunately, going the other way is not as easy.

▶ EXAMPLE 2:   Use the ordered basis from Example 1 to define coordinates for $\mathbb{R}^2$. What is the coordinate vector for the point $X = [4, 5]^t$?

Solution: Let the coordinate vector be $[x', y']^t$. From Formula (1) we have

$$\begin{bmatrix}4\\5\end{bmatrix} = x'\begin{bmatrix}1\\0\end{bmatrix} + y'\begin{bmatrix}1\\1\end{bmatrix} = \begin{bmatrix}1&1\\0&1\end{bmatrix}\begin{bmatrix}x'\\y'\end{bmatrix}$$

We solve this system by inverting the 2×2 matrix on the right and multiplying $[4, 5]^t$ by it. We get

$$\begin{bmatrix}x'\\y'\end{bmatrix} = \begin{bmatrix}1&-1\\0&1\end{bmatrix}\begin{bmatrix}4\\5\end{bmatrix} = \begin{bmatrix}-1\\5\end{bmatrix}$$

The matrix

$$Q = \begin{bmatrix}1&1\\0&1\end{bmatrix}$$

is called the "point matrix" because multiplication of the coordinate vector by it produces the point with these coordinates. The inverse of the point matrix is called the "coordinate matrix" because multiplication by it produces the coordinate vector. ◀

In Example 1, the Q_1 and Q_2 axes have different units. To explain this, note that the point with coordinates $[0, 1]^t$ is just

$$0Q_1 + Q_2 = [1, 1]^t$$

From the distance formula in $\mathbb{R}^2$, the distance from this point to the origin is $\sqrt{2}$. Thus, each unit of distance on the Q_2 axis is really $\sqrt{2}$ units in real terms. This, of course, is because the vector Q_2 has length $\sqrt{2}$. Clearly, Q_1 has length 1, so one unit of length along the Q_1 axis is also one unit in "real terms." In general, a basis is said to be "normal" if each basis vector has length one. Normal bases are significant because for points on the axes, the coordinates correspond to distance from the origin.

Certainly, the most useful coordinates are those defined by perpendicular axes. Determining whether the axes are perpendicular is easy. Recall that two vectors $X = [x_1, x_2]^t$ and $Y = [y_1, y_2]^t$ are perpendicular if and only if $X \cdot Y = 0$ where

$$X \cdot Y = x_1 y_1 + x_2 y_2$$

The axes will be perpendicular if and only if the basis vectors are perpendicular. Such bases are called "orthogonal" bases. (Orthogonal is a synonym for "perpendicular.") Orthogonal bases are especially nice because there is a technique that allows us to compute coordinates without inverting the point matrix.

▶ EXAMPLE 3:   Show that the vectors $Q_1 = [2, -3]^t$ and $Q_2 = [3, 2]^t$ form an orthogonal basis for $\mathbb{R}^2$. Compute the coordinates of the point $X = [4, -3]^t$ with respect to the ordered basis formed by Q_1 and Q_2.

Solution: We compute

$$Q_1 \cdot Q_2 = 2 \cdot 3 + (-3) \cdot 2 = 6 - 6 = 0$$

showing the orthogonality.

The coordinates of X are defined by the equation

$$X = x' Q_1 + y' Q_2$$

We dot both sides of this equation with Q_1:

$$X \cdot Q_1 = (x' Q_1 + y' Q_2) \cdot Q_1 = x' Q_1 \cdot Q_1 + y' Q_2 \cdot Q_1 = x' Q_1 \cdot Q_1$$

Hence

$$x' = \frac{X \cdot Q_1}{Q_1 \cdot Q_1} = \frac{4 \cdot 2 + (-3) \cdot (-3)}{2^2 + (-3)^2} = \frac{17}{13}$$

Similarly

$$y' = \frac{X \cdot Q_2}{Q_2 \cdot Q_2} = \frac{4 \cdot 3 + (-3) \cdot 2}{2^2 + (-3)^2} = \frac{6}{13}$$

As a check on our work, we compute

$$\frac{17}{13}[2, -3]^t + \frac{6}{13}[3, 2]^t = [4, -3]^t$$

as expected.   ◀

The Q_i in Example 3 are orthogonal, but not normal. In fact, if $X = [x_1, x_2]'$, as indicated in Figure 2, then its length is

$$|X| = \sqrt{x_1^2 + x_2^2} \tag{2}$$

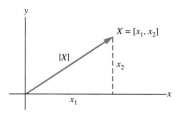

Figure 2

Thus

$$|Q_1| = \sqrt{(-3)^2 + 2^2} = \sqrt{13}$$

Similarly, $|Q_2| = \sqrt{13}$.

A basis consisting of vectors that are both orthogonal and normal is called "orthonormal." Orthonormal bases describe perpendicular axes, with the standard units of measure on each axis. Thus, for example, if you rotate the natural axes by some angle, you produce an orthonormal system. We can always convert an orthogonal basis into an orthonormal basis by dividing each basis vector by its length. (This process is called normalization.) If, then, we normalize the Q_i above, we obtain the orthonormal basis

$$P_1 = [\frac{2}{\sqrt{13}}, \frac{-3}{\sqrt{13}}]' \text{ and } P_2 = [\frac{3}{\sqrt{13}}, \frac{2}{\sqrt{13}}]'$$

Geometrically, normalization amounts to changing the units of measure on the axes.

The ideas just discussed extend far beyond $\mathbb{R}^2$. Let $\mathcal{V}$ be an n-dimensional vector space and let $Q_1, Q_2, \ldots, Q_n$ be a basis for $\mathcal{V}$, listed in a particular order. Since the Q_i span $\mathcal{V}$, any X in $\mathcal{V}$ may be written as a linear combination of the Q_i:

$$X = x_1' Q_1 + x_2' Q_2 + \cdots + x_n' Q_n \tag{3}$$

There is, in fact, only one way to express X as a linear combination of the X_i. To see this, suppose that

$$X = y_1 Q_1 + y_2 Q_2 + \cdots + y_n Q_n$$

Subtracting these two equalities, we see that

$$\mathbf{0} = (x_1' - y_1)Q_1 + (x_2' - y_2)Q_2 + \cdots + (x_n' - y_n)Q_n$$

Since the Q_i are independent, all of the coefficients in this linear combination must be zero, showing that for all i, $x_i' = y_i$, as claimed.

The vector $X' = [x_1', x_2', \ldots, x_n']^t$ is the **coordinate vector** for X with respect to the ordered basis defined by the Q_i, and the numbers x_i' are the **coordinates** for X.

Notice that if $\mathcal{V} = \mathbb{R}^n$, then the Q_i are column vectors. Hence, equation (3) may be written as a matrix product as

$$X = QX'$$

where $Q = [Q_1, Q_2, \ldots, Q_n]$. As above, Q is called the **point matrix**. This matrix is invertible, since it has rank n. (The columns are linearly independent since the Q_i are assumed to be a basis for $\mathbb{R}^n$.) The inverse matrix to Q is called the **coordinate matrix**; multiplication of points in $\mathbb{R}^n$ by the coordinate matrix produces the coordinate vector of the point.

▶ **EXAMPLE 4:**   Let coordinates in $\mathbb{R}^3$ be defined by the ordered basis formed by the vectors

$$Q_1 = [1, 1, 0]^t \quad Q_2 = [1, -1, 1]^t \quad Q_3 = [1, -1, -2]^t$$

Calculate the coordinates of the point $X = [1, 1, 1]^t$. Show that the coordinates do express X as a linear combination of the given basis. What are the point and coordinate matrices for this basis?

Solution: The coordinates are defined by the equality

$$\begin{bmatrix} 1 \\ 1 \\ 1 \end{bmatrix} = x' \begin{bmatrix} 1 \\ 1 \\ 0 \end{bmatrix} + y' \begin{bmatrix} 1 \\ -1 \\ 1 \end{bmatrix} + z' \begin{bmatrix} 1 \\ -1 \\ -2 \end{bmatrix} = \begin{bmatrix} 1 & 1 & 1 \\ 1 & -1 & -1 \\ 0 & 1 & -2 \end{bmatrix} \begin{bmatrix} x' \\ y' \\ z' \end{bmatrix}$$

The 3×3 matrix in the above formula is the point matrix Q.

We solve the above system by computing $C = Q^{-1}$. This is the coordinate matrix for the given basis. We find

$$C = \frac{1}{6} \begin{bmatrix} 3 & 3 & 0 \\ 2 & -2 & 2 \\ 1 & -1 & -2 \end{bmatrix}$$

The coordinate vector of X is $X' = CX = [1, \frac{1}{3}, -\frac{1}{3}]'$. As a check of our computations, we see that

$$Q_1 + \frac{1}{3}Q_2 - \frac{1}{3}Q_3 = [1, 1, 0]' + \frac{1}{3}[1, -1, 1]' - \frac{1}{3}[1, -1, -2]' = [1, 1, 1]' = X \quad \blacktriangleleft$$

The concepts of orthogonality and orthonormality extend readily to $\mathbb{R}^n$. If $X = [x_1, x_2, \ldots, x_n]'$ and $Y = [y_1, y_2, \ldots, y_n]'$ are two vectors in $\mathbb{R}^n$, then we define

$$X \cdot Y = x_1 y_1 + x_2 y_2 + \cdots + x_n y_n$$

and

$$|X| = \sqrt{X \cdot X} = \sqrt{x_1^2 + x_2^2 + \cdots + x_n^2}$$

Thus, for example

$$[1, 2, 3, 4]' \cdot [2, -4, 5, 3]' = 2 - 8 + 15 + 12 = 21$$

and

$$|[1, 2, 3, 4]'| = \sqrt{1 + 2^2 + 3^2 + 4^2} = \sqrt{30}$$

There is a surprisingly useful description of the dot product in terms of matrix multiplication. Specifically, note the matrix product formula

$$\begin{bmatrix} x_1 & x_2 & \cdots & x_n \end{bmatrix} \begin{bmatrix} y_1 \\ y_2 \\ \vdots \\ y_n \end{bmatrix} = x_1 y_1 + x_2 y_2 + \cdots + x_n y_n$$

This formula says that for X and Y in $\mathbb{R}^n$,

$$X' Y = X \cdot Y \tag{4}$$

A direct consequence of this is the following:

Dot Product Theorem.   *Let X, Y, and Z be vectors in $\mathbb{R}^n$. Let c be a scalar. Then*

(a) $X \cdot Y = Y \cdot X$ *(commutative law)*
(b) $Z \cdot (X + Y) = Z \cdot X + Z \cdot Y$ *(additive law)*
(c) $(cX) \cdot Y = c(X \cdot Y) = X \cdot (cY)$ *(scalar law)*
(d) $|X|^2 = X \cdot X$ ■

As in $\mathbb{R}^2$, we say that two vectors X and Y in $\mathbb{R}^n$ are **perpendicular** if $X \cdot Y = 0$. A set of non-zero vectors is **orthogonal** if each vector in the set is perpendicular to every other vector in the set. A set of vectors is **normal** if each vector in the set has length 1. A set that is both orthogonal and normal is, once again, **orthonormal**. The technique we used in Example 3 applies to orthogonal bases in any dimension.

▶ **EXAMPLE 5:**   Show that the basis from Example 4 is orthogonal. Use the orthogonality to redo Example 4.

Solution: To prove orthogonality we must show that each basis vector is perpendicular to every other basis vector. Thus, we compute

$$Q_1 \cdot Q_2 = [1, 1, 0]^t \cdot [1, -1, 1]^t = 1 \cdot 1 + 1 \cdot (-1) + 0 \cdot 1 = 0$$

Similarly, we compute $Q_1 \cdot Q_3 = 0$ and $Q_2 \cdot Q_3 = 0$, showing the orthogonality.
The coordinates of $X = [1, 1, 1]^t$ are defined by the equation

$$X = x'Q_1 + y'Q_2 + z'Q_3$$

Dotting both sides of this equation with Q_1, we see

$$\begin{aligned} X \cdot Q_1 &= (x'Q_1 + y'Q_2 + z'Q_3) \cdot Q_1 \\ &= x'Q_1 \cdot Q_1 + y'Q_2 \cdot Q_1 + z'Q_3 \cdot Q_1 = x'Q_1 \cdot Q_1 \end{aligned}$$

The last equality follows from the perpendicularity of Q_1 with Q_2 and Q_3. We arrive at the formula

$$x' = \frac{X \cdot Q_1}{Q_1 \cdot Q_1} = \frac{1 + 1 + 0}{1^2 + 1^2} = 1$$

Similarly,

$$y' = \frac{X \cdot Q_2}{Q_2 \cdot Q_2} = \frac{1 - 1 + 1}{1^2 + (-1)^2 + 1^2} = \frac{1}{3}$$

and

$$z' = \frac{X \cdot Q_3}{Q_3 \cdot Q_3} = \frac{1 - 1 - 2}{1^2 + 1^2 + 2^2} = -\frac{1}{3}$$

This is, of course, in agreement with what we found in Example 3.   ◀

The technique we used in the last two examples is important enough to be called a theorem.

Theorem 1.   *Let* $S = \{Q_1, Q_2, \ldots, Q_n\}$ *be an orthogonal basis for* $\mathbb{R}^n$ *and let* $X \in \mathbb{R}^n$. *Then*

$$X = x_1 Q_1 + x_2 Q_2 + \cdots + x_n Q_n$$

where

$$x_i = \frac{X \cdot Q_i}{Q_i \cdot Q_i} \qquad \blacksquare$$

Notice that in Example 4 it was given that the Q_i formed a basis. We, of course, could have checked this by showing that the Q_i were independent. (We need to show only independence, since $\mathbb{R}^3$ is three-dimensional.) However, we saw in Example 5 that the Q_i are mutually perpendicular. It seems apparent that they are independent— how could, say, Q_3 be a linear combination of Q_1 and Q_2 if it is perpendicular to the plane they span? (See Figure 3.) Proving independence involves nothing more than repeating the technique used in Example 5. Explicitly, consider the dependence equation

$$x_1 Q_1 + x_2 Q_2 + x_3 Q_3 = \mathbf{0}$$

We dot both sides with Q_1, producing

$$x_1 Q_1 \cdot Q_1 + x_2 Q_2 \cdot Q_1 + x_3 Q_3 \cdot Q_1 = 0$$

which yields

$$x_1 Q_1 \cdot Q_1 = 0$$

since $Q_i \cdot Q_j = 0$ for $i \neq j$. We conclude that $x_1 = 0$ since $Q_1 \cdot Q_1 \neq 0$. Similarly, dotting by Q_2 and Q_3 proves that x_2 and x_3 both equal 0.

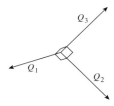

Three mutually perpendicular vectors in 3-space

Figure 3

This example demonstrates another general principle. A set $\{Q_1, Q_2, \ldots, Q_k\}$ of vectors in $\mathbb{R}^n$ is said to be an orthogonal set if none of the Q_i are zero and each Q_i is perpendicular to all of the other Q_j. Using the dot product just as before, we can prove that *orthogonal sets of vectors are independent.* Thus, *in $\mathbb{R}^n$, there can exist at most n mutually perpendicular, non-zero vectors.*

EXERCISES

(1) The next three exercises demonstrate that an appropriate choice of coordinates can simplify the form of some equations. The issue choosing coordinates is discussed in Section 6.4.

(a) Sketch the graph of the curve $y = 1/x$. Use the lines $y = x$ and $y = -x$ as new coordinate axes. Show that relative to these new axes, the curve is describable by a formula of the form

$$\frac{(x')^2}{a^2} - \frac{(y')^2}{b^2} = c$$

where a, b, and c are constants. Note that this proves that $y = 1/x$ represents a hyperbola.

[*Note:* To do this exercise, you will need to choose a vector X on the line $y = x$ and a vector Y on the line $y = -x$. Let $[x', y']^t$ be the coordinate vector for $[x, y]^t$ relative to this basis and compute expressions for x and y in terms of x' and y'. If you substitute these expressions into the equation $xy = 1$, you should obtain a formula of the desired form.]

(b) Figure 4 is a rough sketch of the graph of the equation $x^2 + y^2 + xy = 1$. It appears to represent an ellipse with its major axis along the line $y = -x$ and its minor axis along the line $y = x$. If we use these lines as coordinate axes, what will be the equation of this curve with respect to these coordinates? [*Hint:* See the note for part (a).]

(c) A curve is described by the equation $13x^2 - 8xy + 7y^2 = 5$. Use the basis $X = [1, 2]^t$ and $Y = [-2, 1]^t$ to define new coordinates. What is the formula for this curve in these coordinates? Draw a graph of this curve. What is the "width" of the figure? [*Hint:* See the note for part (a).]

(2) For each of the following ordered bases for $\mathbb{R}^3$, find the point matrix and the coordinate matrix. Let $X = [1, 2, 3]^t$. For each basis, find the coordinates of X with respect to the given basis. Check your answer by showing that your coordinates really do express X as a linear combination of the basis elements.

(a) $S = \{[1, 1, 1]^t, [0, 1, 1]^t, [0, 0, 1]\}$

(b) $S = \{[1, -2, 1]^t, [2, 3, 2]^t, [1, 1, 0]^t\}$

(c) $S = \{[1, 3, 2]^t, [-1, 1, -1]^t, [5, 1, -4]^t\}$

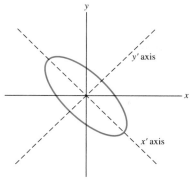

Figure 4

(3) One (and only one) of the bases in Exercise 2 is orthogonal.

(a) Which one is it?
(b) Use the technique from Example 5 to compute the coordinates of $X = [1, 2, 3]'$.
(c) Use the technique from Example 5 to compute the coordinates of $X = [x, y, x]'$.
(d) Use your answer to write down the coordinate matrix for this basis. (Remember that multiplication of X by the coordinate matrix computes the coordinate vector.)

(4) Normalize the basis from Example 4 in the text so as to obtain an orthonormal basis P_1, P_2, and P_3.

(a) What is the point matrix for this basis?
(b) Use the technique from Example 5 to compute the coordinates of $X = [x, y, x]'$.
(c) Use your answer from part (b) to find the coordinate matrix. (Remember that multiplication of X by the coordinate matrix computes the coordinate vector.)
(d) How do the answers to parts (a) and (c) relate to each other? Do you see a general principle at work? If so, what? Can you prove it?

(5) Show that the following vectors form an orthogonal basis for $\mathbb{R}^4$. Find the coordinates for the point $X = [1, 2, -1, -3]$ with respect to this basis.

$$[2, -1, -1, -1]', \ [1, 3, 3, -4]', \ [1, 1, 0, 1]', \ [1, -2, 3, 1]'$$

(6) Show that the three vectors below form an orthogonal set in $\mathbb{R}^4$. Find a fourth vector Q_4 such that $\{Q_1, Q_2, Q_3, Q_4\}$ forms an orthogonal basis in $\mathbb{R}^4$.

$$Q_1 = [1, 1, 1, 1]', \ Q_2 = [1, -2, 1, 0]', \ Q_3 = [1, 1, 1, -3]'$$

To what extent is Q_4 unique?

(7) Let $\{Q_1, Q_2, Q_3\}$ be a set of three orthogonal vectors in $\mathbb{R}^4$. Prove that there is a fourth vector Q_4 such that $\{Q_1, Q_2, Q_3, Q_4\}$ forms an orthogonal basis for $\mathbb{R}^4$. To what extent is Q_4 unique? Prove your answer. What if the vectors were in $\mathbb{R}^5$? Would Q_4 exist? To what extent would it be unique?

(8) In the text we showed that any orthogonal set of three vectors in $\mathbb{R}^n$ is linearly independent. Prove that any orthogonal set of four vectors is independent.

(9) In the text we only proved Theorem 1 for $\mathbb{R}^3$. Prove it for $\mathbb{R}^4$. Then prove it for $\mathbb{R}^n$.

(10) We have seen previously that the polynomials 1, x, and x^2 form a basis for the space P_2 of polynomials of degree less than or equal to 2. When we write the general element of P_2 as

$$p(x) = a + bx + cx^2$$

we are expressing the general element of P_2 as a linear combination of the basis elements. It follows that the coordinate vector for $p(x)$ with respect to this basis in the stated order is $[a, b, c]^t$. Thus, for example, the coordinate vector for $p(x) = 2 + 3x + x^2$ with respect to this ordered basis is $[2, 3, 1]^t$. Conversely, the element of P_2 with coordinates $[3, -5, 2]^t$ is $3 - 5x + 2x^2$.

We can use coordinate vectors to convert questions about polynomials into questions about vectors. For example, suppose that we want to know if the polynomials $p(x) = 3 + x + x^2, q(x) = 1 - 4x + 5x^2$, and $r(x) = 7 + 11x - 7x^2$ form a dependent set. We note that the corresponding coordinate vectors are dependent since

$$[7, 11, -7]^t = 3[3, 1, 1]^t - 2[1, -4, 5]^t$$

This suggests that $r(x) = 3p(x) - 2q(x)$. The reader can verify that this is indeed the case.

In this exercise, you are given dependent sets of polynomials. In each case, write one of the coordinate vectors as a linear combination of other coordinate vectors. (You may need to use the technique of Example 3, Section 2.1.) Finally, express the corresponding polynomial as a linear combination of the other polynomials. Use the ordered basis formed by $1, x, \ldots, x^n$ to define the coordinates for P_n.

(a) $\{12x + 14x^2, 1 + 2x + 3x^3, 2 - 2x - x^2\}$
(b) $\{1 - 4x + 4x^2 + 4x^3, 2 - x + 2x^2 + x^3, 3 + 2x - 2x^3\}$
(c) $\{1 + 2x + 3x^3, 2 + x + x^2 + x^3, 1 + 4x + 3x^2 - 3x^3, 3 + 15x + 8x^2 - 4x^3, 3 - 11x - 9x^2 + 18x^3\}$ [*Hint*: The first part of this exercise was done for you in Example 3, Section 2.2]

(11) Let $\mathcal{V}$ be an n-dimensional vector space and let $Q_1, Q_2, \ldots, Q_n$ form an ordered basis for $\mathcal{V}$. Let X, Y, and Z be elements of $\mathcal{V}$ and suppose that the coordinate

vectors for these elements with respect to the stated basis are respectively X', Y', and Z'. Suppose that Z' is a linear combination of X' and Y'. Prove that Z is a linear combination of X and Y.

Remark: Exercise 11 is the mathematical justification for the technique demonstrated in Exercise 10 above.

(12) We have seen that the polynomials 1, $x - 1$, and $(x - 1)^2$ also form a basis for P_2. We can use this basis in the stated order to define coordinates for P_2 as well. Computing coordinates with respect to this basis requires writing a given polynomial as a linear combination of the basis elements. For example, since $(a + b)^2 = a^2 + 2ab + b^2$, we see that

$$x^2 = (1 + (x - 1))^2 = 1 + 2(x - 1) + (x - 1)^2$$

Thus, the coordinate vector for x^2 with respect to this basis is $[1, 2, 1]^t$. Compute the coordinate vectors for the following polynomials with respect to the ordered basis formed by 1, $x - 1$, and $(x - 1)^2$.

(a) x
(b) $1 + 3x + x^2$
(c) $5 - 3x + 2x^2$

(13) The polynomials 1, $x + 3$, and $(x + 3)^2$ also form an ordered basis for P_2. Compute the coordinates of the polynomials in Exercise 12 with respect to this basis.

ON LINE

The following MATLAB commands generate a piece of graph paper for the coordinates in $\mathbb{R}^2$ defined by the basis $Q_1 = [1, 2]^t$ and $Q_2 = [-1, 1]^t$. The "grid on" command causes it to be overlaid on top of a rectangular grid. Begin by initializing the coordinate axes with the commands:

```
cla; hold on; axis equal; grid on;
Q1=[1;2]; Q2=[-1; 1];
A=[5*Q1,-5*Q1] ; B=[5*Q2,-5*Q2]
plot(A(1,:),A(2,:),'r'); plot(B(1,:),B(2,:),'r');
for n=-5:5,
  plot(A(1,:)+n*Q2(1),A(2,:)+n*Q2(2),'r:');
  plot(B(1,:)+n*Q1(1),B(2,:)+n*Q1(2),'r:');
end
```

(1) Modify the above sequence of code to produce a sheet of graph paper for the coordinates defined by the basis $Q_1 = [1, 2]^t$ and $Q_2 = [1, -2]^t$.

(2) The curve defined by the equation $x^2 - y^2/4 = 1$ represents the hyperbola of Figure 5. Show that this hyperbola is given by the equation $x'y' = 4$ in coordinates relative to the basis from Exercise 1.

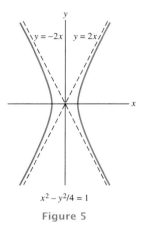

$$x^2 - y^2/4 = 1$$

Figure 5

(3) The following sequence of commands plots the right side of the hyperbola of Figure 5. (Note the period before the caret.)

```
x=.8:.1:5; y=4*(x.^(-1)); C=Q1*x+Q2*y;
plot(C(1,:),C(2,:));
```

Execute the above sequence of commands to produce a plot of the hyperbola. Explain what each command does. To help you in understanding what these commands do, get MATLAB to print out both x and y. How would you describe x? How does y relate to x? What is the size of the matrix Q1$*x$? How does each column of this matrix relate to x and Q1? The "plot" command plots the first row of Q1$*x$+Q2$*y$ against the second. Why does this produce the right side of Figure 5?

(4) Pick a point (reader's choice) on the graph of the hyperbola from Exercise 3. Use your graph to estimate both the new and natural coordinates of this point. Call these coordinates $[x', y']^t$ and $[x, y]^t$ respectively. Show that (approximately) $x'y' = 4$ and $x^2 - y^2/4 = 1$. Repeat for another point on the graph.

4.2   PROJECTIONS: THE GRAM–SCHMIDT PROCESS

Let us imagine that we are studying a particle that is moving in a direction defined by a vector Q, and that there is a force acting on the particle defined by a vector B. Often in such problems it is useful to think of the force B as the sum of two forces B_1 and B_0 where B_0 acts in the direction of the motion and B_1 is perpendicular to the direction of motion, as in Figure 1. The vector B_0 is called the **projection of B onto Q** and B_1 is referred to as the **complementary vector**.

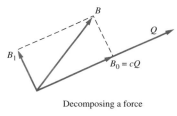

Decomposing a force

Figure 1

Computing the projection vector and the corresponding complementary vector is not hard. Assume that we choose our coordinates so that the particle is at the origin. Since B_0 points along Q, there is a scalar c such that

$$B_0 = c\,Q$$

What we need to know is how to choose c so that $B - cQ$ is perpendicular to Q. But, two vectors are perpendicular if and only if their dot product is 0. Hence

$$0 = (B - cQ) \cdot Q = B \cdot Q - c(Q \cdot Q)$$

Solving for c yields

$$c = \frac{B \cdot Q}{Q \cdot Q}$$

and thus

$$B_0 = \frac{B \cdot Q}{Q \cdot Q}\,Q \qquad (1)$$

We may then find B_1 simply from $B_1 = B - B_0$.

▶ **EXAMPLE 1:**   Find the projection of the vector $B = [1, 2, 3]^t$ onto $Q = [1, -1, 1]^t$. Also find the complementary vector B_1 and check that B_1 is perpendicular to Q.

Solution: According to formula (1),

$$B_0 = \frac{B \cdot Q}{Q \cdot Q} Q = \frac{1 - 2 + 3}{1 + 1 + 1}[1, -1, 1]^t = \left[\frac{2}{3}, -\frac{2}{3}, \frac{2}{3}\right]^t$$

The complementary vector is

$$B_1 = B - B_0 = \left[\frac{1}{3}, \frac{8}{3}, \frac{7}{3}\right]^t$$

It is easily checked that $B_1 \cdot Q = 0$.   ◀

In Example 1, we projected a vector onto a line determined by another vector. We can also project vectors onto planes, as indicated in Figure 2. In this case, we wish to resolve our vector into a sum of two vectors, one lying in the plane and one perpendicular to the plane (i.e., perpendicular to every vector in the plane). More generally, given a subspace $\mathcal{W}$ of $\mathbb{R}^n$, and a vector $B \in \mathbb{R}^n$, we can hope to find a point $B_0 \in \mathcal{W}$ such that $B_1 = B - B_0$ is perpendicular to every vector in $\mathcal{W}$.

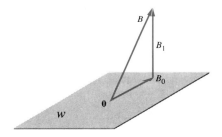

The projection of B onto $\mathcal{W}$

Figure 2

We shall show that B_0 exists and is unique. (We will present a different proof in Section 4.5.)

Our starting point will be an orthogonal basis for $\mathcal{W}$. By this we mean a basis $\{Q_1, Q_2, \ldots, Q_m\}$ such that

$$Q_i \cdot Q_j = 0 \qquad (i \neq j) \tag{2}$$

Since B_0 (if it exists) lies in $\mathcal{W}$, there are constants c_i such that

$$B_0 = c_1 Q_1 + c_2 Q_2 + \cdots + c_m Q_m$$

As in the case of a line, the question is how to choose the c_i so that $B - B_0$ is perpendicular to $\mathcal{W}$. Notice, however, that the dot product of Q_i with any other Q_j is zero. Hence

$$B_0 \cdot Q_i = c_1 Q_1 \cdot Q_i + c_2 Q_2 \cdot Q_i + \cdots + c_m Q_m \cdot Q_i = c_i Q_i \cdot Q_i$$

Thus

$$(B - B_0) \cdot Q_i = B \cdot Q_i - B_0 \cdot Q_i = B \cdot Q_i - c_i Q_i \cdot Q_i$$

The vector $B - B_0$ will be perpendicular to $\mathcal{W}$ if and only if the expression above is zero for all i, which is the same as

$$c_i = \frac{B \cdot Q_i}{Q_i \cdot Q_i}$$

Thus, we have proven the following theorem, which tells us how to find projections.

Fourier Theorem.   *Let $\{Q_1, Q_2, \ldots, Q_m\}$ be an orthogonal basis for a subspace $\mathcal{W}$ of $\mathbb{R}^n$ and let $B_0 \in \mathbb{R}^n$. Then the vector B_0 in formula (3) is the unique vector in $\mathcal{W}$ such that $B_1 = B - B_0$ is perpendicular to $\mathcal{W}$.*

$$B_0 = \frac{B \cdot Q_1}{Q_1 \cdot Q_1} Q_1 + \frac{B \cdot Q_2}{Q_2 \cdot Q_2} Q_2 + \ldots + \frac{B \cdot Q_m}{Q_m \cdot Q_m} Q_m \qquad (3)$$

■

The vector B_0 in the Fourier theorem is called the **projection of B onto $\mathcal{W}$** and is denoted

$$B_0 = \text{proj}_{\mathcal{W}}(B)$$

The formula for B_0 in the Fourier theorem should look familiar. If the Q_i were an orthogonal basis for all of $\mathbb{R}^n$, this formula would be exactly the expression of B in terms of the basis introduced in Section 4.1. The computations involved in doing specific examples are almost identical with those done in Section 4.1.

The shadow cast by an object when the sun is directly overhead can be modeled as the orthogonal projection of the object onto the plane of the earth.

▶ **EXAMPLE 2:** Let W be the subspace of $\mathbb{R}^3$ spanned by the vectors Q_1 and Q_2:

$$Q_1 = [1, -1, 1]^t \text{ and } Q_2 = [1, -1, -2]^t.$$

Show that these vectors define an orthogonal basis for W and compute the projection B_0 of $B = [1, 1, 1]^t$ onto this subspace. Show that $B - B_0$ is perpendicular to both Q_1 and Q_2.

Solution: It is easily seen that $Q_1 \cdot Q_2 = 0$; hence the Q_i form an orthogonal basis for W. From formula (3), we can write

$$B_0 = \frac{B \cdot Q_1}{Q_1 \cdot Q_1} Q_1 + \frac{B \cdot Q_2}{Q_2 \cdot Q_2} Q_2 = \frac{1}{3} Q_1 + \frac{-2}{6} Q_2 = [0, 0, 1]^t.$$

As a check, we compute that

$$B - B_0 = [1, 1, 1]^t - [0, 0, 1]^t = [1, 1, 0]^t$$

This is clearly orthogonal to both Q_1 and Q_2. ◀

Remark: The reader should compare the work done in this example with the work in Example 5 of Section 4.1.

To compute projections using formula (3), we need an orthogonal basis for $\mathcal{W}$. Unfortunately, most of the bases that occur in linear algebra are not orthogonal (e.g., row spaces, column spaces, spanning vectors for homogeneous systems). Fortunately, there is a process for converting nonorthogonal bases into orthogonal bases, the so-called Gram–Schmidt process.

Suppose that we are given a basis $\{A_1, A_2, \ldots, A_n\}$ for a subspace $\mathcal{W}$ of $\mathbb{R}^n$. We will "straighten out" each of the A_i, one at a time, to produce an orthogonal basis $\{Q_1, Q_2, \ldots, Q_n\}$ for $\mathcal{W}$.

The first step is simple enough. We set

$$Q_1 = A_1$$

This is our first "new" basis element.

To define the next "new" basis element, let B_2 be the projection of A_2 onto Q_1 and let

$$Q_2 = A_2 - B_2$$

From formula (3) [or formula (1)]

$$B_2 = \frac{A_2 \cdot Q_1}{Q_1 \cdot Q_1} Q_1$$

It is clear that Q_2 is perpendicular to Q_1 (Figure 3, left). We think of Q_2 as having been obtained by somehow "straightening" A_2.

We claim that Q_1 and Q_2 span the same subspace as A_1 and A_2. In fact, we need only show that A_1 and A_2 are combinations of the Q_i, since the Q_i are clearly combinations of the A_i. But $A_1 = Q_1$ and $A_2 = Q_2 + B_2$. B_2 is, by definition, a combination of the Q_i.

Our next step, as Figure 3 (right) indicates, is to take the projection of A_3 onto the plane spanned by Q_1 and Q_2. We call this element B_3. We then define

$$Q_3 = A_3 - B_3$$

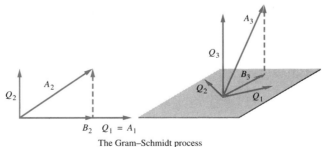

The Gram–Schmidt process

Figure 3

From the Fourier theorem, we write

$$B_3 = \frac{A_3 \cdot Q_1}{Q_1 \cdot Q_1} Q_1 + \frac{A_3 \cdot Q_2}{Q_2 \cdot Q_2} Q_2$$

Again, Q_3 is perpendicular to Q_2 and Q_1 and these three vectors span the same set as does $\{A_1, A_2, A_3\}$. (Note that $A_3 = Q_3 + B_3$.) As before, we think of Q_3 as being obtained by "straightening" A_3.

We may continue this process until we run out of basis elements for $\mathcal{W}$. Specifically, suppose that we have defined $Q_1, Q_2, \ldots, Q_k$ so that the Q_i are mutually orthogonal and span the same subspace as $A_1, A_2, \ldots, A_k$. To define the next Q, we let B_{k+1} be the projection of A_{k+1} onto the span of the Q_i. This amounts to setting

$$B_{k+1} = \frac{A_{k+1} \cdot Q_1}{Q_1 \cdot Q_1} Q_1 + \frac{A_{k+1} \cdot Q_2}{Q_2 \cdot Q_2} Q_2 + \cdots + \frac{A_{k+1} \cdot Q_k}{Q_k \cdot Q_k} Q_k$$

We then set

$$Q_{k+1} = A_{k+1} - B_{k+1}$$

It is clear that, in this manner, we produce an orthogonal basis $\{Q_1, Q_2, \ldots, Q_n\}$ that spans $\mathcal{W}$. In particular, we have proved the following theorem.

Theorem 1.   *Every subspace of $\mathbb{R}^n$ has an orthogonal basis.*   ∎

▶ EXAMPLE 3:   Find an orthogonal basis for the solution set to

$$2x + y + 3z - w = 0$$

Solution: The general solution to this system is

$$\left[-\frac{1}{2}, 1, 0, 0\right]^t y + \left[-\frac{3}{2}, 0, 1, 0\right]^t z + \left[\frac{1}{2}, 0, 0, 1\right]^t w$$ ◀

To avoid fractions, we multiply each basis element by 2, producing the basis

$$A_1 = [-1, 2, 0, 0]^t \quad A_2 = [-3, 0, 2, 0]^t \quad A_3 = [1, 0, 0, 2]^t.$$

Then, from the formulas above

$$Q_1 = A_1 = [-1, 2, 0, 0]^t$$
$$B_2 = \frac{A_2 \cdot Q_1}{Q_1 \cdot Q_1} Q_1 = \tfrac{3}{5}[-1, 2, 0, 0]^t$$
$$Q_2 = A_2 - B_2 = [-3, 0, 2, 0]^t - \tfrac{3}{5}[-1, 2, 0, 0]^t = \tfrac{2}{5}[-6, -3, 5, 0]^t$$

And

$$B_3 = \frac{A_3 \cdot Q_1}{Q_1 \cdot Q_1} Q_1 + \frac{A_3 \cdot Q_2}{Q_2 \cdot Q_2} Q_2$$

$$= \left(-\tfrac{1}{5}\right) Q_1 + \left(-\tfrac{3}{14}\right) Q_2 = \tfrac{1}{7}[5, -1, -3, 0]^t$$

$$Q_3 = A_3 - B_3 = \tfrac{1}{7}[2, 1, 3, 14]^t$$

As in Section 4.1, an orthogonal basis in which every basis element has length one is called **orthonormal**. Since we may always normalize an orthogonal basis by dividing each basis element by its length, it follows from Theorem 1 that every subspace of $\mathbb{R}^n$ has an orthonormal basis.

Orthonormal bases can help us to understand the geometry of subspaces of $\mathbb{R}^n$. To explain this, let $\mathcal{W}$ be a 2-dimensional subspace of $\mathbb{R}^3$. We picture $\mathcal{W}$ as a plane through the origin. We choose an orthonormal basis for $\mathcal{W}$ consisting of two vectors Q_1 and Q_2 which we use to define coordinates for $\mathcal{W}$ (Figure 4). Every point X in $\mathcal{W}$ is defined by its coordinate vector $[x', y']^t$ and every such pair of numbers defines a unique point in $\mathcal{W}$. This allows us to think of $\mathcal{W}$ as being, in some sense, equal to $\mathbb{R}^2$.

We can even use formulas that are valid for $\mathbb{R}^2$ to do computations in $\mathcal{W}$. For example, we claim that we can use the $\mathbb{R}^2$ formula for vector addition to add elements of $\mathcal{W}$. To see this, suppose that X and Y are elements of $\mathcal{W}$ whose coordinate vectors are respectively $[x_1, y_1]^t$ and $[x_2, y_2]^t$. Then

$$X = x_1 Q_1 + y_1 Q_2$$
$$Y = x_2 Q_1 + y_2 Q_2$$

Hence,

$$X + Y = (x_1 + x_2) Q_1 + (y_1 + y_2) Q_2$$

This shows that the coordinate vector for $X + Y$ is

$$[x_1 + x_2, y_1 + y_2]^t = [x_1, y_1]^t + [x_2, y_2]^t$$

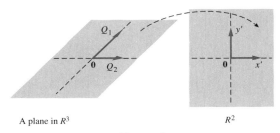

A plane in R^3 R^2

Figure 4

Thus, we do indeed add elements of W by adding their $\mathbb{R}^2$ coordinate vectors.

We can also use the $\mathbb{R}^2$ formula for the dot product to compute dot products in W. To explain this, let X and Y be as in the previous paragraph. Then, using properties of the dot product, we note that

$$X \cdot Y = (x_1 Q_1 + y_1 Q) \cdot (x_2 Q_1 + y_2 Q_2)$$
$$= x_1 y_2 (Q_1 \cdot Q_1) + (x_1 y_2 + x_2 y_1) Q_1 \cdot Q_2 + y_1 y_2 (Q_2 \cdot Q_2)$$

The second term on the right is zero since Q_1 and Q_2 are perpendicular. Also, since the Q_i are normal, $Q_i \cdot Q_i = |Q_i|^2 = 1$. Hence, the preceding formula reduces to

$$X \cdot Y = x_1 y_1 + x_2 y_2$$

which is exactly the $\mathbb{R}^2$ dot product formula.

Notice that one consequence of the computation just done is that for X as above

$$|X| = \sqrt{X \cdot X} = \sqrt{x_1 x_1 + y_1 y_1} = \sqrt{x_1^2 + y_1^2}$$

Thus, the $\mathbb{R}^2$ distance formula also works for W.

Very similar comments apply to any subspace of $\mathbb{R}^n$. In the following theorem, part (a) says that we may think of W as equal to $\mathbb{R}^k$, part (b) says that addition and scalar multiplication in W are described by the corresponding formulas in $\mathbb{R}^k$ and part (c) says that we may compute lengths and dot products in W using the corresponding formulas for $\mathbb{R}^k$.

Theorem 2. Let W be a subspace of $\mathbb{R}^n$ and let $Q_1, Q_2, \ldots, Q_k$ form an ordered, orthonormal basis for W. For each X in W, let X' denote its skewed coordinate vector with respect to the given basis. (Note that X' is an element of $\mathbb{R}^k$.) Then

(a) *The transformation that transforms each element of W to its skewed coordinate vector in $\mathbb{R}^k$ is a one-to-one transformation of W onto $\mathbb{R}^k$.*

(b) *For all X and Y in W and all scalars c,*

$$(X + Y)' = X' + Y'$$
$$(cX)' = cX'$$

(c) *For all X and Y in W,*

$$X \cdot Y = X' \cdot Y'$$
$$|X| = |X'|$$

Proof: For (a), we note that saying that the coordinate vector for X is $[x_1', x_2', \ldots, x_k']^t$ is equivalent with the equality

$$X = x_1' Q_1 + x_2' Q_2 + \cdots + x_k' Q_k$$

It follows immediately that each vector in $\mathbb{R}^k$ is the coordinate vector for one, and only one, X in $\mathcal{W}$, which proves (a).

The proof of (b) is very similar to the 2-dimensional case and is left to the reader.

For (c), let Y be another element of $\mathcal{W}$. Assume that the coordinate vector for Y is $Y' = [y_1', y_2', \ldots, y_k']^t$ so that

$$Y = y_1' Q_1 + y_2' Q_2 + \cdots + y_k' Q_k$$

Notice that since $Q_i \cdot Q_j = 0$ for $i \neq j$,

$$Q_i \cdot Y = y_i' Q_i \cdot Q_i = y_i' |Q_i|^2 = y_i'.$$

Hence

$$
\begin{aligned}
X \cdot Y &= (x_1' Q_1 + x_2' Q_2 + \cdots + x_k' Q_k) \cdot Y \\
&= x_1'(Q_1 \cdot Y) + x_2'(Q_2 \cdot Y) + \cdots + x_k'(Q_k \cdot Y) \\
&= x_1' y_1' + x_2' y_2' + \cdots + x_k' y_k' = X' \cdot Y'
\end{aligned}
$$

The fact that $|X| = |X'|$ follows just as in the two dimensional case. ∎

Remark: Parts (a) and (b) of the above theorem say that the transformation that transforms X to X' is a one-to-one linear transformation of $\mathcal{W}$ onto $\mathbb{R}^k$. In general, two vector spaces $\mathcal{V}$ and $\mathcal{W}$ are said to be **isomorphic** if there is a one-to-one linear transformation that transforms $\mathcal{V}$ onto $\mathcal{W}$. Such a transformation is said to be an **isomorphism**. If $\mathcal{V}$ is isomorphic with $\mathcal{W}$, then we may at times think of $\mathcal{V}$ as being equal to $\mathcal{W}$ in much the same sense that we think of a two-dimensional subspace of $\mathbb{R}^3$ as being equal to $\mathbb{R}^2$.

Notice that the proofs of both (a) and (b) only used the fact that the Q_i formed a basis for $\mathcal{W}$. Thus, (a) and (b) would be true for any k-dimensional vector space. Thus, we see that *any k-dimensional vector space is isomorphic with $\mathbb{R}^k$*.

One aspect of the Gram–Schmidt process should be emphasized. If we start with a basis A_i and apply Gram–Schmidt producing a basis Q_i, then $A_1 = Q_1$, A_2 is a linear combination of Q_1 and Q_2 and, in general, A_j is a linear combination of the first j Q's. Thus, in Example 3, we could write:

$$
\begin{aligned}
A_1 &= Q_1 \\
A_2 &= B_2 + Q_2 = \frac{3}{5} Q_1 + Q_2 \\
A_3 &= B_3 + Q_3 = \frac{-1}{5} Q_1 + \frac{-3}{14} Q_2 + Q_3
\end{aligned}
$$

We may express these three equalities with a single matrix equality as

$$[A_1, A_2, A_3] = [Q_1, Q_2, Q_3] \begin{bmatrix} 1 & \frac{3}{5} & -\frac{1}{5} \\ 0 & 1 & -\frac{3}{14} \\ 0 & 0 & 1 \end{bmatrix}$$

Symbolically, we could write this as $A = QR$ where A has the A_i as columns, Q has the Q_i as columns and R is the 3×3 upper triangular matrix on the right. The triangularity of R is a consequence of the property that each A_j is a linear combination of the first j Q_i's.

In general, we could begin with an $m \times n$ matrix A with independent columns and apply the Gram–Schmidt process to the columns of A, producing vectors Q_i. The fact that each A_j is expressible as a linear combination of the first j of the Q_i says exactly that $A = QR$ where R is upper triangular and Q has the Q_j as columns. We could, in fact, have normalized the Q_i to have length one before expressing the A_j as a linear combination of the Q_j. This yields the famous "QR Factorization":

QR Factorization Theorem.   *Let A be an $m \times n$ matrix with independent columns. Then we may write $A = QR$ where Q is an $m \times n$ matrix with orthonormal columns and R is an $n \times n$ upper-triangular matrix.*   ∎

EXERCISES

(1) Show that the following set of vectors is an orthogonal set in $\mathbb{R}^3$. Find the orthogonal projection of $[1, 2, 2]^t$ onto the subspace spanned by these elements.

$$S = \{[-1, 1, -1]^t, \ [1, 3, 2]^t\}$$

(2) Find the orthogonal projection X_0 of $X = [x, y, z]^t$ onto the subspace spanned by the set S in Exercise 1. Find a matrix P such that $X_0 = PX$. Show that $P^2 = P$. Why is this to be expected?

(3) Show that the following vectors form an orthogonal set in $\mathbb{R}^4$. Find the orthogonal projection of $[1, 2, -1, -3]$ onto the subspace spanned by these elements. You should compare the work done here with that done for Exercise 5 in Section 4.1.

$$[2, -1, -1, -1]^t, \ [1, 3, 3, -4]^t, \ [1, 1, 0, 1]^t$$

(4) Show that the three vectors below form an orthogonal set in $\mathbb{R}^4$. Find the projection of $X = [1, 0, 1, 0]$ onto the subspace spanned by these elements. What does your answer tell you about X?

$$Q_1 = [1, 1, 1, 1]^t, \ Q_2 = [1, -2, 1, 0]^t, \ Q_3 = [1, 1, 1, -3]^t$$

(5) Use the Gram–Schmidt process to find an orthogonal basis for the subspaces spanned by the following sets of vectors.

(a) $\{[0, 1, 1]^t, [1, 1, 1]^t\}$

(b) $\{[1, 2, 1, 1]^t, [-2, 1, 1, -1]^t, [1, 2, 1, 3]^t\}$

(6) Compute the orthogonal projection of $[1, 2, 3]^t$ onto the subspace spanned by the vectors in Exercise 5a and of $[1, 2, 3, 4]^t$ onto the subspace spanned by the vectors in Exercise 5b.

(7) Below, you are given two sets of vectors S_1 and S_2 in $\mathbb{R}^4$.

$$S_1 = \{[1, 2, 1, -1]^t, [0, -1, 1, -1]^t\}$$
$$S_2 = \{[1, 1, 2, -2]^t, [3, 13, -4, 4]^t\}$$

(a) Show that both S_1 and S_2 are orthogonal sets.

(b) Show that S_1 and S_2 both span the *same* subspace of $\mathbb{R}^4$. Call this subspace $\mathcal{W}$. [*Hint:* Express each element of S_2 as a linear combination of elements of S_1.]

(c) Let $X = [x, y, z, w]^t$. Use formula (3) from the text to compute the projection of X onto $\mathcal{W}$ using the basis S_1.

(d) Let $X = [x, y, z, w]^t$. Use formula (3) from the text to compute the projection of X onto $\mathcal{W}$ using the basis S_2. If you work carefully, the answer should be the same as you found for part (c). Why?

(8) For each of the following matrices A (i) find an orthogonal basis for the subspace spanned by the columns of the matrix, (ii) write the columns of A as linear combinations of the basis elements, and (iii) use your results to write $A = QN$ where the columns of Q are perpendicular to each other and N is upper triangular and has ones on its diagonal.

(a)
$$A = \begin{bmatrix} 1 & 1 \\ 1 & 2 \\ 0 & 1 \\ 1 & 0 \end{bmatrix}$$

(b)
$$A = \begin{bmatrix} 2 & 2 & 1 \\ 3 & 3 & -1 \\ 1 & -1 & 0 \end{bmatrix}$$

(9) Let $\mathcal{W}$ be the solution set to the system $AX = \mathbf{0}$ where A is the following matrix.

$$A = \begin{bmatrix} 1 & 3 & 1 & -1 \\ 2 & 6 & 0 & 1 \\ 4 & 12 & 2 & -1 \end{bmatrix}$$

(a) Find an orthogonal basis for $\mathcal{W}$.

(b) Find the projection Q_o of the vector $X = [1, 1, 1, 1]^t$ onto this subspace.

(c) Let $Q_1 = X - Q_o$. Show that $AX = AQ_1$.

(10) Suppose that S is some set of vectors in $\mathbb{R}^n$. We define $S^\perp$ to be the set of all vectors $W \in \mathbb{R}^n$ such that $X \cdot W = 0$ for all $X \in S$. Let

$$S = \{[1, 3, 1, -1]', \ [2, 6, 0, 1]', \ [4, 12, 2, -1]'\}$$

Find an orthogonal basis for $S^\perp$.

(11) For the set S in the Exercise 10, find an orthogonal basis for the subspace $\mathcal{W}$ spanned by S. Show that the basis found in this problem, together with the basis found in Exercise 10, constitutes an orthogonal basis for all of $\mathbb{R}^4$. [*Note:* Results from Section 4.1 may be used to prove that this set of vectors is linearly independent. Which results?]

(12) Let S be a set of vectors in $\mathbb{R}^n$ and let $S^\perp$ be defined as in Exercise 4. Prove that $S^\perp$ is a subspace.

(13) Let $\mathcal{W}$ be a subspace of $\mathbb{R}^n$. Use the Fourier theorem to show that every vector $B \in \mathbb{R}^n$ may be written uniquely in the form $B = B_1 + B_0$ where $B_0 \in \mathcal{W}$ and $B_1 \in \mathcal{W}^\perp$.

(14) Let $\mathcal{W}$ be a subspace of $\mathbb{R}^n$. Let $\{Q_1, Q_2, \ldots, Q_k\}$ be an orthogonal basis for $\mathcal{W}$ and let $\{A_1, A_2, \ldots, A_m\}$ be an orthogonal basis for $\mathcal{W}^\perp$. Show that the set

$$T = \{Q_1, Q_2, \ldots, Q_k, A_1, A_2, \ldots, A_m\}$$

is an orthogonal set. Prove that this set spans $\mathbb{R}^n$, hence is a basis. How does it follow that $\dim(\mathcal{W}) + \dim(\mathcal{W}^\perp) = n$.

(15) Let $\mathcal{W}$ be a subspace of $\mathbb{R}^n$ and let $\{Q_1, Q_2, \ldots, Q_k\}$ be an orthogonal basis for $\mathcal{W}$. Let $X \in \mathcal{W}$. Without using any of the theorems in this section, prove that

$$X = x_1 Q_1 + x_2 Q_2 + \cdots + x_k Q_k$$

where

$$x_i = \frac{X \cdot Q_i}{Q_i \cdot Q_i}$$

For your proof, reason as in the proof of Theorem 1 in Section 4.1. The difference between what you are given here and Theorem 1 is that in the theorem, the Q_i are a basis for $\mathbb{R}^n$ and here the Q_i are a basis for $\mathcal{W}$.

(16) Let $\{Q_1, Q_2\}$ be an ordered orthogonal (but not orthonormal) basis for some subspace $\mathcal{W}$ of $\mathbb{R}^n$. Let X and Y be elements of $\mathcal{W}$ whose coordinate vectors with respect to these bases are $X' = [x_1, y_1]'$ and $Y' = [x_2, y_2]'$. Prove that

$$X \cdot Y = x_1 y_1 |Q_1|^2 + x_2 y_2 |Q_2|^2$$

What is the corresponding formula for $|X|^2$?

(17) Generalize the result of Exercise 16 to arbitrary subspaces of $\mathbb{R}^n$ and prove your theorem.

ON LINE

You are working for an engineering firm and your boss insists that you find one single solution to the following system:

$$2x + 3y + 4z + 3w = 12.9$$
$$4x + 7y - 6z - 8w = -7.1$$
$$6x + 10y - 2z - 5w = 5.9$$

You object, noting that

(a) The system is clearly inconsistent: the sum of the first two equations contradicts the third.

(b) You need at least four equations to uniquely determine four unknowns. Even if the system were solvable, you couldn't produce just one solution.

The boss won't take no for an answer. Concerning objection (a), you are told that the system was obtained from measured data and any inconsistencies must be due to experimental error. Indeed, if any one of the constants on the right sides of the equations were modified by 0.1 unit in the appropriate direction, the system would be consistent.

Concerning objection (b), the boss says, "Do the best you can. We will pass this data on to our customers because they wouldn't know what to do with multiple answers."

After some thought, you realize that projections can help with the inconsistency problem. The above system can be written in vector format as

$$x \begin{bmatrix} 2 \\ 4 \\ 6 \end{bmatrix} + y \begin{bmatrix} 3 \\ 7 \\ 10 \end{bmatrix} + z \begin{bmatrix} 4 \\ -2 \\ -2 \end{bmatrix} + w \begin{bmatrix} 3 \\ -8 \\ -5 \end{bmatrix} = \begin{bmatrix} 12.9 \\ -7.1 \\ 5.9 \end{bmatrix}$$

You realize that this system would be solvable if the vector on the right of the equality were in the space spanned by the four vectors on the left. Using MATLAB's "rank" command, you quickly compute that the rank of the following matrix A is 2, showing that these four vectors in fact span a plane (call it $\mathcal{W}$).

$$A = \begin{bmatrix} 2 & 3 & 4 & 3 \\ 4 & 7 & -6 & -8 \\ 6 & 10 & -2 & -5 \end{bmatrix}$$

Your idea is to let P be the projection of $B = [12.9, -7.1, 5.9]^t$ onto $\mathcal{W}$. Since the system is so nearly consistent, P should be very close to B. Furthermore, the system $AX = P$ should certainly be solvable, and one of the solutions should be what the boss is looking for.

Point (b) will require some further thought. However, you eventually come up with an idea, which is developed in Exercises 1–5.

ON LINE EXERCISES

(1) Begin by computing an orthogonal basis for the span of the columns of A. You could use the Gram–Schmidt process.... Instead, however, try typing "help orth" in MATLAB. Once you have found your orthogonal basis, use the Fourier theorem from the text to compute the projection BO. Finally, use "rref([A,BO])" to find all solutions to $AX = BO$. Express your solution in parametric form. [*Hint:* Let $W = \text{orth}(A)$. Then "BO'∗W(:,i)" computes the dot product of BO with the ith basis element.]

(2) Concerning your objection (b), your first thought is that maybe you could just report the translation vector as the solution. But there is nothing special about the translation vector. The translation theorem says that the general solution can be expressed using any particular solution instead of the translation vector.

Your next idea, however, is a good one. Let T be the translation vector and let TO be its projection to the nullspace of A. Let $X = T - TO$. Then X is what you will report to your boss. Why is X a solution?

Find X. For some help on computing X, enter "help null" at the MATLAB prompt.

(3) Try computing X starting with some solution other than T. You should get the same X. Why does it work out this way?

Remark: It can be shown that your X is the solution of minimal length.

(4) Although your solution was ingenious, MATLAB is way ahead of you. In MATLAB, let $B = [12.9, -7.1, 5.9]^t$ and ask MATLAB to compute $A\backslash B$. This solution is called the "pseudo-inverse" solution. We could devote a whole chapter to it alone!

(5) After giving the boss your answer, you delete all your data except for A and X. A month later the customer calls, saying, "We know that there must be other solutions. Will you please provide us with the general solution?"

You are able to provide the desired information immediately, simply by executing a single MATLAB command. How?

4.3   FOURIER SERIES: SCALAR PRODUCT SPACES

Imagine that you have been hired by a company that makes music synthesizers. You are assigned to design a model that produces a rasping noise. Being expert in sound design, you know that sound is produced by vibrating air. For simplicity, let us assume that your synthesizer will produce only longitudinal waves—that is, the particles of air vibrate back and forth in the direction of the motion of the wave. Such a wave may be described by giving the displacement (forward or back) of the typical particle as a function of time. You also know that the wave graphed in Figure 1 produces an excellent rasp. (The time units are not in seconds. To be audible, your rasp would need to oscillate hundreds or even thousands of times a second. We assume that our time units are such that one oscillation of the rasp takes two units of time to complete.)

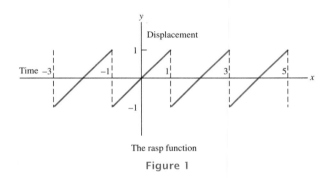

The rasp function

Figure 1

For $-1 < x \leq 1$, the rasp is defined by the function

$$f(x) = x$$

Since the rasp is periodic of period 2, it also satisfies

$$f(x + 2) = f(x)$$

These two conditions uniquely determine the rasp for all x.

You also know where to buy computer chips that produce "pure tones." A pure tone is a sound that consists of a single note. Mathematically, pure tones are described by sine (or cosine) functions. For example, the function $q(x) = 2 \sin 3\pi x$ produces the wave graphed in Figure 2.

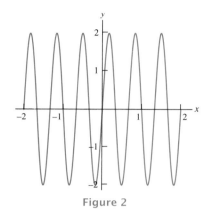

The maximum height (amplitude) of this wave is 2, which is the coefficient multiplying the sine function. The wave has three complete cycles between -1 and 1. In general, a function of the form

$$q_n(x) = \sin n\pi x$$

describes a wave with amplitude 1, having n cycles between -1 and 1.

To build your synthesizer, you buy a chip that produces the pure tones described by the following functions:

$$q_1(x) = \sin \pi x \quad q_2(x) = \sin 2\pi x \quad q_3(x) = \sin 3\pi x \quad q_4(x) = \sin 4\pi x \qquad (1)$$

Your idea is to use this chip to produce four pure tones, each with a different amplitude. You will then mix these tones to produce a sound that, hopefully, will at least approximate the sound of a rasp. In physical terms, your problem is determining what the amplitudes would be to best imitate the rasp.

Mathematically, mixing simply means adding the graphs together. Thus, in mathematical terms, what you want is constants c_1, c_2, c_3, and c_4 (the amplitudes) such that the function

$$f_o = c_1 q_1 + c_2 q_2 + c_3 q_3 + c_4 q_4 \qquad (2)$$

approximates the rasp function f as closely as possible.

Notice that both the rasp function f and the pure tone functions q_k are periodic of period 2. This means that if f_o approximates f over the interval $-1 < x \leq 1$, the approximation will hold for all x. Thus, we shall think of our functions as being defined over the interval $(-1, 1]$.

To solve our approximation problem we shall use a quite remarkable line of reasoning. Let us imagine, for the moment, that instead of being functions, f, q_1, q_2, q_3, and q_4 are vectors in $\mathbb{R}^n$. Let $\mathcal{W}$ be the subspace spanned by q_1, q_2, q_3, and q_4. The "vector" $f_o = c_1 q_1 + c_2 q_2 + c_3 q_3 + c_4 q_4$ is then an element of $\mathcal{W}$. The constants c_i should be chosen so that f_o is as close to f as possible while still lying in $\mathcal{W}$. Thus, f_o should be the projection of f onto $\mathcal{W}$ (Figure 3).

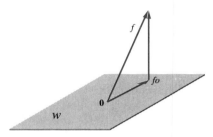

The projection of B onto W

Figure 3

This, of course, is nonsense, since f is not a vector in $\mathbb{R}^n$; f is a function. There are, however, ways in which functions are like vectors. A vector $[x_1, x_2, x_3]^t$ is uniquely defined by giving each of its entries x_1, x_2, and x_3. A function f would be uniquely defined by giving each of its values $f(x)$. We add vectors by adding their entries; the third entry of $X + Y$ is $x_3 + y_3$. We add functions by adding their values; the value of $f + g$ at $x = 3$ is $f(3) + g(3)$. To multiply a vector by a scalar, we multiply each entry by the scalar. To multiply a function by a scalar, we multiply each value by the scalar. Just as all the entries of the zero vector equal zero, there is a zero function (denoted **0**) having all its values equal to zero. In fact, the set of bounded, continuous functions defined over the interval $(-1, 1]$ is a vector space that contains both the rasp function f and the pure tone functions q_n. We shall call this space $\mathcal{F}((-1, 1])$.

There is also a "dot product" for functions. For vectors, the dot product of two vectors is the sum of the products of their entries. For functions, the corresponding concept is the integral of the products of the functional values. For functions, we prefer not to denote the dot product by means of the symbol $f \cdot g$, which is too easily confused with the usual notion of product of functions. Instead, we use the symbol (f, g). We call this the **scalar product** of the functions. Thus, for f and g in $\mathcal{F}((-1, 1])$, we define

$$(f, g) = \int_{-1}^{1} f(x)g(x)\, dx \tag{3}$$

▶ **EXAMPLE 1:**   Compute (q_1, q_2) where q_1 and q_2 are as in formula (1).

Solution: From formula (3), we write

$$(q_1, q_2) = \int_{-1}^{1} (\sin \pi x)(\sin 2\pi x)\, dx$$

This integral is easily evaluated using the identity

$$\sin A \sin B = \frac{\cos(A + B) - \cos(A - B)}{2}$$

We get

$$\left(\frac{-\sin 3\pi x}{6\pi} + \frac{\sin \pi x}{2\pi} \right) \Bigg|_{-1}^{1} = 0.$$

◀

When dealing with vectors in $\mathbb{R}^n$, two vectors whose dot products are zero are called orthogonal. We use the same terminology for functions: two functions f and g are orthogonal if $(f, g) = 0$. Thus, according to Example 1, the functions q_1 and q_2 are orthogonal. In fact, it is not hard to prove the following proposition, where the q_n are defined in the paragraph preceding formula (1).

Proposition 1.   *The functions q_n defined above form an orthogonal set of functions in that $(q_i, q_j) = 0$ for all $i \neq j$. Furthermore, $(q_n, q_n) = 1$ for all n.*   ■

The scalar product in formula (3) has properties very similar to the dot product. Specifically, it is easily seen that the following is true:

Scalar Product Theorem.   *Let f, g, and h be functions in $\mathcal{F}((-1, 1])$. Let c be a scalar. Then*

(a) $(f, g) = (g, f)$ *(commutative law)*
(b) $(f + g, h) = (f, h) + (g, h)$ *(additive law)*
(c) $(cf, g) = c(f, g) = (f, cg)$ *(scalar law)*
(d) $(f, f) > 0$ *for $f \neq \mathbf{0}$ (positivity).*   ■

Now, let us recall that our goal is to find constants c_i such that the function f_o of formula (2) approximates f as closely as possible. As we have seen, if f and q_i were vectors, the answer would be found by letting f_o be the projection of f onto the subspace spanned by the q_i. The q_i, of course, are not vectors in $\mathbb{R}^n$. What if we were to ignore this fact and use the Fourier theorem from Section 4.2, replacing dot products with scalar products? This seems, in a way, reasonable in that the q_i do form an orthogonal set of functions. What we find is

$$f_o = \frac{(f, q_1)}{(q_1, q_1)} q_1 + \frac{(f, q_2)}{(q_2, q_2)} q_2 + \frac{(f, q_3)}{(q_3, q_3)} q_3 + \frac{(f, q_4)}{(q_4, q_4)} q_4$$

We noted earlier that $(q_k, q_k) = 1$. The terms (f, q_k) are computed by integration as before. Since $f(x) = x$ for $-1 < x \le 1$,

$$c_k = (f, q_k) = \int_{-1}^{1} x \sin k\pi x \, dx = \frac{2(-1)^{k+1}}{k\pi} \tag{4}$$

(We used integration by parts.) Hence, our candidate for f_o is:

$$f_o = \frac{2}{\pi} q_1 - \frac{2}{2\pi} q_2 + \frac{2}{3\pi} q_3 - \frac{2}{4\pi} q_4.$$

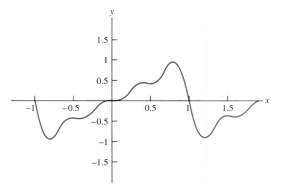

Figure 4

Figure 4 shows the graph of f_o. The graph makes it clear that f_o does a reasonable (but far from perfect) job of approximating a rasp. If we want a better approximation, we must use more pure tones. For example, if we used 10 pure tones, then [from formula (4)], we would have

$$f_o = \sum_{1}^{10} \frac{2(-1)^{k+1}}{k\pi} \sin k\pi x$$

Our new approximation has the graph shown in Figure 5.

Is there any sense in which our original f_o is the closest approximation to f possible using the four pure tones? Answering this question is, once again, a matter of thinking about f as a vector. If X is a vector, we define the magnitude $|X|$ of X to be $\sqrt{X \cdot X}$. We use the analogous definition for functions, except we usually denote the magnitude of a function by the symbol $\|f\|$ to prevent confusion of the magnitude of f with the absolute value of f. Thus, for $f \in \mathcal{F}((-1, 1])$, we define

$$\|f\| = \sqrt{(f, f)}$$

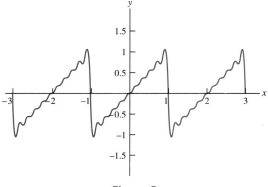

Figure 5

In formulas

$$\|f\| = \sqrt{\int_{-1}^{1} (f(x))^2 \, dx}$$

This is very similar to vectors; the magnitude of a vector is the square root of the sum of the squares of its entries. The magnitude of a function is the square root of the integral of the squares of its values.

In what sense does does $\|f\|$ represent the "size" of the function f? In the exercises you will prove the following proposition:

Proposition 2. *For all $f \in \mathcal{F}((-1, 1])$,*

$$\int_{-1}^{1} |f(x)| \, dx \leq \sqrt{2}\|f\|$$ ■

The integral on the left side of this inequality represents the total area under the graph of f. The proposition says that if $\|f\|$ is small, then f will have very little area under its graph. If f and g belong to $\mathcal{F}((-1, 1])$, then we define the "distance" between f and g to be $\|f - g\|$. If the distance between f and g is small, then the graphs of f and g will be close in the sense that there will be very little area between the graphs (Figure 6).

This discussion is meant to lead to the following theorem. This result is called the Fourier Sine Theorem after Jean Fourier, the discoverer of this type of analysis, and the coefficients c_i are called the Fourier coefficients. The proof will be discussed in the exercises, in the context of general scalar product spaces.

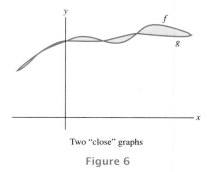

Two "close" graphs

Figure 6

Fourier Sine Theorem. *Let f belong to $\mathcal{F}((-1, 1])$ and let c_i be scalars for $i = 1, 2, \ldots, n$. Let*

$$f_o = c_1 q_1 + c_2 q_2 + \cdots + c_n q_n$$

where

$$q_k(x) = \sin k\pi x$$

Then, setting $c_k = (f, q_k)$ yields the best approximation to f by functions of the form f_o in the sense that this choice of constants minimizes $\|f - f_o\|$. ∎

Remark: The approximations to waves by pure tones that we have been discussing have a remarkable amount of physical reality. We showed that it is possible to use appropriate combinations of pure tones to find better and better approximations to our rasp. It turns out that in real life, waves, such as our rasp, behave as if they somehow were produced by combining an infinite number of pure tones precisely as we have described. For example, suppose that we fed our rasp (the real one, not an approximation) into an audio system. All audio systems, no matter how good, can only reproduce sounds whose pitch lies within a certain range—a tone whose pitch is either too high or too low will not be reproduced. This means that some of the pure tones used in producing the rasp will not be heard. If the amplifier reproduces, say, the first four pure tones perfectly, while completely attenuating all others, the sound we hear would be precisely described by the graph in Figure 4. In essence, our amplifier projects us down to the subspace spanned by the first four pure tones.

The method we used to solve our synthesizer problem occurs in many different contexts. In general, if $\mathcal{V}$ is a vector space, then a **scalar product** on $\mathcal{V}$ is a function $(\ ,\ )$ that takes pairs of vectors from $\mathcal{V}$ and produces real numbers and satisfies the properties (a)–(d) of the scalar product theorem. A vector space on which a scalar product is given is called a **scalar product space**. Thus, for example, $\mathbb{R}^n$ is a scalar product space with the dot product as a scalar product. Our space $\mathcal{F}((-1, 1])$ is a scalar product space with the scalar product defined by formula (3).

Almost all the concepts already discussed above are meaningful for any scalar product space. For example, if V is an element of $\mathcal{V}$, then we define

$$\|V\| = \sqrt{(V, V)}$$

A set $Q_1, Q_2, \ldots, Q_n$ of elements of $\mathcal{V}$ is an orthogonal set if each of the Q_i is non-zero and $(Q_i, Q_j) = 0$ for all $i \neq j$. The projection of V to the span of the Q_i is (from the Fourier theorem in Section 4.2)

$$V_o = \frac{(V, Q_1)}{(Q_1, Q_1)} Q_1 + \frac{(V, Q_2)}{(Q_2, Q_2)} Q_2 + \frac{(V, Q_3)}{(Q_3, Q_3)} Q_3 + \cdots + \frac{(V, Q_n)}{(Q_n, Q_n)} Q_n$$

The coefficients

$$c_i = \frac{(V, Q_i)}{(Q_i, Q_i)} \tag{5}$$

are called the Fourier coefficients of V. Just as in the proof of the Fourier theorem, it follows that V_o is the closest point to V in the span of the Q_i. The exercises present further examples.

EXERCISES

(1) Prove vector space properties (c), (e), (h), (i), and (j), Section 1.2, for $\mathcal{F}((-1, 1])$. As an example of how your proof should go, we shall prove property (b) (commutativity) for you:

Let f and g be two functions in $\mathcal{F}((-1, 1])$. Then, for all $-1 < x \leq 1$,

$$(f + g)(x) = f(x) + g(x) = g(x) + f(x) = (g + f)(x)$$

Since $(f + g)(x) = (g + f)(x)$ for all such x, it follows that $f + g = g + f$, as claimed.

(2) Let $\mathcal{F}(\mathbb{R})$ be the vector space of all continuous, real-valued functions defined on all of $\mathbb{R}$. The following sets of functions are linearly dependent in $\mathcal{F}(\mathbb{R})$. Show this by expressing one of them as a linear combination of the others. (You may need to look up the definitions of the sinh and cosh functions as well as some trig identities in a calculus book.)

(a) $\{e^x, e^{-x}, \sinh x\}$
(b) $\{\sinh x, \cosh x, e^{-x}\}$
(c) $\{\cos(2x), \sin^2 x, \cos^2 x\}$

(d) $\{\cos(2x), 1, \cos^2 x\}$
(e) $\{\sin x, \sin(x + \frac{\pi}{4}), \cos(x + \frac{\pi}{6})\}$
(f) $\{(x + 3)^2, 1, x, x^2\}$
(g) $\{x^2 + 3x + 3, x + 1, 2x^2\}$

(3) Prove the scalar product theorem from the text.

(4) Prove Proposition 1 from the text.

(5) The synthesizer company wants you to build a synthesizer to produce the wave shown in Figure 7. For $-1 \leq x \leq 1$, this wave is defined by the function $f(x) = x^3$. Find the best approximation to f using the functions q_n as in the Fourier sine theorem.

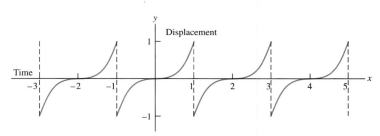

The cubic rasp function

Figure 7

(6) Now you are to build a synthesizer to produce the wave shown in Figure 8. For $-1 \leq x \leq 1$, this wave is defined by the function $f(x) = x^2$. Find the best approximation to f using 20 function q_n as in the Fourier sine theorem. [*Note:* The integrals can be evaluated with almost no work at all if you observe that for all n, $g(x) = x^2 \sin n\pi x$ is an *odd* function.]

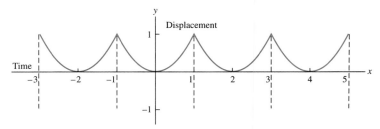

The quadratic rasp function

Figure 8

(7) Exercise 6 says that it is not possible to approximate an even function such as $f(x) = x^2$ over the interval $-1 \le x \le 1$ by sine functions. This is because the sine functions are odd and any linear combination of odd functions will be odd. This observation suggests the use of cosine functions.

    (a) Let $p_n(x) = \cos n\pi x$ for $n = 0, 1, 2, \ldots$. Show that the p_n form an orthogonal set of functions in $\mathcal{F}((-1, 1])$.

    (b) Compute (p_n, p_n) for all n (including $n = 0$.)

    (c) Compute the Fourier coefficients for f relative to the p_n.

(8) Finally, the synthesizer company wants a model that will produce the sawtooth wave shown in Figure 9. This wave is defined by the function $g(x) = |x|$ for $-1 \le x \le 1$. Compute the Fourier coefficients for g relative to the p_n from Exercise 7.

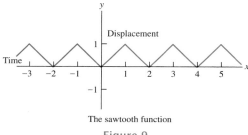

The sawtooth function

Figure 9

(9) For a function that is neither odd nor even, we must use both sine and cosine functions to obtain a good approximation. Consider the wave graphed in Figure 10. This wave is described by the function

$$h(x) = x \quad 0 \le x \le 1$$
$$h(x) = 0 \quad -1 < x \le 0$$

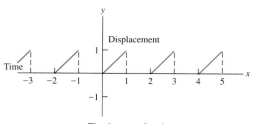

The slow rasp function

Figure 10

(a) Show that the functions, q_i, together with the p_i, form an orthogonal set of functions in $\mathcal{F}((-1, 1])$.
(b) Compute the Fourier coefficients of h with respect to this orthogonal set. [*Note*: If you completed Exercise 8, this part can be done with essentially no work. Specifically, let f be the rasp function from the text and let g be the sawtooth function from Exercise 8. How does the function $f + g$ relate to the function in the present problem?]

(10) Show that the functions 1, x, and x^2 are *not* orthogonal with respect to the scalar product from formula (3). Use the Gram–Schmidt process to find an orthogonal set of functions that is contained in their span. Use your answer to compute the projection of x^3 onto this span.

> **Remark:** The three orthogonal functions you found were the first three Legendre polynomials. In general, the Legendre polynomials are what you would get by applying the Gram–Schmidt process to the functions $1, x, x^2, \ldots, x^n$.

(11) The assumption that our waves have period 2 is rather artificial. In real life, we would want to assume that our waves repeat every T seconds. (T is the "period" of the wave.) The corresponding pure sine tones would be $q_n(x) = \sin \frac{2\pi n x}{T}$. The relevant scalar product space is $\mathcal{F}((-\frac{T}{2}, \frac{T}{2}])$ with the scalar product

$$(f, g) = \int_{-T/2}^{T/2} f(x)g(x)\, dx$$

Suppose that $T = 0.002$ and that $f(x) = x$ for $-0.001 < x \leq 0.001$. Compute the corresponding Fourier coefficients. [*Warning*: $(q_n, q_n) \neq 1$]

(12) A function of the form $p(x) = a_n x^n + a_{n-1} x^{n-1} + \cdots + a_0$ is said to be a polynomial of degree less than or equal to n. (It has degree n if $a_n \neq 0$.) The space of all polynomials of degree less than or equal to n is denoted P_n. Prove that P_n is a subspace of $\mathcal{F}(\mathbb{R})$.

(13) The general element of P_2 is describable in the form $f(x) = ax^2 + bx + c$ for a, b, and c real numbers. We define a scalar product on P_2 by setting

$$(f, g) = f(0)g(0) + f(1)g(1) + f(2)g(2)$$

Thus, for example, we can write:

$$(x^2 + 1, 2x) = (0^2 + 1)(2 \cdot 0) + (1^2 + 1)(2 \cdot 1) + (2^2 + 1)(2 \cdot 2) = 24 \quad (6)$$

(a) Compute $(x^2 + x - 2, x^2 + 1)$.
(b) Prove that $(\ ,\ )$ is a scalar product on P_2.
(c) Let $q_1(x) = (x - 1)(x - 2)$, $q_2(x) = x(x - 2)$ and $q_3(x) = x(x - 1)$. Show that this set of functions forms an orthogonal set of functions in $\mathcal{V}$.

(d) Compute the Fourier coefficients c_i for $f(x) = x^2$ with respect to the orthogonal set in part (c). Show by direct computation that $x^2 = c_1 q_1(x) + c_2 q_2(x) + c_3 q_3(x)$ for all x.

(e) We may use formula (6) to define a product on $\mathcal{F}(\mathbb{R})$. Why is this not a scalar product? Let $f \in \mathcal{F}(\mathbb{R})$ and consider the function

$$f_o = c_1 q_1 + c_2 q_2 + c_3 q_3$$

where the c_i are as in formula (5). Show that f_o is a polynomial function that satisfies $f_o(0) = f(0)$, $f_o(1) = f(1)$, and $f_o(2) = f(2)$.

This problem demonstrates the method of Lagrange interpolation. We have succeeded in finding (quite simply) a polynomial that equals f at three specific points, $x = 0$, $x = 1$, and $x = 2$. One can do the same for any finite number of x values $a_1, a_2, \ldots, a_n$. The ideas are similar.

(14) Is the set of all polynomial functions (regardless of degree) a subspace of $\mathcal{F}(\mathbb{R})$? What about the set of all polynomial functions with integral coefficients?

(15) Use the idea from the following proof that the functions e^x and xe^x are independent elements of $\mathcal{F}(\mathbb{R})$ to prove that the functions 1, x, x^2, and x^3 are independent. (You will need to differentiate more than once.)

Consider the dependency equation

$$c_1 e^x + c_2 x e^x = \mathbf{0}$$

It is important to realize that $\mathbf{0}$ represents the zero function. Thus, from this equation, we may conclude that *for all x,*

$$c_1 e^x + c_2 x e^x = 0 \tag{7}$$

We differentiate on both sides:

$$c_1 e^x + c_2 (e^x + x e^x) = 0 \tag{8}$$

Setting $x = 0$ in each equation:

$$
\begin{aligned}
c_1 &= 0 \\
c_1 + c_2 &= 0
\end{aligned}
$$

This yields $c_1 = c_2 = 0$, proving independence.

(16) Explain why 1, x, x^2, and x^3 span P_3. What is the dimension of P_3?

(17) What is the dimension of P_n? Prove your answer.

(18) Let $\mathcal{P}$ be the space of all polynomial functions (regardless of degree). What is its dimension?

(19) Define a product on $\mathbb{R}^2$ by

$$([x_1, y_1]', [x_2, y_2]') = 2x_1 x_2 + 3y_1 y_2$$

Prove that this defines a scalar product.

(20) Let B be an invertible $n \times n$ matrix and let $A = B' B$. Define a product on $\mathbb{R}^n$ by

$$(X, Y) = X^t A Y$$

Prove that this defines a scalar product. [*Hint:* Use formula (4) in Section 4.1.]

(21) Let $S = \{V_1, V_2, \ldots, V_n\}$ be an orthogonal set of vectors in a scalar product space. Prove that S is linearly independent. For your proof, see the discussion following Theorem 1 in Section 4.1.

(22) Let $\mathcal{V}$ be a scalar product space and let V and W be elements of $\mathcal{V}$. Prove that $||V + W||^2 = ||V||^2 + ||W||^2$ if and only if V and W are orthogonal. Explain why this result is called the Pythagorean theorem.

(23) Let $\mathcal{V}$ be a scalar product space and let V and W be elements of $\mathcal{V}$. Prove that $||V - W||^2 + ||V + W||^2 = 2(||V||^2 + ||W||^2)$. What is the geometric interpretation of this result?

(24) Let notation be as in formula (5) and the immediately preceding paragraph.

(a) Let $B = V - V_o$. Prove that B is perpendicular to each of the elements $V_1, V_2, \ldots, V_n$. [*Hint:* See the comments preceding the Fourier theorem in Section 4.2.] How does it follow that B is perpendicular to the span $\mathcal{W}$ of the V_i?

(b) Prove that B is the closest element of $\mathcal{W}$ to V in the sense that $||V - (V_o + Y)|| > ||V - V_o||$ for all non-zero Y in $\mathcal{W}$. [*Hint:* Use part (a) along with Exercise 22 to show that $||V - (V_o + Y)||^2 = ||B||^2 + || - Y||^2$.]

(25) Let $S = \{V_1, V_2, \ldots, V_n\}$ be an orthogonal set of vectors in a scalar product space. Let $W = c_1 V_1 + c_2 V_2 + \cdots + c_n V_n$ where the c_i are scalars. Prove that

$$||W||^2 = |c_1|^2 ||V_1||^2 + |c_2|^2 ||V_2||^2 + \cdots + |c_n|^2 ||V_n||^2$$

(26) Let $\mathcal{V}$ be a finite-dimensional vector space that is also a scalar product space. Prove that $\mathcal{V}$ has an orthogonal basis. [*Hint:* Look in Section 4.2.]

(27) In this problem, you are asked to prove the following theorem:

Cauchy-Schwarz Theorem.   Let $\mathcal{V}$ be a scalar product space. Then, for all non-zero elements V and W in $\mathcal{V}$, we have

$$-1 \le \frac{(V, W)}{||V|| \, ||W||} \le 1 \qquad \blacksquare$$

For your proof, let $U = V/||V|| + W/||W||$ and note that $(U, U) > 0$. This should yield one of the desired inequalities. How do you get the other?

Remark: This theorem has a rather simple geometric interpretation. Recall that if X and Y are vectors in $\mathbb{R}^3$, then the angle θ between them satisfies

$$\cos \theta = \frac{X \cdot Y}{|X| \, |Y|}$$

The Cauchy–Schwarz theorem says that $-1 \le \cos \theta \le 1$.

The Cauchy–Schwarz theorem allows us to define the angle between any pair of non-zero vectors in our space; we simply define the angle θ between u and v to be

$$\theta = \arccos \left(\frac{(V, W)}{||V|| \, ||W||} \right)$$

The Cauchy–Schwarz Theorem is needed to guarantee that the fraction is in the domain of the arccos.

(28) Let f and g belong to $\mathcal{F}((-1, 1])$. Use Exercise 27 to prove that

$$\left(\int_{-1}^{1} f(x)g(x)dx \right)^2 \le \left(\int_{-1}^{1} (f(x))^2 dx \right) \left(\int_{-1}^{1} (g(x))^2 dx \right)$$

(29) Use Exercise 28 to prove Proposition 2 from the text. For your proof, use $|f(x)|$ in place of $f(x)$ and make an appropriate choice for $g(x)$.

ON LINE

The following "program" can be used to generate the graphs of the Fourier sine series for the rasp from the text. Enter each line in MATLAB *exactly as shown*. Note in particular the period after the "y" in the last line.

```
hold on; axis([-1,4,-2.5,2.5]); grid on;
x=-1:.01:1; t=-1:.01:4;
n=1; sm=0; y=x;
b=trapz(x,y.*sin(n*pi*x)); sm=b*sin(n*pi*t)+sm; n=n+1;
```

The first time you enter the last line, you compute the first Fourier approximation to the rasp. [The command "trapz(t,y.*sin(n*pi*t));" tells MATLAB to integrate $x \sin(\pi x)$

over the interval $-1 \leq t \leq 1$ using a trapezoidal approximation.] If you press the up-arrow key and re-enter this line you will compute the second Fourier approximation. (Note that the last term in this line advances the value of n each time the line is entered.)

To plot your approximation and compare it with the rasp, enter the following line. The rasp will plot in red over three periods.

cla; plot(t,sm); plot(x,y,'r'); plot(x+2,y,'r'); plot(x+4,y,'r');

ON LINE EXERCISES

(1) Plot the first, fourth, and tenth approximations to the rasp. (If you need to restart the program, just reenter the second-to-last line. If you have closed the Figure window, you will also need to reenter the first line to reinitialize the graph.)

(2) Use the up-arrow key to move up to the next to last line and change the expression "y=x;" to "y=-1*(x<0)+1*(x>=0);". This represents a function $y(x)$ that equals -1 if $x < 0$ and equals 1 if $x \geq 0$. Compute and graph the best approximations to y using 4, 8, and 20 sine functions.

The wave you are approximating is called a "square wave." Notice the "ear-like" peaks on your graph at the discontinuities of the wave. These peaks are referred to as the "Gibbs phenomenon." They are quite pronounced, even after twenty terms of the Fourier series. Their existence shows that it takes a very high fidelity amplifier to accurately reproduce a square wave. For this reason, square waves are sometimes used to test the fidelity of an amplifier.

(3) Use the up-arrow key to move up to the next to last line in the program and change the expression "y=x;" to "y=(1+x.^2).^(-1);" (Again, note the placement of the periods.) Compute and graph the fourth approximation to this function. Explain why your graphs don't agree. (See Exercises 6 and 7 on page 225.)

(4) Modify the above program to approximate for the function in Exercise 3 using cosine functions. Plot the approximations using one, four, and eight cosine functions. [*Note:* Be sure to reenter the line that defines y before running your program as this line also initializes n and sm.]

(5) You might have wondered about the placement of the periods in the above program. To understand their meaning, let A=[1,2,3,4,5] and ask MATLAB to compute the following quantities (one at a time): "A*A", "B=A.*A", and "C=B.^(1/2)". Try the same operations on a 2×3 matrix of your choice. Describe what you think these operations do.

This type of multiplication of matrices is called "Hadamard multiplication." It is necessary in our program because x is a vector. (Ask MATLAB to display x by entering "x" at the prompt. What size is x? What do its entries represent?) The quantities "y", "sin(n*pi*x)", and "sm" are also vectors while "b" is a scalar.

4.4 ORTHOGONAL MATRICES

In Section 3.1, we saw that rotation about the origin by a fixed angle θ in $\mathbb{R}^2$ defines a linear transformation. This transformation is described by the matrix

$$R_\theta = \begin{bmatrix} \cos\theta & -\sin\theta \\ \sin\theta & \cos\theta \end{bmatrix} \tag{1}$$

One of the most striking features of rotation is that it does not change lengths of vectors; if X is rotated into X', then X and X' have the same length (Figure 1).

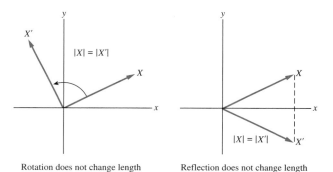

Rotation does not change length   Reflection does not change length

Figure 1

Another example of a linear transformation that does not change length is the transformation defined by the matrix

$$M = \begin{bmatrix} 1 & 0 \\ 0 & -1 \end{bmatrix}$$

Multiplication by M transforms $[x, y]^t$ into $[x, -y]^t$. Geometrically, this describes a reflection in the x axis. We shall show that the only linear transformations of $\mathbb{R}^2$ into itself that do not change length are combinations of reflections and rotations.

There also exist transformations of $\mathbb{R}^3$ into itself that do not change length. For example, multiplication by

$$R_\theta^z = \begin{bmatrix} \cos\theta & -\sin\theta & 0 \\ \sin\theta & \cos\theta & 0 \\ 0 & 0 & 1 \end{bmatrix} \tag{2}$$

rotates by θ radians around the z axis (Figure 2). (Notice that multiplication of $[x, y, z]^t$ by this matrix multiplies $[x, y]^t$ by the two-dimensional rotation matrix without changing z.)

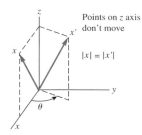

Rotation about z axis

Figure 2

This certainly should not change lengths of vectors. Similarly, multiplication by either of the following matrices should not change vectors' lengths, since each describes rotation about a different axis in $\mathbb{R}^3$.

$$R_\theta^x = \begin{bmatrix} 1 & 0 & 0 \\ 0 & \cos\theta & -\sin\theta \\ 0 & \sin\theta & \cos\theta \end{bmatrix} \quad R_\theta^y = \begin{bmatrix} \cos\theta & 0 & -\sin\theta \\ 0 & 1 & 0 \\ \sin\theta & 0 & \cos\theta \end{bmatrix} \tag{3}$$

Orthogonal matrices are used to describe rotations in three dimensions.

In general, an $n \times n$ matrix that does not change lengths is called **orthogonal**. More formally:

Definition.   An $n \times n$ matrix A is orthogonal if

$$|AX| = |X|$$

for all $X \in \mathbb{R}^n$.   ■

We think of general orthogonal matrices as being analogous to rotations and reflections in $\mathbb{R}^2$ and $\mathbb{R}^3$.

Remark: Recall that points in $\mathbb{R}^n$ represent vectors whose initial point is at the origin. Thus, our definition says only that orthogonal matrices preserve the lengths of such vectors. However, it follows that orthogonal matrices in fact preserve the lengths of all vectors. To see this, let X and Y be points in $\mathbb{R}^n$. Multiplication by A transforms the vector from X to Y into the vector from AX to AY. The length of this latter vector is

$$|AX - AY| = |A(X - Y)| = |X - Y|$$

which is the length of the former vector.

In $\mathbb{R}^2$, rotation preserves angles as well as lengths. Specifically, if X and Y are two vectors that get rotated into X' and Y', the angle between X and Y will be the same as that between X' and Y' (Figure 3).

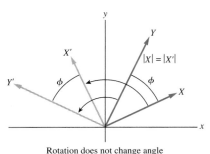

Rotation does not change angle

Figure 3

There is a sense in which general orthogonal matrices also preserve angles. To explain this, consider two vectors X and Y in $\mathbb{R}^2$. The angle ϕ between X and Y is determined by

$$\cos \phi = \frac{X \cdot Y}{|X| \, |Y|} \qquad (4)$$

(See formula (6), Section 1.1).

Now, suppose that X and Y were transformed into X' and Y' under some orthogonal transformation of $\mathbb{R}^2$. The angle between X' and Y' would be described by

$$\cos \phi' = \frac{X' \cdot Y'}{|X'| |Y'|}$$

Since $|X| = |X'|$ and $|Y| = |Y'|$, we see that these two expressions are equal if and only if $X \cdot Y = X' \cdot Y'$. The following theorem implies that this equality is true for any orthogonal matrix.

Preservation of Angles Theorem. *Suppose that A is an orthogonal $n \times n$ matrix. Then, for all X and Y in $\mathbb{R}^n$,*

$$A X \cdot A Y = X \cdot Y$$

Proof: The following formula, which expresses the dot product in terms of length, is the key to our proof:

$$2(X \cdot Y) = |X|^2 + |Y|^2 - |X - Y|^2 \tag{5}$$

Granted this formula, our theorem follows directly from the fact that multiplication by A preserves lengths:

$$2(AX \cdot AY) = |AX|^2 + |AY|^2 - |AX - AY|^2 = |X|^2 + |Y|^2 - |X - Y|^2 = 2(X \cdot Y)$$

Formula (5) follows from the following computation:

$$|X - Y|^2 = (X - Y) \cdot (X - Y) = X \cdot (X - Y) - Y \cdot (X - Y)$$
$$= X \cdot X - X \cdot Y - Y \cdot X + Y \cdot Y = |X|^2 + |Y|^2 - 2X \cdot Y \qquad \blacksquare$$

In the 2×2 case, the orthogonal matrices are easy to describe. Let A be some 2×2 orthogonal matrix and let $I_1 = [1, 0]^t$ and $I_2 = [0, 1]^t$. The vectors I_1 and I_2 have length one and are perpendicular to each other. The vectors $A_1 = AI_1$ and $A_2 = AI_2$ are the columns of A. From the preservation of angles theorem, A_1 and A_2 are perpendicular to each other. Since A is orthogonal, these vectors also have length one. In particular, the A_i represent points on the unit circle. Thus, we may write

$$A_1 = \begin{bmatrix} \cos \theta \\ \sin \theta \end{bmatrix}$$

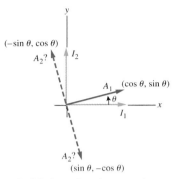

2 x 2 Orthogonal matrices as rotations

Figure 4

where θ is the angle between A_1 and the x axis. (See Figure 4.) Since A_2 is perpendicular to A_1, there are only two possibilities for A_2:

$$A_2 = \pm \begin{bmatrix} -\sin\theta \\ \cos\theta \end{bmatrix}$$

In the first case

$$A = \begin{bmatrix} \cos\theta & -\sin\theta \\ \sin\theta & \cos\theta \end{bmatrix}$$

which is just the rotation matrix R_θ.

In the second case,

$$A = \begin{bmatrix} \cos\theta & \sin\theta \\ \sin\theta & -\cos\theta \end{bmatrix} = \begin{bmatrix} \cos\theta & -\sin\theta \\ \sin\theta & \cos\theta \end{bmatrix} \begin{bmatrix} 1 & 0 \\ 0 & -1 \end{bmatrix}$$

This is $R_\theta M$ where M is the reflection matrix on page 224. Thus, all orthogonal 2×2 matrices describe combinations of rotations and reflections.

In higher dimensions, the description of the general orthogonal transformation is less simple. However, some of what we have said carries over to the general case.

Proposition 1. *The columns of any $n \times n$ orthogonal matrix A are mutually perpendicular and have length one.*

Proof: Let $\{I_1, I_2, \ldots, I_n\}$ be the standard basis for $\mathbb{R}^n$. Then the ith column of A is $A_i = AI_i$. Our proposition follows from the preservation of angles theorem together with the observation that the I_k are mutually perpendicular and have length one.   ∎

Remark: Mathematical notation is not always consistent or logical. Proposition 1 says that the columns of an *orthogonal matrix* form an orthonormal set of vectors. One would think that such matrices would be called "orthonormal" matrices and that orthogonal matrices would have perpendicular columns that are not necessarily of length one. Unfortunately, this is not "standard" terminology.

There is a very useful way of stating Proposition 1 using matrix multiplication. Let A be an $n \times n$ orthogonal matrix and consider $B = A^t A$. Each entry of B is a row of A^t times a column of A. But the rows of A^t are just columns of A, flipped. Thus, each entry of B is a product of two columns of A. More explicitly

$$B_{ij} = A_i \cdot A_j$$

Since the columns of A are mutually perpendicular, $B_{ij} = 0$ if $i \neq j$. If $i = j$, then $A_i \cdot A_i = |A_i|^2 = 1$. Hence $B_{ii} = 1$. This exactly describes the identity matrix. Thus, Proposition 1 may be summarized in the single matrix equality

$$A^t A = I$$

Hence (using Theorem 2 in Section 3.4), $A^t = A^{-1}$. This makes orthogonal matrices wonderful: to invert them, just take the transpose!

Actually, Proposition 1 characterizes orthogonal matrices.

Theorem 1. *Let A be an $n \times n$ matrix. Then A will be orthogonal if and only if either of the following statements holds:*

(a) $A^t A = I$.

(b) *The columns of A are mutually perpendicular and have length one.*

Proof: We noted that both (a) and (b) hold for orthogonal matrices and that (a) and (b) are equivalent statements. Thus, we need only prove that if $A^t A = I$, then A is orthogonal. Thus, we must prove that multiplication by A does not change lengths. Using formula (4) from Section 4.1, we see that

$$|AX|^2 = AX \cdot AX = (AX)^t AX = X^t A^t AX = X^t I X = X^t X = X \cdot X = |X|^2$$

Thus, A preserves length. ∎

Theorem 1 offers us two techniques for deciding whether a given matrix is orthogonal: we can check to see if $A^t A = I$, or we can check the columns to see if they are perpendicular and length one. Which is easier depends on the context, as the next examples show.

▶ EXAMPLE 1:   Prove that the matrix A is orthogonal.

$$A = \frac{1}{2} \begin{bmatrix} 1 & \sqrt{2} & 1 \\ -\sqrt{2} & 0 & \sqrt{2} \\ 1 & -\sqrt{2} & 1 \end{bmatrix}$$

Solution: It is easily checked that $A^t A = I$.   ◀

▶ EXAMPLE 2:   Find numbers a, b, c, and d such that the matrix A is orthogonal.

$$A = d \begin{bmatrix} 2 & -1 & a \\ -1 & 2 & b \\ 2 & 2 & c \end{bmatrix}$$

Solution: The first column of A is $A_1 = [2d, -d, 2d]^t$. Its length is

$$\sqrt{(2d)^2 + (-d)^2 + (2d)^2} = 3|d|$$

Since the length must be one, we see that $d = \pm 1/3$. This also makes A_2 have length one.

   We note that $A_1 \cdot A_2 = 0$. This is good, since otherwise, the problem would not be solvable. For the third column to be perpendicular to the first two, we require

$$2a - b + 2c = 0$$
$$-a + 2b + 2c = 0$$

The general solution is $[a, b, c]^t = [-2, -2, 1]^t c$. When we multiply this column by $d = \pm 1/3$, we must get a column of length 1. This forces $c = \pm 1$, yielding $a = b = -2(\pm 1)$.   ◀

EXERCISES

(1) Let $X = [4, 2\sqrt{2}, 6]^t$. Show by direct computation that $|AX| = |X|$ where A is as in Example 1.

(2) Show that the matrices M and R_θ defined at the beginning of this section are orthogonal using Theorem 1.

(3) In the text, we used geometric intuition to justify the orthogonality of the rotation matrices from formulas (2) and (3). Use Theorem 1 to prove that they are orthogonal.

(4) Find a constant c such that cA is orthogonal where A is as shown.

$$A = \begin{bmatrix} 2 & 3 & 6 \\ 6 & 2 & -3 \\ 3 & -6 & 2 \end{bmatrix}$$

(5) Check your work in Exercise 4 by choosing a specific vector X and showing (by direct computation) that $|cAX| = |X|$. Choose your X so that none of its entries are zero.

(6) Let A and B be $n \times n$ orthogonal matrices. Prove that AB is orthogonal by showing that for all $X \in \mathbb{R}^n$, $|(AB)X| = |X|$.

(7) Redo Exercise 6 by showing that $(AB)^t(AB) = I$.

(8) Prove that the inverse of an orthogonal matrix is orthogonal.

(9) Suppose that an $n \times n$ matrix A is such that its rows, considered as vectors in $\mathbb{R}^n$, are of length one and are mutually perpendicular. Prove that A is orthogonal. [*Hint:* Consider A^t.]

(10) Find numbers a, b, c, and d such that the matrix A below is orthogonal.

$$A = \begin{bmatrix} \frac{1}{2} & \frac{1}{2} & -\frac{1}{\sqrt{2}} & a \\ \frac{1}{2} & -\frac{1}{2} & 0 & b \\ \frac{1}{2} & \frac{1}{2} & -\frac{1}{\sqrt{2}} & c \\ \frac{1}{2} & -\frac{1}{2} & 0 & d \end{bmatrix}$$

(11) Prove that an $n \times n$ matrix is orthogonal if and only if its columns form an orthonormal basis for $\mathbb{R}^n$.

(12) Is it possible to find a 3×2 matrix with orthonormal *rows*? Explain.

(13) Give an example of a 3×2 matrix A with all entries non-zero that has orthonormal columns. Compute AA^t and A^tA. Which is the identity? Prove that the similar product equals I for any A that has orthonormal columns.

ON LINE

Let R_θ^x and R_θ^y be as defined on page 245 and let

$$A = R_{\pi/6}^x R_{\pi/4}^y$$

Since the product of two orthogonal matrices is orthogonal, A is orthogonal. The purpose of this set of exercises is to demonstrate that A defines a rotation around a fixed axis by a particular angle as in Figure 5.

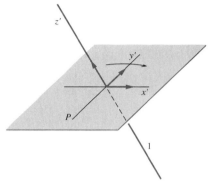

Figure 5

Points on this axis remain fixed under the rotation. Thus, if X is on the axis of rotation, it will satisfy $AX = X$. Notice that this equation is equivalent with

$$(A - I)X = 0$$

ON LINE EXERCISES

(1) Use MATLAB to compute A.

(2) We can find an X on the axis of rotation with the command

```
X=null(A-eye(3))
```

Note that "eye(n)" is MATLAB's notation for the $n \times n$ identity matrix and "null" produces an orthonormal basis for the nullspace of the matrix.

(3) Plot the line segment from $-X$ to X. For this set "W=[X,-X]" and enter:

```
plot3(W(1,:),W(2,:),W(3,:),'r')
```

This tells MATLAB to plot the columns of W, (in red) using the first row as the x coordinates, the second row as the y coordinates, and the third row as z coordinates. MATLAB connects the columns with a line.

(4) The plane P through the origin perpendicular to X is called the "plane of rotation." Since this plane contains the origin, it is a subspace of $\mathbb{R}^3$. It consists of all vectors Y such that $X \cdot Y = 0$. Since $X \cdot Y = X^t Y$, we can find an orthonormal basis for $\mathcal{W}$ by entering "Q=null(X')" into MATLAB. The columns of Q form the basis. To "see" this basis, we must plot the line segment from the first column of Q to the origin and the line segment from the origin to the second column of Q. We begin by entering "Z=[0;0;0]" followed by

```
P=[Q(:,1),Z,Q(:,2)]
hold on; axis equal;
plot3(P(1,:),P(2,:), P(3,:),'y')
```

From this angle, the basis elements appear neither orthogonal nor normal. To select a better viewing angle, enter "view([1,3,10])". This causes us to look down toward the origin in the direction of the vector [1, 3, 10]. Now it at least is plausible that the basis is orthonormal.

(5) We expect that multiplication by A will rotate elements in P around the axis of rotation by a fixed angle. To test this, enter the following line. Explain what you see.

```
P=A*P;plot3(P(1,:),P(2,:),P(3,:),'y')
```

(6) Enter the line given in Exercise 5, 15 times and explain what you see. (Use the up-arrow to avoid retyping.)

(7) Determine the angle of the rotation in P. [*Hint:* Use formula (4).]

4.5   LEAST SQUARES

We shall begin with a story. The consulting firm that employs you as a statistician has been hired by a large midwestern university to help make admissions decisions. Each applicant provides the university with three pieces of information: the score on the SAT exam, the score on the ACT exam and a high school GPA (0–4 scale). The college wishes to know what weight to put on each of these numbers.

You begin by collecting data from the previous year's freshman class. In addition to the admissions data, you collect the student's current (college) GPA (0–4 scale). A partial listing of your data might look like

SAT	ACT	GPA	C-GPA
600	30	3.0	3.2
500	28	2.9	3.0
750	35	3.9	3.5
650	30	3.5	3.5
550	25	2.8	3.2
800	35	3.7	3.7

Ideally, you would like numbers x_1, x_2, and x_3 such that, for all students,

$$(SAT)x_1 + (ACT)x_2 + (GPA)x_3 = C\text{-}GPA$$

These numbers would tell you exactly what weight to put on each each piece of data. Finding such numbers would be equivalent to solving the system $AX = B$ where $X = [x_1, x_2, x_3]'$,

$$A = \begin{bmatrix} 600 & 30 & 3.0 \\ 500 & 28 & 2.9 \\ 750 & 35 & 3.9 \\ 650 & 30 & 3.5 \\ 550 & 25 & 2.8 \\ 800 & 35 & 3.7 \end{bmatrix} \text{ and } B = \begin{bmatrix} 3.2 \\ 3.0 \\ 3.5 \\ 3.5 \\ 3.2 \\ 3.7 \end{bmatrix}$$

Statistically, it is highly unlikely that such numbers exist: we do not expect to be able to predict a student's college GPA with certainty on the basis of admissions data. Thus, we feel confident that the system $AX = B$ is inconsistent. More mathematically, the set of B for which $AX = B$ is solvable is the column space of A. This subspace is at most three-dimensional, since A has only three columns. The system is solvable only if B happens to lie in this subspace. The probability that a randomly selected vector in $\mathbb{R}^6$ lies in a given three-dimensional subspace is practically nil. (Think about the probability of a randomly selected vector in $\mathbb{R}^3$ lying in some given plane.)

The mathematical interpretation, however, suggests a "geometric" solution to our problem. Since it is hard to picture six dimensions, we shall draw a three-dimensional picture. However, it will be seen that our three-dimensional intuition is an accurate guide to six-dimensional geometry. Our game plan may be described as follows:

Let $\mathcal{W}$ be the span of the columns of A. We think of $\mathcal{W}$ as a plane in space and B as some vector lying outside of this plane, as in Figure 1. We drop a perpendicular from B to $\mathcal{W}$, producing the point B_0. The equation $AX = B_0$ is definitely solvable, since $\mathcal{W}$ is the column space. We solve this equation for X. Our geometric intuition suggests that B_0 should be the closest point in $\mathcal{W}$ to B. Thus, while $AX \neq B$, it will at least be as close to equaling B as possible.

In fulfilling our game plan our first question is, of course, How do we find B_0? This vector is defined by two properties:

(a) $B_0 \in \mathcal{W}$.
(b) $B_1 = B - B_0$ is perpendicular to $\mathcal{W}$.

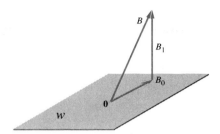

The projection of B onto W

Figure 1

Since $\mathcal{W}$ is spanned by the columns of A, the second condition means that the dot product of B_1 with each column of A must be zero. Explicitly, if the columns of A are A_1, A_2, and A_3, then

$$A_1^t B_1 = 0$$
$$A_2^t B_1 = 0$$
$$A_3^t B_1 = 0$$

The vectors A_i^t are the rows of A^t. *Thus, we see that B_1 is perpendicular to the column space of A if and only if*

$$A^t B_1 = \mathbf{0} \qquad\qquad (1)$$

The condition that $B_0 \in \mathcal{W}$ is equivalent to $B_0 = AX$ for some $X \in \mathbb{R}^3$. Thus, conditions (a) and (b) above may be combined together into the single equation

$$A^t(B - AX) = \mathbf{0}$$

Note that this equation is equivalent with

$$A^t B = A^t A X$$

(This equation is called the **normal equation**.) The least squares theorem below proves the existence of an X satisfying this equation and Theorem 1 proves that $B_0 = AX$ is the closest point in $\mathcal{W}$ to B. What these theorems say is quite remarkable. We seem to have multiplied both sides of the equation $B = AX$ (which has no solutions) by A^t. Suddenly, we have a unique solution and this solution is, in some sense, the best possible answer! Specifically, the X we find by this process makes $|B - AX|$ as small as possible. More explicitly, if $B = [b_1, b_2, \ldots, b_n]^t$ and $AX = [y_1, y_2, \ldots, y_n]^t$, then the X we find has the property that

$$(b_1 - y_1)^2 + (b_2 - y_2)^2 + \cdots + (b_n - y_n)^2$$

is as small as possible. This explains the term "least squares."

Granted all this, we can solve our consulting problem for the university. We compute (using MATLAB)

$$A^t A = \begin{bmatrix} 2537500 & 119500 & 12950.0 \\ 119500 & 5659 & 612.2 \\ 12950.0 & 612.2 & 66.40 \end{bmatrix} \text{ and } A^t B = \begin{bmatrix} 13040.0 \\ 617.0 \\ 66.85 \end{bmatrix}$$

Next, we ask our computer to solve $A^t A X = A^t B$. We get:

$$X = \begin{bmatrix} -0.0002 \\ 0.0461 \\ 0.6227 \end{bmatrix}$$

As a check on our work, we compute

$$B_0 = AX = \begin{bmatrix} 3.1251 \\ 2.9917 \\ 3.8846 \\ 3.4260 \\ 2.7806 \\ 3.7495 \end{bmatrix}$$

This matrix represents the "expected GPA" of each student on the basis of the admissions data. It compares favorably with the actual GPA, although the third student did not do quite as well as we might have expected, while the fourth was definitely an overachiever.

Remark: You might have noticed that the first coefficient of X was negative (-0.00021). This suggests that there is a slight *negative* correlation between SAT scores and GPA. This, of course, is nonsense! This is an artifact of the data (made up by the author). For the specific data set given, there is in fact very little correlation between the SAT and the final grade. This certainly would not be true for real data. This demonstrates an important point: a good mathematician always looks at the output of any calculation and asks if it makes sense.

Now, let us prove the least squares theorem. We should remark that the assumption that A has rank n is equivalent with saying that the columns of A are independent. This condition would be satisfied in most applications. We certainly would not, for example, expect to be able to express the GPA column of the data matrix as a linear combination of the ACT and SAT columns–this would say that we can predict with certainty a student's high school GPA solely on the basis of ACT and SAT scores. Actually, the only part of the conclusion that requires the rank assumption is the uniqueness of X. The proof, however, is somewhat easier with this assumption in place. The general case is discussed in the exercises.

Least Squares Theorem.   Let A be an m × n matrix that has rank n. Then for all $B \in \mathbb{R}^m$, there exists a unique $X \in \mathbb{R}^n$ such that

$$A^t B = (A^t A) X$$

The vector $B_0 = AX$ is the unique vector in the column space of A with the property that $B - B_0$ is perpendicular to the column space of A.

Proof: The matrix A^t is of size $n \times m$ while A is $m \times n$. Hence $A^t A$ is $n \times n$. If we can show that this matrix is invertible, then both the existence and uniqueness of X will follow.

For invertibility, it suffices to show that the nullspace is zero. Thus, suppose that

$$A^t A X = \mathbf{0}$$

From the reasoning that lead up to formula (1), we see that AX is perpendicular to the column space of A. But, AX is also in the column space of A. Hence

$$0 = (AX) \cdot (AX) = |AX|^2$$

The only vector with zero length is the zero vector. Thus $AX = \mathbf{0}$. Since the columns of A are assumed to be independent, this implies that $X = \mathbf{0}$, as desired. ∎

The least squares theorem tells us how to find B_0. It does not, however, tell us that B_0 really is the closest point in $\mathcal{W}$ to B. At the moment, all we know about B_0 is that it is the "foot" of the perpendicular from B to $\mathcal{W}$; i.e. $B_1 = B - B_0$ is perpendicular to every vector in $\mathcal{W}$. To prove this, we need to state the least squares theorem more geometrically.

For any subspace $\mathcal{W}$ of $\mathbb{R}^n$, we define $\mathcal{W}^\perp$ to be the set of all vectors V such that $V \cdot W = 0$ for all $W \in \mathcal{W}$. It is not difficult to show that $\mathcal{W}^\perp$ is a subspace of $\mathbb{R}^n$. Notice in Figure 2 that in $\mathbb{R}^3$, if $\mathcal{W}$ is a plane through $\mathbf{0}$, then $\mathcal{W}^\perp$ is the line through $\mathbf{0}$ perpendicular to the plane.

Figure 2 also indicates that any $B \in \mathbb{R}^3$ may be written in the form $B = B_0 + B_1$ where $B_0 \in \mathcal{W}$ and $B_1 \in \mathcal{W}^\perp$. Notice that in this picture, since B_0 and B_1 are perpendicular, it follows from the Pythagorean theorem that

$$|B|^2 = |B_0|^2 + |B_1|^2 \tag{2}$$

The geometric content of Theorem 1 is that this is all true in any dimension.

Theorem 1.   *Let $\mathcal{W}$ be any subspace of $\mathbb{R}^n$ and let $B \in \mathbb{R}^n$. Then there exist uniquely determined vectors $B_0 \in \mathcal{W}$ and $B_1 \in \mathcal{W}^\perp$ such that $B = B_0 + B_1$. Furthermore, formula (2) holds, and B_0 is the point in $\mathcal{W}$ that is closest to B.*

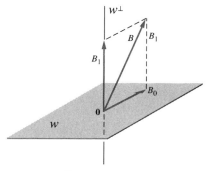

The orthogonal complement

Figure 2

Proof: Let $A_1, A_2, \ldots, A_k$ be a basis for $\mathcal{W}$. Then $\mathcal{W}$ is the column space of the matrix $A = [A_1, A_2, \ldots, A_k]$. The existence and uniqueness of B_0 and B_1 follows directly from the least squares theorem.

To prove formula (2), notice that since B_1 is perpendicular to $\mathcal{W}$, $B_1 \cdot B_0 = 0$. Hence

$$|B|^2 = (B_1 + B_0) \cdot (B_1 + B_0) = B_1 \cdot B_1 + 2(B_1 \cdot B_0) + B_0 \cdot B_0 = |B_1|^2 + |B_0|^2$$

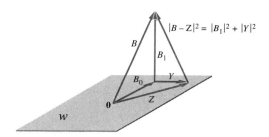

Figure 3

Finally, to prove that B_0 is the closest point in $\mathcal{W}$ to B, let Z be another point in $\mathcal{W}$ and set $Y = Z - B_0$ so that $Z = B_0 + Y$ as in Figure 3. Then

$$B - Z = B - B_0 - Y = B_1 - Y$$

Hence, from formula (2),

$$|B - Z|^2 = |B_1 - Y|^2 = |B_1|^2 + |-Y|^2 = |B_1|^2 + |Y|^2$$

(Note that $-Y \in \mathcal{W}$ and $B_1 \in \mathcal{W}^{\perp}$.) It follows that the distance is least when $Y = \mathbf{0}$, which is the same as $Z = B_0$, as desired.   ■

The vector B_0 in the least squares theorem is called the **orthogonal projection** of B onto $\mathcal{W}$. It is denoted

$$B_o = \text{proj}_{\mathcal{W}}(B)$$

Every subspace of $\mathbb{R}^n$ has a basis. Hence, every subspace is the column space of some matrix. It follows that we may use the least squares theorem to compute orthogonal projections onto subspaces, as Example 1 demonstrates.

▶ **EXAMPLE 1:** Let $\mathcal{W}$ be the subspace of $\mathbb{R}^4$ spanned by $A_1 = [1, 2, 1, 1]^t$ and $A_2 = [1, 0, 1, 0]^t$. Compute $\text{proj}_{\mathcal{W}}(B)$ for $B = [1, 2, 3, 4]^t$.

Solution: The subspace in question is the column space of the matrix A:

$$A = \begin{bmatrix} 1 & 1 \\ 2 & 0 \\ 1 & 1 \\ 1 & 0 \end{bmatrix}$$

From the above discussion, $\text{proj}_{\mathcal{W}}(B) = B_0 = AX$ where X is found by solving the normal equation $A^t A X = A^t B$. Explicitly, as the reader may check,

$$A^t A = \begin{bmatrix} 7 & 2 \\ 2 & 2 \end{bmatrix} \qquad A^t B = \begin{bmatrix} 12 \\ 4 \end{bmatrix}$$

We compute X as

$$X = \begin{bmatrix} 7 & 2 \\ 2 & 2 \end{bmatrix}^{-1} \begin{bmatrix} 12 \\ 4 \end{bmatrix} = \begin{bmatrix} 0.2 & -0.2 \\ -0.2 & 0.7 \end{bmatrix} \begin{bmatrix} 12 \\ 4 \end{bmatrix} = \begin{bmatrix} 1.6 \\ 0.4 \end{bmatrix}$$

Finally,

$$\text{proj}_{\mathcal{W}}(B) = B_0 = AX = \begin{bmatrix} 1 & 1 \\ 2 & 0 \\ 1 & 1 \\ 1 & 0 \end{bmatrix} \begin{bmatrix} 1.6 \\ 0.4 \end{bmatrix} = \begin{bmatrix} 2.0 \\ 3.2 \\ 2.0 \\ 1.6 \end{bmatrix}$$

We may check our work by checking that $B_1 = B - B_0 = [-1, -1.2, 1, 2.4]^t$ is perpendicular to both A_1 and A_2. It is easily seen that this is valid.   ◀

Abstractly, what we did in Example 1 was to solve the normal equations for X by inverting $A^t A$. What we found was

$$X = (A^t A)^{-1} A^t B$$

Next, we said

$$B_0 = AX = A(A^t A)^{-1} A^t B$$

Thus, the work done in Example 1 can be summarized in Theorem 2.

Theorem 2.   Let W be a subspace of $\mathbb{R}^n$ and let $A_1, A_2, \ldots, A_k$ form a basis for W. Then for all $B \in \mathbb{R}^n$

$$\text{proj}_W(B) = A(A^t A)^{-1} A^t B$$

where $A = [A_1, A_2, \ldots, A_k]$.   ■

We shall close this section with another (very typical) example of the least squares technique.

▶ EXAMPLE 2:   Let us imagine that we are studying a physical system that gets hotter over time. Let us also suppose that we expect a linear relationship between time and temperature. That is, we expect time and temperature to be related by a formula of the form

$$T = a + bt$$

where T is temperature (in degrees Celcius), t is time (in seconds), and a and b are unknown physical constants. We wish to do an experiment to determine approximate values for the constants a and b. We allow our system to get hot and measure the temperature at various times t. The following data summarizes our findings:

t	0.5	1.1	1.5	2.1	2.3
T	32.0	33.0	34.2	35.1	35.7

A very simple-minded solution to this problem would be to plot our data on a sheet of graph paper, as shown in Figure 4.

Since we expect a linear relationship, all these points should lie on a single straight line. The slope of this line will be b, and the intercept (not shown in Figure 4) is a. Unfortunately, when we plot our data, we discover that the points do not lie on a single line. This is only to be expected, since our measurements are subject to experimental error. On the other hand, it appears that the points are "approximately" colinear. We can

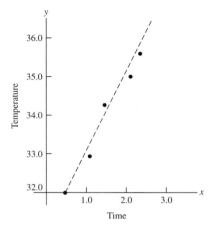

Figure 4

find good approximations to a and b by constructing a single line that best fits the data. But how do we find this line? We could simply take out a ruler and draw in a line that seems "pretty good." This approach, however, is rather imprecise and inaccurate. It is much better to use linear algebra. Specifically, if the points did all lie on a single straight line, each of these five equations below would be satisfied

$$32.0 = a + 0.5b$$
$$33.0 = a + 1.1b$$
$$34.2 = a + 1.5b$$
$$35.1 = a + 2.1b$$
$$35.7 = a + 2.3b$$

As a matrix equation, this system may be written $B = AX$ where $X = [a, b]^t$, $B = [32.0, 33.0, 34.2, 35.1, 35.7]^t$, and

$$A = \begin{bmatrix} 1 & 0.5 \\ 1 & 1.1 \\ 1 & 1.5 \\ 1 & 2.1 \\ 1 & 2.3 \end{bmatrix}$$

Since the points do not all lie on a line, we know that this equation does not have a solution. However, we can proceed to find an approximate solution just as before. Specifically, we replace the system $B = AX$ by the system $A^t B = A^t AX$ and solve this system for $X = [a, b]^t$. We shall leave the computation of X as an exercise. ◀

Remark: This kind of analysis of data is also called **regression analysis**, since one of the early applications of least squares was to genetics, to study the well-known phenomenon that children of abnormally tall or abnormally short parents tend to be more normal in height than their parents. In more technical language, the children's height tends to "regress toward the mean."

EXERCISES

(1) Compute X in Example 2 by solving the normal equation. Then compute $B_0 = AX$. The answer should agree closely with B. Show that $B - B_0$ really is perpendicular to the columns of A.

(2) Suppose that in the context of Example 2, it is proposed that our data would be better described by an equation of the form

$$T = a + b \ln t$$

Use the techniques of this section to find approximate values for a and b. How would you go about deciding which equation better fits the given data: the logarithmic equation proposed in this exercise or the linear equation proposed in Example 1? [*Hint:* Think in terms of distance in $\mathbb{R}^5$.]

(3) Suppose that in the context of Example 2, it is proposed that our data would be better described by an equation of the form

$$T = a + bt + ct^2$$

   (a) Explain how you would use the techniques of this section to find approximate values for a, b, and c. Specifically, in the normal equation $A^t B = A^t A X$, what are appropriate choices for A and B?
   (b) If you have appropriate software available, compute the solution to the normal equation. Note the size of the constant c. What is your interpretation of this result?

(4) The data in the chart below is the estimated population of the United States (in millions), rounded to three digits, from 1980–1995.[1] Your goal in this exercise is to predict the U.S. population in the year 2010.

Year	1980	1985	1990	1995
Population	227	237	249	262

---

[1] *Source: U.S. Census Bureau.*

For this, let t denote "years after 1980" and I represent increase in population over the 1985 level (see the chart below). Use the method of least squares to find constants a and b such that I is approximately equal to $at + b$. Then use your formula to predict the 2010 population.

Years after 1980	0	5	10	15
Increase over 1980	0	10	22	35

(5) In population studies, exponential models are much more commonly used than linear models. This means that we hope to find constants a and b such that the population P is given approximately by the equation $P = ae^{bt}$. To convert this into a linear equation, we take the natural logarithm of both sides, producing

$$\ln P = \ln a + bt$$

Use the method of least squares to find values for $\ln a$ and b that make this equation approximate the data from Exercise 4. You can let t denote "years after 1980" but you should use the actual population for P, not the increase in population over 1980. Finally, use the formula $P = e^{bt}$ to predict the population in the year 2010.

(6) Let W be the plane $2x - y + 3z = 0$. Compute the orthogonal projection of $B = [1, 2, 3]$ onto this plane. Use your answer to compute the distance from B to this plane.

(7) Let W be the subspace of $\mathbb{R}^5$ spanned by $A_1 = [1, 2, 0, 1, 0]^t$ and $A_2 = [1, 1, 1, 1, 1]^t$. Compute the orthogonal projection B_0 of $B = [1, 0, 0, 0, 0]^t$ onto W. Show by direct calculation that $B_1 = B - B_0$ is perpendicular to A_1 and A_2.

(8) The matrix $P = A(A^tA)^{-1}A^t$ from Theorem 2 is called the **projection matrix**.

   (a) Compute P for the subspace W in Exercise 7. Then compute P^2. If you are careful, it should be the case that $P = P^2$. Why, geometrically, should this not be a surprise?
   (b) Prove algebraically that if P is a projection matrix, then $P^2 = P$.

(9) Exercise 15 in Section 3.4 proved that $(AB)^{-1} = B^{-1}A^{-1}$. Using this property, we can prove that the matrix P from Exercise 8 must be the identity matrix. Specifically

$$P = A(A^tA)^{-1}A^t = AA^{-1}(A^t)^{-1}A^t = II = I$$

But P did not turn out to be I. What's wrong?

(10) Let W be the subspace of $\mathbb{R}^5$ spanned by $C_1 = [0, 1, -1, 0, -1]^t$ and $C_2 = [2, 3, 1, 2, 1]^t$. Compute the projection matrix P for W. If you work carefully,

it should be the same as the P found in Exercise 8. Why, geometrically, should this not surprise you? Why, algebraically, should this not surprise you? *Hint:*

$$\begin{bmatrix} 0 & 2 \\ 1 & 3 \\ -1 & 1 \\ 0 & 2 \\ -1 & 1 \end{bmatrix} = \begin{bmatrix} 1 & 1 \\ 2 & 1 \\ 0 & 1 \\ 1 & 1 \\ 0 & 1 \end{bmatrix} \begin{bmatrix} 1 & 1 \\ -1 & 1 \end{bmatrix}$$

(11) Let W be a subspace of $\mathbb{R}^n$ and suppose that $A_1, A_2, \ldots, A_m$ and $C_1, C_2, \ldots, C_m$ are both bases for W. Prove algebraically that

$$A(A^t A)^{-1} A^t = C(C^t C)^{-1} C^t$$

where $A = [A_1, A_2, \ldots, A_m]$ and $C = [C_1, C_2, \ldots, C_m]$.

(12) If W is a k-dimensional subspace of $\mathbb{R}^n$ and P is the corresponding projection matrix, what is the rank of P? [*Hint:* What is the image of the corresponding transformation?] What is the nullspace of P? (It was discussed in this section.)

(13) In the consulting example given in the text, check that $B_1 = B - B_0$ is really perpendicular to the columns of A.

(14) In our consulting example, we used the fact that if Y is perpendicular to $A_1, A_2,$ and A_3, then Y is perpendicular to their span. Prove it!

(15) In this exercise, we investigate how the least squares technique is affected by dependencies among the columns of A. For this exercise let

$$A = \begin{bmatrix} 1 & 2 & 3 \\ 3 & 1 & 4 \\ -2 & 1 & -1 \end{bmatrix}$$

Note that the columns of A are dependent.

(a) Compute $C = A^t A$. Explain why, on general principles, the rank of C can be at most 2. Explicitly exhibit one row of C as a combination of the others. This means that $A^t A$ cannot be invertible. This suggests that we may not be able to solve the normal equation for X. But read on. . . .

(b) Let $B = [1, 2, 3]^t$. Write the augmented matrix for the system $CX = A^t B$. Express one row of this matrix as a linear combination of the others. Is the corresponding system of equations consistent? How many free variables will it have?

(c) The consistency of the system in part (b) is too improbable to be a coincidence. The matrices Y for which $A^t A X = Y$ is solvable comprise the column space of $A^t A$. What the calculations in part (b) suggest is that for any B in $\mathbb{R}^m$ and any $m \times n$ matrix A, $A^t B$ belongs to the column space of $A^t A$. To prove this, note that from Theorem 1, we may write $B = B_1 + B_0$ where B_0 belongs to the column space of A and B_1 is perpendicular to the column space. Use this to prove that $A^t B = A^t B_0$. How does this prove that $A^t B$ is in the column space of $A^t A$?

(d) Suppose now that X is any solution to $A^t A X = A^t B$. Explain why $B - AX$ is perpendicular to the column space of A. How does it follow that $AX = B_0$?

Remark: To summarize the conclusion of this exercise, if A is any $m \times n$ matrix, then the normal equation is consistent, regardless of the rank of A, and any solution X will be a least squares solution in the sense that $AX = B_0$. If the rank of A is less than n, however, X will not be unique.

(16) In Exercise 15, we noted that if A has rank k, then rank $A^t A$ is at most k. Prove that in fact it is exactly k. The "easy" way to do this is to prove that A and $A^t A$ have the same nullspace. If you read the text carefully, you should find the equality of the nullspaces proved for you.

(17) For any subset S of $\mathbb{R}^n$, we define $S^\perp$ to be the set of all vectors V such that $V \cdot X = 0$ for all $X \in S$.

(a) Suppose that S is a line through $\mathbf{0}$ in $\mathbb{R}^2$. Indicate with a graph what $S^\perp$ is. [*Hint*: Points in $\mathbb{R}^n$ represent vectors that begin at $\mathbf{0}$.] What is $(S^\perp)^\perp$? What about $((S^\perp)^\perp)^\perp$?

(b) Suppose that S is a pair of independent vectors in $\mathbb{R}^3$. Answer the questions posed in part (a).

(c) Suppose that S is a plane through $\mathbf{0}$ in $\mathbb{R}^3$. Answer the questions posed in part (a).

(d) Suppose that S is a line through $\mathbf{0}$ in $\mathbb{R}^3$. Answer the questions posed in part (a).

(18) The purpose of Exercise 17 was to convince you that if $\mathcal{W}$ is a subspace of either $\mathbb{R}^2$ or $\mathbb{R}^3$, then $\mathcal{W} = (\mathcal{W}^\perp)^\perp$. Now you are to prove that this is true in any dimension. Thus, let $\mathcal{W}$ be a subspace of $\mathbb{R}^n$.

(a) Let $\mathcal{W}$ be spanned by the columns $A_1, A_2, \ldots, A_k$ and let $A = [A_1, A_2, \ldots, A_k]$ be the matrix having the A_i as columns. Then $\mathcal{W}$ is the column space of A. According to the text, $\mathcal{W}^\perp$ is the nullspace of A^t. Use Theorem 1 (Section 3.3) to prove that

$$\dim(\mathcal{W}) + \dim(\mathcal{W}^\perp) = n$$

(b) Prove that every element of $\mathcal{W}$ belongs to $(\mathcal{W}^\perp)^\perp$.

(c) Use part (a) to prove that $\mathcal{W}$ and $(\mathcal{W}^\perp)^\perp$ have the same dimension. How does part (b) then prove that these spaces are equal?

(19) Prove that in $\mathbb{R}^n$, the only vector orthogonal to itself is the zero vector. Use this to prove that for any subspace $\mathcal{W}$ of $\mathbb{R}^n$, $\mathcal{W} \cap \mathcal{W}^\perp = \{\mathbf{0}\}$. (Recall that if A and B are sets, then $A \cap B$ is the set of points that belong to both A and B.)

(20) Let $\mathcal{W}$ be a subspace of $\mathbb{R}^n$. Let $B \in (\mathcal{W}^\perp)^\perp$. Use Theorem 1 to write $B = B_1 + B_0$ where $B_0 \in \mathcal{W}$ and $B_1 \in \mathcal{W}^\perp$. Then, $B - B_0$ belongs to $\mathcal{W}^\perp$. Prove that $B - B_0$ also belongs to $(\mathcal{W}^\perp)^\perp$. Explain how this proves that $B = B_0$. This proves that $(\mathcal{W}^\perp)^\perp \subset \mathcal{W}$ and hence that $(\mathcal{W}^\perp)^\perp = \mathcal{W}$.

ON LINE

(1) Imagine that you are an astronomer investigating the orbit of a newly discovered asteroid. You want to determine (a) how close the asteroid will come to the sun and (b) how far away from the sun the asteroid will get. To solve your problem, you will make use of the following facts.

(a) Asteroids' orbits are approximately elliptical, with the sun as one focus.
(b) In polar coordinates, an ellipse with one focus at the origin can be described by a formula of the following form where a, b, and c are constants.

$$r = \frac{c}{1 + a\sin\theta + b\cos\theta}$$

You have also collected the data tabulated below where r is the distance from the sun in millions of miles and θ is the angle in radians between the vector from the sun to the asteroid and a fixed axis through the sun. This data is, of course, subject to experimental error.

Your strategy is to use the given data to find values of a, b, and c that cause the formula to agree as closely as possible with the given data. This will involve setting up a system of linear equations and solving the normal equation. (Give the augmented matrices for both the original system and the normal equation.) You will then graph the given formula and measure the desired data from the graph. Good luck!

θ	0.00	0.60	1.80	1.40	2.10	3.20	5.40
r	329.27	313.80	319.49	310.91	327.88	374.91	367.49

Remark: Once you have found values of a, b, and c, you will need to plot the orbit of the asteroid. For this, you should create a vector of θ values and a vector of the corresponding r values. Thus, you might enter "T=0:.1:2*pi" followed by

```
r=c*(1+a*sin T+b*cos T).^(-1)
```

(Note the period.) Then you enter "polar(T,r)" to construct the plot.

The asteroid field.

(2) MATLAB has a way of solving least squares problems that is easier (and better) than solving the normal equation. If the (inconsistent) system is expressed in the form $AX = B$, then the least squares solution is $A \backslash B$. Try it!

4   CHAPTER SUMMARY

In this chapter, our emphasis was on the concepts of distance and angle. We began in Section 4.1 by studying the role of distance and angle in defining the coordinates one uses in analytic geometry. We saw that we could define new coordinates using non-perpendicular axes with different units on each axis.

Coordinates are defined by bases. Specifically, we use a basis which contains one vector on each of the new axes. The coordinates of a point are the coefficients used in expressing the point as a linear combination of the basis elements. The transformation between the coordinates and the usual coordinates (the *natural coordinates*) is accomplished by multiplying the coordinate vector by the *point matrix*, which is the matrix whose columns are formed by the basis elements. The inverse of the point matrix is the *coordinate matrix*. Multiplication of points by the coordinate matrix produces the new coordinate vector of the point. When the basis elements are mutually perpendicular (i.e. when the basis is *orthogonal*), we can use Theorem 1 in Section 4.1 (the "dot product trick") to find the coordinates.

This idea of describing a space using coordinates applies to many vector spaces other than $\mathbb{R}^n$. For example, we saw in some of the exercises for Section 4.1 that coordinates could be used to reduce questions about polynomial functions to questions about vectors. We also saw in Section 4.2 that by introducing the right coordinates, we could think if any subspace of $\mathbb{R}^n$ as being just $\mathbb{R}^k$ for some k (i.e. every subspace of $\mathbb{R}^n$ is *isomorphic* with $\mathbb{R}^k$ for some k). This could be done in such a way that the concepts of distance and angle in the subspace "match" those in $\mathbb{R}^k$ (Theorem 2, Section 4.2).

In this context, the "right" coordinates are those defined by an *orthonormal* basis. (An orthonormal basis for a subspace is a basis consisting of mutually perpendicular vectors of length one.) It is a fundamental result (Theorem 1 in Section 4.2) that every subspace of $\mathbb{R}^n$ has an orthonormal basis. This theorem is a direct consequence of the Gram–Schmidt process, which is a technique for transforming non-orthogonal bases into orthogonal bases. The Gram–Schmidt process also produces the *QR factorization* of matrices.

One important application of orthogonal bases for subspaces is to computing the *projection* of a vector onto the subspace (the Fourier theorem from Section 4.2). (The projection of a vector onto a subspace is the closest point in the subspace to the vector.) In Section 4.5, we saw different ways of computing projections (the least-squares theorem) as well as an application of the notion to the study of inconsistent systems of equations. Specifically, given an inconsistent system $AX = B$ where A in $m \times n$, we can let B_0 be the projection of B onto the column space of A. The system $AX = B_0$ is solvable and any solution X comes as close to solving $AX = B$ as possible in the sense that the distance from AX to B in $\mathbb{R}^m$ is as small as possible. Remarkably, X may be found by solving the system $A^t A X = A^t B$ (the *normal equations*). We may then find B_0 from $AX = B_0$.

In Section 4.5 we saw another remarkable application of the ideas of projections and orthogonal bases. In this section, we noted that one could define a *scalar product* for functions which is similar to the dot product of vectors. This notion is so similar, in fact, that it can be used to compute projections in spaces of functions in just the same way as in $\mathbb{R}^n$. We used these ideas to solve the problem of approximating given wave forms by combinations of "pure tones" (sine and cosines). In the exercises, we saw many examples of vector spaces on which one could define meaningful scalar products. (A scalar product on a vector space is a scalar valued product which satisfies the properties listed in the scalar product theorem in Section 4.3. A vector space together with a scalar product is called a *scalar product space*).

Another important topic from this chapter was *orthogonal matrices* (Section 4.4). We know that in $\mathbb{R}^2$ rotating a vector counterclockwise about the origin by some fixed angle (say $\pi/4$) does not change the length of the vector. An $n \times n$ matrix A is *orthogonal* if multiplication of vectors by A does not change their lengths. Orthogonal matrices are characterized by the lovely property that their inverse equals their transpose ($A^t A = I$). They are also characterized by the fact that their columns are mutually orthogonal and have length one. In $\mathbb{R}^2$ and $\mathbb{R}^3$, the trasformation defined by an orthogonal matrix can be expressed as a product of rotations and reflections.

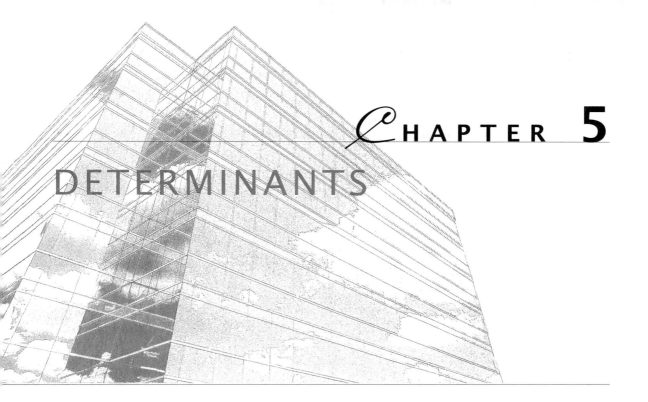

CHAPTER 5

DETERMINANTS

5.1 DETERMINANTS

You may have wondered why we need to row-reduce whenever we compute the inverse of a matrix. Should we not be able to write down a general formula for the inverse and simply plug in our specific numbers? The answer is, yes, formulas do exist. Using them for large matrices, however, usually requires more work than simply doing the row reduction! The formula for a 2×2 matrix is, nevertheless, very simple:

$$\begin{bmatrix} a & b \\ c & d \end{bmatrix}^{-1} = \frac{1}{ad - cb} \begin{bmatrix} d & -b \\ -c & a \end{bmatrix}$$

The factor $\delta = ad - cb$, which occurs in the formula for A^{-1}, is called the determinant of A. We denote it by "det A." Thus

$$\det \begin{bmatrix} a & b \\ c & d \end{bmatrix} = ad - cb$$

Hence

$$\det \begin{bmatrix} 1 & 2 \\ 3 & 4 \end{bmatrix} = (1 \cdot 4) - (3 \cdot 2) = -2$$

It is also common to denote the determinant of a matrix by replacing the square brackets with straight lines. Thus, we define

$$\begin{vmatrix} a & b \\ c & d \end{vmatrix} = ad - cb$$

In Section 5.3, we give a formula for the inverse of the general $n \times n$ matrix A that has the form

$$A^{-1} = \frac{1}{\delta} B$$

where δ is a number which, once again, is called the determinant of A and is denoted "det A." Just as in the 2×2 case, it will turn out that A is invertible if and only if $\delta \neq 0$. (We prove this fact in Section 5.2.) As commented above, using formulas to compute matrix inverses is usually very inefficient. However, the fact that a matrix is invertible if and only if the determinant is non-zero is one of the most important facts in linear algebra. In particular, it will become essential when we study eigenvalues in Chapter 6.

Our first goal is to describe how to compute determinants for matrices larger than 2×2. The determinant of a 3×3 matrix is computable in terms of determinants of 2×2 matrices. Specifically, we define

$$\begin{vmatrix} a_{11} & a_{12} & a_{13} \\ a_{21} & a_{22} & a_{23} \\ a_{31} & a_{32} & a_{33} \end{vmatrix} = a_{11} \begin{vmatrix} a_{22} & a_{23} \\ a_{32} & a_{33} \end{vmatrix} - a_{12} \begin{vmatrix} a_{21} & a_{23} \\ a_{31} & a_{33} \end{vmatrix} + a_{13} \begin{vmatrix} a_{21} & a_{22} \\ a_{31} & a_{32} \end{vmatrix} \quad (1)$$

(Note the use of the vertical lines to denote determinants.)

Notice that in this formula, the first 2×2 matrix on the right is what we get if we eliminate the row and column containing a_{11} from the original 3×3 matrix:

$$\begin{bmatrix} a_{11} & a_{12} & a_{13} \\ a_{21} & a_{22} & a_{23} \\ a_{31} & a_{32} & a_{33} \end{bmatrix} = \begin{bmatrix} a_{22} & a_{23} \\ a_{32} & a_{33} \end{bmatrix}$$

In formula (1), the determinant of this matrix is multiplied by a_{11}. Similarly, the term which is multiplied by a_{12} is the determinant of the 2×2 matrix obtained by eliminating the row and column containing a_{12}, and the term which is multiplied by a_{13} is the determinant of the 2×2 matrix obtained by eliminating the row and column containing a_{13}. We add the first product, subtract the second, and add the third.

▶ EXAMPLE 1:   Compute det A, where

$$A = \begin{bmatrix} 2 & 0 & 1 \\ 6 & 1 & 7 \\ 4 & 5 & 9 \end{bmatrix}$$

Solution: From formula (3):

$$\begin{vmatrix} 2 & 0 & 1 \\ 6 & 1 & 7 \\ 4 & 5 & 9 \end{vmatrix} = 2 \det \begin{bmatrix} 2 & 0 & 1 \\ 6 & 1 & 7 \\ 4 & 5 & 9 \end{bmatrix} - 0 \det \begin{bmatrix} 2 & 0 & 1 \\ 6 & 1 & 7 \\ 4 & 5 & 9 \end{bmatrix} + 1 \det \begin{bmatrix} 2 & 0 & 1 \\ 6 & 1 & 7 \\ 4 & 5 & 9 \end{bmatrix}$$

$$= 2 \begin{vmatrix} 1 & 7 \\ 5 & 9 \end{vmatrix} - 0 \begin{vmatrix} 6 & 7 \\ 4 & 9 \end{vmatrix} + 1 \begin{vmatrix} 6 & 1 \\ 4 & 5 \end{vmatrix}$$

$$= 2[(1 \cdot 9) - (5 \cdot 7)] - 0 + [(6 \cdot 5) - (4 \cdot 1)] = -26 \quad \blacktriangleleft$$

The definition of the determinant of an $n \times n$ matrix is similar. If A is any $n \times n$ matrix, we let A_{ij} denote the matrix obtained from A by deleting the ith row and jth column. We then define the determinant of A inductively by the formula

$$\det A = a_{11} \det A_{11} - a_{12} \det A_{12} + a_{13} \det A_{13} + \cdots + (-1)^{n-1} a_{1n} \det A_{1n} \quad (2)$$

Thus, as in the 3×3 case, we multiply each entry in the first row by the determinant of the matrix obtained by eliminating the row and column containing the entry. Also, as in the 3×3 case, we alternately add and subtract the terms. This method of computing determinants is called either the **Laplace expansion** or the **cofactor expansion**

▶ **EXAMPLE 2:** Compute

$$\begin{vmatrix} 1 & 2 & 3 & 4 \\ 5 & 6 & 7 & 8 \\ 9 & 10 & 11 & 12 \\ 13 & 14 & 15 & 16 \end{vmatrix}$$

Solution: We compute:

$$1 \begin{vmatrix} 6 & 7 & 8 \\ 10 & 11 & 12 \\ 14 & 15 & 16 \end{vmatrix} - 2 \begin{vmatrix} 5 & 7 & 8 \\ 9 & 11 & 12 \\ 13 & 15 & 16 \end{vmatrix} + 3 \begin{vmatrix} 5 & 6 & 8 \\ 9 & 10 & 12 \\ 13 & 14 & 16 \end{vmatrix} - 4 \begin{vmatrix} 5 & 6 & 7 \\ 9 & 10 & 11 \\ 13 & 14 & 15 \end{vmatrix}$$

Each of these 3×3 determinants works out to be zero. For example,

$$\begin{vmatrix} 6 & 7 & 8 \\ 10 & 11 & 12 \\ 14 & 15 & 16 \end{vmatrix} = 6 \begin{vmatrix} 11 & 12 \\ 15 & 16 \end{vmatrix} - 7 \begin{vmatrix} 10 & 12 \\ 14 & 16 \end{vmatrix} + 8 \begin{vmatrix} 10 & 11 \\ 14 & 15 \end{vmatrix}$$

$$= 6[(11 \cdot 16) - (15 \cdot 12)] - 7[(10 \cdot 16) - (14 \cdot 12)]$$

$$+ 8[(10 \cdot 15) - (11 \cdot 14)] = 0$$

Thus, the answer is zero. ◀

Formulas (1) and (2) make it appear that the first row plays a special role in computing determinants. Actually, the following theorem tells us that we can compute the determinant by applying the exact same process to any row, except if we use an even numbered row (such as row 2), we must multiply the final answer by -1. (This is indicated by the factor -1^{i+1} in the formula.) This is equivalent with saying that we should *subtract* the first product, add the second, subtract the third, etc. We will not prove this theorem as its proof would involve a substantial digression. (In the statement of this result, recall that A_{ij} denotes the matrix obtained by deleting the ith row and jth column from A.)

Theorem 1.   Let A be an $n \times n$ matrix. Then, for each i,

$$\det A = (-1)^{i+1}(a_{i1} \det A_{i1} - a_{i2} \det A_{i2} + a_{i3} \det A_{i3} + \cdots + (-1)^{n-1} a_{in} \det A_{in}) \quad (3)$$

▶ EXAMPLE 3:   Redo Example 1 by expanding along the second row.

Solution: Since we are using the second row, we begin by subtracting the first product and alternate the signs thereafter. Thus

$$\begin{bmatrix} 2 & 0 & 1 \\ 6 & 1 & 7 \\ 4 & 5 & 9 \end{bmatrix} = -6 \det \begin{bmatrix} 2 & 0 & 1 \\ 6 & 1 & 7 \\ 4 & 5 & 9 \end{bmatrix} + 1 \det \begin{bmatrix} 2 & 0 & 1 \\ 6 & 1 & 7 \\ 4 & 5 & 9 \end{bmatrix} - 7 \det \begin{bmatrix} 2 & 0 & 1 \\ 6 & 1 & 7 \\ 4 & 5 & 9 \end{bmatrix}$$

$$= -6 \begin{vmatrix} 0 & 1 \\ 5 & 9 \end{vmatrix} + 1 \begin{vmatrix} 2 & 1 \\ 4 & 9 \end{vmatrix} - 7 \begin{vmatrix} 2 & 0 \\ 4 & 5 \end{vmatrix}$$

$$= 30 + 14 - 70 = -26 \quad ◀$$

The determinant has many beautiful properties. One of the nicest is the following. Again, we will not prove this property.

Theorem 2.   For any $n \times n$ matrix A, $\det A = \det A^t$.

It follows that we can compute $\det A$ by applying formula (3) to A^t. Specifically,

$$\det A = (-1)^{i+1}(a_{1i} \det(A^t)_{i1} - a_{2i} \det(A^t)_{i2} + a_{3i} \det(A^t)_{i3}$$
$$+ \cdots + (-1)^{n-1} a_{ni} \det(A^t)_{in})$$
$$= (-1)^{i+1}(a_{1i} \det A_{1i} - a_{2i} \det A_{2i} + a_{3i} \det A_{3i} + \cdots + (-1)^{n-1} a_{ni} \det A_{ni})$$
$$(4)$$

(Note that the (i, j) entry of A^t is a_{ji}.)

This formula tells us that we may expand determinants using columns in much the same manner as with rows.

▶ EXAMPLE 4:   Compute det A by expanding along any row or column.

$$A = \begin{bmatrix} 1 & 0 & 2 & 0.7 \\ 1 & 0 & 3 & 0 \\ 17 & 5 & -37 & 0.002 \\ 1 & 0 & 6 & -9 \end{bmatrix}$$

Solution: Since the second column has only one non-zero term, we use it. Since we are expanding along an even-numbered column, the first term is subtracted.

$$\begin{vmatrix} 1 & 0 & 2 & 0.7 \\ 1 & 0 & 3 & 0 \\ 17 & 5 & -37 & 0.002 \\ 1 & 0 & 6 & -9 \end{vmatrix} = -0 \det A_{12} + 0 \det A_{22} - 5 \det A_{32} + 0 \det A_{42}$$

$$= -5 \begin{vmatrix} 1 & 2 & 0.7 \\ 1 & 3 & 0 \\ 1 & 6 & -9 \end{vmatrix}$$

We expand this 3×3 determinant along the second row.

$$-5 \left(-1 \begin{vmatrix} 2 & 0.7 \\ 6 & -9 \end{vmatrix} + 3 \begin{vmatrix} 1 & 0.7 \\ 1 & -9 \end{vmatrix} \right) = 34.5 \qquad ◀$$

Recall that a matrix is said to be upper triangular if all of the entries below the main diagonal are zero. Computing the determinant of an upper triangular matrix is simple.

▶ EXAMPLE 5:   Find the determinant of

$$A = \begin{bmatrix} 2 & 7 & -1 & 16 \\ 0 & 3 & -2 & 9 \\ 0 & 0 & 4 & 8 \\ 0 & 0 & 0 & 6 \end{bmatrix}$$

Solution: Expanding determinants along the first column, we see that

$$
\begin{vmatrix} 2 & 7 & -1 & 16 \\ 0 & 3 & -2 & 9 \\ 0 & 0 & 4 & 8 \\ 0 & 0 & 0 & 6 \end{vmatrix} = 2 \begin{vmatrix} 3 & -2 & 9 \\ 0 & 4 & 8 \\ 0 & 0 & 6 \end{vmatrix}
$$

$$
= 2 \cdot 3 \begin{vmatrix} 4 & 8 \\ 0 & 6 \end{vmatrix}
$$

$$
= 2 \cdot 3 \cdot 4 \cdot 6 = 144
$$

Thus, the determinant is just the product of the entries on the main diagonal.   ◄

Following the same reasoning as in the above example, it is not difficult to prove that *the determinant of an upper triangular matrix is the product of the entries on the main diagonal.*

In Section 5.2 we will describe some vastly more efficient methods for computing determinants than presented so far. These techniques are based on some general properties of the determinant which we describe below.

To explain our first property, consider the following two matrices:

$$
A = \begin{bmatrix} 1 & 2 & 1 \\ 1 & 2 & 3 \\ 4 & 3 & 4 \end{bmatrix} \quad B = \begin{bmatrix} 1 & 2 & 1 \\ 2 & 4 & 6 \\ 4 & 3 & 4 \end{bmatrix}
$$

Note that B differs from A only in that its second row is twice the second row of A. Expanding both determinants along the second row shows that

$$
\det A = -1 \cdot \begin{vmatrix} 2 & 1 \\ 3 & 4 \end{vmatrix} + 2 \cdot \begin{vmatrix} 1 & 1 \\ 4 & 4 \end{vmatrix} - 3 \cdot \begin{vmatrix} 1 & 2 \\ 4 & 3 \end{vmatrix}
$$

$$
\det B = -2 \cdot \begin{vmatrix} 2 & 1 \\ 3 & 4 \end{vmatrix} + 4 \cdot \begin{vmatrix} 1 & 1 \\ 4 & 4 \end{vmatrix} - 6 \cdot \begin{vmatrix} 1 & 2 \\ 4 & 3 \end{vmatrix}
$$

$$
= 2 \left(-1 \cdot \begin{vmatrix} 2 & 1 \\ 3 & 4 \end{vmatrix} + 2 \cdot \begin{vmatrix} 1 & 1 \\ 4 & 4 \end{vmatrix} - 3 \cdot \begin{vmatrix} 1 & 2 \\ 4 & 3 \end{vmatrix} \right) = 2 \det A
$$

The same argument used in the above example can be used to prove the following property.

Scalar Property.   *Suppose the $n \times n$ matrix B is obtained from A by multiplying each element in one particular row by some scalar k. Then $\det B = k \det A$.*   ▉

An interesting consequence of this property is that in general, $\det cA \neq c \det A$. For example, suppose that

$$A = \begin{bmatrix} 1 & 2 & 3 \\ 2 & 1 & 1 \\ 0 & 0 & 2 \end{bmatrix}$$

Then

$$\det(2A) = \begin{vmatrix} 2 & 4 & 6 \\ 4 & 2 & 2 \\ 0 & 0 & 4 \end{vmatrix} = 2 \begin{vmatrix} 1 & 2 & 3 \\ 4 & 2 & 2 \\ 0 & 0 & 4 \end{vmatrix} = 4 \begin{vmatrix} 1 & 2 & 3 \\ 2 & 1 & 1 \\ 0 & 0 & 4 \end{vmatrix} = 8 \begin{vmatrix} 1 & 2 & 3 \\ 2 & 1 & 1 \\ 0 & 0 & 2 \end{vmatrix} = 8 \det A$$

In general, if A is $n \times n$,

$$\det(cA) = c^n \det A$$

Our next property is the "additive property." We will prove it at the end of this section.

Additive Property.   *Let U, V, and A_i be $1 \times n$ row matrices where $i = 2, 3, \ldots, n$. Then*

$$\det \begin{bmatrix} U + V \\ A_2 \\ A_3 \\ \vdots \\ A_n \end{bmatrix} = \det \begin{bmatrix} V \\ A_2 \\ A_3 \\ \vdots \\ A_n \end{bmatrix} + \det \begin{bmatrix} U \\ A_2 \\ A_3 \\ \vdots \\ A_n \end{bmatrix}$$

Similar equalities hold for other rows.   ∎

▶ EXAMPLE 6:   Given that

$$\begin{vmatrix} 1 & 2 & 3 \\ 0 & 2 & 2 \\ 1 & 0 & 0 \end{vmatrix} = -2$$

compute

$$\begin{vmatrix} 1 & 2 & 3 \\ 0 & 2 & 2 \\ 1 & 0 & 5 \end{vmatrix}$$

Solution: We shall apply the additive property to the third row. We note that

$$[1, 0, 5] = [1, 0, 0] + [0, 0, 5]$$

Hence, from additivity,

$$
\begin{vmatrix} 1 & 2 & 3 \\ 0 & 2 & 2 \\ 1 & 0 & 5 \end{vmatrix} = \begin{vmatrix} 1 & 2 & 3 \\ 0 & 2 & 2 \\ 1 & 0 & 0 \end{vmatrix} + \begin{vmatrix} 1 & 2 & 3 \\ 0 & 2 & 2 \\ 0 & 0 & 5 \end{vmatrix} = -2 + 10 = 8
$$

since the matrix on the right is triangular.   ◀

Our final property is the interchange property. We outline a proof in Exercise 8.

Interchange Property.   *Suppose that the $n \times n$ matrix B is obtained from A by interchanging two rows. Then $\det B = -\det A$.*   ■

▶ **EXAMPLE 7:**   Without computing determinants, how are the determinants of A and B related?

$$
A = \begin{bmatrix} 1 & 7 & 6 \\ 2 & 3 & 4 \\ 1 & 1 & 1 \end{bmatrix} \qquad B = \begin{bmatrix} 2 & 3 & 4 \\ 3 & 3 & 3 \\ 1 & 7 & 6 \end{bmatrix}
$$

Solution: We use the scalar and interchange properties on B to conclude that

$$
\det B = 3 \begin{vmatrix} 2 & 3 & 4 \\ 1 & 1 & 1 \\ 1 & 7 & 6 \end{vmatrix} = -3 \begin{vmatrix} 1 & 7 & 6 \\ 1 & 1 & 1 \\ 2 & 3 & 4 \end{vmatrix} = 3 \begin{vmatrix} 1 & 7 & 6 \\ 2 & 3 & 4 \\ 1 & 1 & 1 \end{vmatrix} = 3 \det A.
$$
◀

Proof of the Additive Property.   Let A be the matrix on the left in the statement of the additive property. We expand the determinant of A along the first row, finding that

$$
\begin{aligned}
\det A &= (u_1 + v_1) \det A_{11} - (u_2 + v_2) \det A_{12} + (u_3 + v_3) \det A_{13} \\
&\quad + \cdots + (-1)^{n-1}(u_n + v_n) \det A_{1n} \\
&= u_1 \det A_{11} - u_2 \det A_{12} + u_3 \det A_{13} + \cdots + (-1)^{n-1} u_n \det A_{1n} \\
&\quad + v_1 \det A_{11} - v_2 \det A_{12} + v_3 \det A_{13} + \cdots + (-1)^{n-1} v_n \det A_{1n} \\
&= \det B + \det C
\end{aligned}
$$

where B and C are, respectively, the first and second matrices on the right in the statement of the additive property. The proof for other rows would be similar—we only need to expand along the row in question.   ■

EXERCISES

(1) Compute the following determinants.

(a) $\begin{vmatrix} 1 & 4 \\ -3 & 2 \end{vmatrix}$

(b) $\begin{vmatrix} 1 & 4 \\ 2 & 8 \end{vmatrix}$

(c) $\begin{vmatrix} 2 & 0 & 1 \\ -1 & 1 & 1 \\ -2 & 2 & 2 \end{vmatrix}$

(d) $\begin{vmatrix} 7 & 1 & 1 \\ 0 & a & b \\ 0 & d & c \end{vmatrix}$

(e) $\begin{vmatrix} 0 & 5 & 1 \\ -1 & 1 & 3 \\ -2 & -2 & 2 \end{vmatrix}$

(f) $\begin{vmatrix} 2 & 1 & 1 \\ 5 & 4 & 3 \\ 7 & 5 & 4 \end{vmatrix}$

(g) $\begin{vmatrix} 1 & 0 & 1 \\ 2 & 1 & 1 \\ 3 & 2 & 1 \end{vmatrix}$

(h) $\begin{vmatrix} -3 & 2 & 2 \\ 1 & 4 & 1 \\ 7 & 6 & -2 \end{vmatrix}$

(i) $\begin{vmatrix} 2 & 0 & 2 & 0 \\ 1 & 1 & 1 & 1 \\ 0 & 0 & 3 & 2 \\ 1 & 0 & 0 & 5 \end{vmatrix}$

(j) $\begin{vmatrix} 3 & 1 & 3 & 0 \\ 3 & 1 & 3 & 1 \\ 0 & 0 & 2 & 1 \\ 6 & 3 & 4 & 5 \end{vmatrix}$

(k) $\begin{vmatrix} 3 & 2 & 1 & 4 & -1 \\ 5 & 4 & 3 & 2 & 1 \\ 2 & 1 & 2 & 3 & 1 \\ -3 & 4 & 1 & 6 & 7 \\ 1 & 2 & 3 & 4 & 5 \end{vmatrix}$

(l) $A = \begin{vmatrix} 1 & 0 & 0 & 0 \\ 0 & 1 & 0 & 2 \\ 3 & 0 & 1 & 0 \\ 0 & 1 & 0 & 0 \end{vmatrix}$

(2) In the next section we will prove that the determinant of a matrix is zero if and only if the rows are dependent. Check the validity of this by expressing one row as a combination of the others in each of the parts of Exercise 1 where the determinant is zero.

(3) Write a formula similar to formula (1) that describes the expansion of the determinant of a 3×3 matrix along the second row. Prove that your formula agrees with formula (1).

(4) Use expansion along the third row to express each of the following determinants as a sum of determinants of 4×4 matrices. Without expanding further, explain why $\alpha = 3\beta$. What theorem from the text does this exercise demonstrate?

$$\alpha = \begin{vmatrix} 24 & -13 & 7 & 9 & 5 \\ 11 & 16 & -37 & 99 & 64 \\ 1 & 4 & 2 & 2 & -3 \\ 31 & -42 & 78 & 55 & -3 \\ 62 & 47 & 29 & -14 & -8 \end{vmatrix} \qquad \beta = \begin{vmatrix} 24 & -13 & 7 & 9 & 5 \\ 11 & 16 & -37 & 99 & 64 \\ 3 & 12 & 6 & 6 & -9 \\ 31 & -42 & 78 & 55 & -3 \\ 62 & 47 & 29 & -14 & -8 \end{vmatrix}$$

(5) Use expansion along the third row to express α, δ, and γ as a sum of determinants of 4×4 matrices where α is as in Exercise 4 and δ and γ are as follows. Without expanding further, explain why $\alpha = \delta + \gamma$. What theorem from the text does this exercise demonstrate?

$$\delta = \begin{vmatrix} 24 & -13 & 7 & 9 & 5 \\ 11 & 16 & -37 & 99 & 64 \\ 1 & 1 & 1 & 1 & 7 \\ 31 & -42 & 78 & 55 & -3 \\ 62 & 47 & 29 & -14 & -8 \end{vmatrix} \qquad \gamma = \begin{vmatrix} 24 & -13 & 7 & 9 & 5 \\ 11 & 16 & -37 & 99 & 64 \\ 2 & 5 & 3 & 3 & 4 \\ 31 & -42 & 78 & 55 & -3 \\ 62 & 47 & 29 & -14 & -8 \end{vmatrix}$$

(6)

    (a) Prove that if all of the entries of a 2×2 matrix are integers, then its determinant must be an integer.

    (b) Use part (a) and formula (1) to prove the statement in part (a) for 3×3 matrices.

    (c) Use part (b) to prove the statement in part (a) for 4×4 matrices. If you are familiar with mathematical induction, prove the statement in part (a) for all $n \times n$ matrices.

(7)

    (a) Prove that a 2×2 matrix with two equal rows has zero determinant.

    (b) Use part (a) to prove that a 3×3 matrix with two equal rows has zero determinant. [*Hint:* Expand along the row which is not one of the ones which are assumed equal.]

    (c) Use part (b) to prove the statement in part (a) for 4×4 matrices. If you are familiar with mathematical induction, prove the statement in part (a) for all $n \times n$ matrices.

(8) Use the row interchange property to prove that an $n \times n$ matrix with two equal rows has zero determinant.

(9) Let $X = [a_1, a_2, a_3]^t$ and $Y = [b_1, b_2, b_3]^t$ be vectors in $\mathbb{R}^3$. Let $X \times Y = [c_1, c_2, c_3]^t$ where

$$c_1 = \begin{vmatrix} a_2 & a_3 \\ b_2 & b_3 \end{vmatrix} \qquad c_2 = -\begin{vmatrix} a_1 & a_3 \\ b_1 & b_3 \end{vmatrix} \qquad c_3 = \begin{vmatrix} a_1 & a_2 \\ b_1 & b_2 \end{vmatrix}$$

Prove that

$$X \cdot (X \times Y) = \begin{vmatrix} a_1 & a_2 & a_3 \\ a_1 & a_2 & a_3 \\ b_1 & b_2 & b_3 \end{vmatrix}$$

How does it follow from this and the previous exercise that X is perpendicular to $X \times Y$? (See Section 1.1 for a discussion of dot products of vectors.)
Prove that Y is also perpendicular to $X \times Y$.

(10) Compute the determinant of each of these matrices. The last matrix is assumed to be 100×100.

$$\begin{bmatrix} 1 & 1 \\ 1 & 1 \end{bmatrix} \quad \begin{bmatrix} 1 & 1 & 0 \\ 0 & 1 & 1 \\ 1 & 0 & 1 \end{bmatrix} \quad \begin{bmatrix} 1 & 1 & 0 & 0 \\ 0 & 1 & 1 & 0 \\ 0 & 0 & 1 & 1 \\ 1 & 0 & 0 & 1 \end{bmatrix} \quad \begin{bmatrix} 1 & 1 & 0 & 0 & \cdots & 0 \\ 0 & 1 & 1 & 0 & \cdots & 0 \\ 0 & 0 & 1 & 1 & \cdots & 0 \\ \vdots & \vdots & \vdots & \vdots & & \vdots \\ 1 & 0 & 0 & 0 & \cdots & 1 \end{bmatrix}$$

(11) In this exercise we lead you through a proof of the row interchange property.

(a) Prove the row interchange property for 2×2 matrices.

(b) Use formula (1) *together with part (a)* to prove the row interchange property for exchanging the second and third rows of a general 3×3 matrix.

(c) Write a formula similar to formula (1) that describes expansion of the determinant of a 3×3 matrix along the second row. Use your formula to prove the row interchange property for exchanging the first and third rows of a general 3×3 matrix.

(d) Write a formula similar to formula (1) that describes expansion of the determinant of a 3×3 matrix along the third row. Use your formula to prove the row interchange property for exchanging the first and second rows of a general 3×3 matrix.

(e) Note that parts (b)-(d) prove the row exchange property for 3×3 matrices. Use the validity of the row exchange property for 3×3 matrices to prove it for 4×4 matrices. For your proof, think about the result of expanding a 4×4 matrix along some row other than the pair of rows that are being interchanged.

(f) Granted that the row interchange property is true for 4×4 matrices, explain why it is true for 5×5 matrices. If you are familiar with mathematical induction, construct a mathematical induction proof of the general row exchange property.

ON LINE

Determinants are extremely useful in many contexts. You will, for example, use them constantly when you study eigenvalues and eigenvectors later in the text. In addition, you will see them used to write formulas for the solutions to many applied problems. In particular, determinants are used extensively in the study of differential equations and in the study of advanced calculus. Determinants are also used extensively in studying the mathematical foundations of linear algebra. Computers, however, do not generally use determinants for computations. Much faster and more efficient numerical techniques have been found. Thus, we will not provide any computer exercises for this chapter.

The reader should be aware, however, that MATLAB will compute determinants. The appropriate command is "det(A)". Incidentally, MATLAB uses the methods of the next section to compute determinants rather than the methods already described.

5.2   REDUCTION AND DETERMINANTS

Usually, computers (both human and electronic) do not use the formulas described in Section 5.1 for computing determinants. There are vastly more efficient methods based on

(a) the row interchange property,

(b) the row additive property, and

(c) the row scalar property.

Before describing these techniques, however, we need to note one consequence of these properties. Consider the following matrix.

$$A = \begin{bmatrix} 1 & 2 & 1 & 3 \\ 4 & 2 & 1 & 5 \\ -1 & 5 & 2 & 6 \\ 4 & 2 & 1 & 5 \end{bmatrix}$$

If we interchange the second and fourth rows, the determinant of A will not change because A doesn't change. And yet, from the row interchange property, interchanging two rows must negate the value of the determinant. Zero is the only real number that equals its own negative. Thus, we see that det $A = 0$. In general *any $n \times n$ matrix with two equal rows will have zero determinant.*

This observation leads us to a very important principle. Consider the following matrix A. If we add twice row 1 onto row 3, we get the matrix B:

$$A = \begin{bmatrix} 1 & 2 & 3 \\ 4 & -1 & 7 \\ -2 & -4 & -2 \end{bmatrix} \qquad B = \begin{bmatrix} 1 & 2 & 3 \\ 4 & -1 & 7 \\ 0 & 0 & 4 \end{bmatrix}$$

It turns out that, remarkably, A and B have the same determinant. To explain why, let A_1, A_2, and A_3 denote the rows of A. Then, from the additive and scalar properties

$$\det B = \det \begin{bmatrix} A_1 \\ A_2 \\ A_3 + 2A_1 \end{bmatrix} = \det \begin{bmatrix} A_1 \\ A_2 \\ A_3 \end{bmatrix} + 2\det \begin{bmatrix} A_1 \\ A_2 \\ A_1 \end{bmatrix} = \det \begin{bmatrix} A_1 \\ A_2 \\ A_3 \end{bmatrix} = \det A$$

(Note that a matrix with two equal rows has zero determinant.)

Hence, expanding $\det B$ along the third row, we see that

$$\det A = \det B = 4 \begin{vmatrix} 1 & 2 \\ 4 & -1 \end{vmatrix} = -36$$

The same argument proves the following theorem.

Theorem 1.   *In an* $n \times n$ *matrix A, adding a multiple of one row of A onto a different row will not change the determinant of A.*

Proof: We will only consider the case of adding a multiple of the first row onto another row. The proof for other rows is similar. Thus, let A_i be the rows of A and suppose that B was obtained from A by adding a multiple of A_1 onto A_k. Then

$$\det B = \det \begin{bmatrix} A_1 \\ A_2 \\ \vdots \\ A_k + cA_1 \\ \vdots \\ A_n \end{bmatrix} = \det \begin{bmatrix} A_1 \\ A_2 \\ \vdots \\ A_k \\ \vdots \\ A_n \end{bmatrix} + c\det \begin{bmatrix} A_1 \\ A_2 \\ \vdots \\ A_1 \\ \vdots \\ A_n \end{bmatrix} = \det \begin{bmatrix} A_1 \\ A_2 \\ \vdots \\ A_k \\ \vdots \\ A_n \end{bmatrix} = \det A$$

since, once again, a matrix with two equal rows has determinant 0.   ∎

The above theorem tells us that we can use row reduction to compute determinants.

▶ **EXAMPLE 1:**   Use row reduction methods to compute the determinant of the matrix A:

$$A = \begin{bmatrix} 3 & 6 & 9 & 12 \\ 1 & 2 & 2 & 1 \\ 3 & 5 & 2 & 1 \\ 0 & 2 & 4 & 2 \end{bmatrix}$$

Solution: We begin by using the scalar property on row 1.

$$\det A = 3 \begin{vmatrix} 1 & 2 & 3 & 4 \\ 1 & 2 & 2 & 1 \\ 3 & 5 & 2 & 1 \\ 0 & 2 & 4 & 2 \end{vmatrix} =$$

$$3 \begin{vmatrix} 1 & 2 & 3 & 4 \\ 0 & 0 & -1 & -3 \\ 0 & -1 & -7 & -11 \\ 0 & 2 & 4 & 2 \end{vmatrix} = \quad \textit{(Multiples of row 1 were added to the other rows)}$$

$$-3 \begin{vmatrix} 1 & 2 & 3 & 4 \\ 0 & -1 & -7 & -11 \\ 0 & 0 & -1 & -3 \\ 0 & 2 & 4 & 2 \end{vmatrix} = \quad \textit{(Rows 2 and 3 were interchanged)}$$

$$-3 \begin{vmatrix} 1 & 2 & 3 & 4 \\ 0 & -1 & -7 & -11 \\ 0 & 0 & -1 & -3 \\ 0 & 0 & -10 & -20 \end{vmatrix} = \quad \textit{(A multiple of row 2 was added to row 4)}$$

$$30 \begin{vmatrix} 1 & 2 & 3 & 4 \\ 0 & 1 & 7 & 11 \\ 0 & 0 & 1 & 3 \\ 0 & 0 & 1 & 2 \end{vmatrix} = \quad \textit{(Constants were factored out of rows 2, 3, and 4)}$$

$$30 \begin{vmatrix} 1 & 2 & 3 & 4 \\ 0 & 1 & 7 & 11 \\ 0 & 0 & 1 & 3 \\ 0 & 0 & 0 & -1 \end{vmatrix} = -30 \quad \textit{(A multiple of row 3 was added to row 4)}$$

◀

▶ **EXAMPLE 2:** Compute det *A* where

$$A = \begin{bmatrix} 2 & 1 & 3 \\ 3 & 2 & 1 \\ 7 & 5 & 2 \end{bmatrix}$$

Solution:

$$\det A = 2 \begin{vmatrix} 1 & \frac{1}{2} & \frac{3}{2} \\ 3 & 2 & 1 \\ 7 & 5 & 2 \end{vmatrix}$$

$$= 2 \begin{vmatrix} 1 & \frac{1}{2} & \frac{3}{2} \\ 0 & \frac{1}{2} & -\frac{7}{2} \\ 0 & \frac{3}{2} & -\frac{17}{2} \end{vmatrix}$$

$$= \frac{2}{4} \begin{vmatrix} 1 & \frac{1}{2} & \frac{3}{2} \\ 0 & 1 & -7 \\ 0 & 3 & -17 \end{vmatrix}$$

$$= \frac{1}{2} \begin{vmatrix} 1 & \frac{1}{2} & \frac{3}{2} \\ 0 & 1 & -7 \\ 0 & 0 & 4 \end{vmatrix} = 2$$

◄

We noted in the last section that for any matrix A, $\det A = \det A^t$. It follows that most of the row properties also apply to columns. In particular, there are column interchange, scalar, and additive properties. Thus, we may, if we wish, do "elementary column" operations instead of elementary row operations. For example, by subtracting twice column 1 from column 3, we see

$$\begin{vmatrix} 1 & 0 & 2 \\ 2 & 1 & 4 \\ 3 & 2 & 3 \end{vmatrix} = \begin{vmatrix} 1 & 0 & 0 \\ 2 & 1 & 0 \\ 3 & 2 & -3 \end{vmatrix} = -3$$

We may even mix the two types of operations within the same problem.

Notice that the only elementary row operations that change the determinant of a matrix A are the row interchange property, which multiplies the determinant by -1, and the multiplication of a given row by a non-zero scalar, which multiplies the determinant by the same non-zero scalar. It follows that the determinant of any matrix will be a non-zero multiple of the determinant of its row-reduced form. If A is invertible, then its row-reduced form is I; hence the determinant of A is non-zero. On the other hand, if A is not invertible, its row-reduced form will have a row of zeros. A matrix with a row of zeros has zero determinant. (This follows by expansion along the row in question.) Thus, $\det A = 0$. These comments prove the following crucial theorem.

Theorem 2.   *An $n \times n$ matrix A is invertible if and only if* $\det A \neq 0$.   ∎

The fact that we may use row operations to compute determinants has another, subtler, consequence. The theorem below is significant in that it says that the properties of the determinant function uniquely determine its form. We will give a formal proof of the uniqueness principle at the end of this section. However, the basic idea of the proof is that the value of any such function on a given matrix can be computed using row reduction and this computation will be, step by step, identical with the computation of the determinant. Hence this function must equal the determinant.

Uniqueness Theorem.   *The determinant function is the only function that transforms $n \times n$ matrices into numbers and*

(a) *produces the number 1 from the identity matrix and*

(b) *satisfies the row interchange, the row scalar, and the row additivity properties.*

∎

Uniqueness principles are very important in mathematics. They often help us to derive many deep properties. The following theorem (and its proof) is a beautiful example of this. There is a more computational proof based on factoring A into a product of simpler matrices for which the theorem can be proven directly. We have chosen to present the more "abstract" proof since it stresses the importance of the properties of the determinant. To gain an appreciation for the depth of this result, the reader should attempt to give a direct proof of it in the 2×2 and 3×3 cases.

Product Theorem.   *For all $n \times n$ matrices A and B, $\det(AB) = (\det A)(\det B)$.*

Proof: If $\det B = 0$, then B has rank less than n. It follows that AB also has rank less than n so $\det(AB) = 0$, proving the theorem in this case. Thus, we may assume that $\det B \neq 0$.

For each $n \times n$ matrix A let

$$D(A) = \frac{\det(AB)}{\det B}$$

(We consider B as fixed and A as varying.) Our theorem is equivalent with showing that $D(A) = \det A$ for all A. This will be true if D satisfies the hypotheses of parts (a) and (b) of the uniqueness principle.

For (a), note that

$$D(I) = \frac{\det(IB)}{\det B} = \frac{\det B}{\det B} = 1$$

To prove the row interchange property, recall that in Section 3.2 we proved that AB may be computed as the rows of A times B. In symbols:

$$\begin{bmatrix} A_1 \\ A_2 \\ \vdots \\ A_n \end{bmatrix} B = \begin{bmatrix} A_1 B \\ A_2 B \\ \vdots \\ A_n B \end{bmatrix}$$

It follows that if two rows of A are interchanged, the corresponding rows of AB become exchanged. Thus, $D(A)$ is negated.

Since the row interchange property is true, it suffices to prove the row additivity property for the first row. (Why?) Thus, suppose that the rows of A are denoted A_i and that $A_1 = U + V$ where U and V are row vectors. Then

$$AB = \begin{bmatrix} (U+V)B \\ A_2 B \\ \vdots \\ A_n B \end{bmatrix} = \begin{bmatrix} UB + VB \\ A_2 B \\ \vdots \\ A_n B \end{bmatrix}$$

Hence, from the row additivity of the determinant

$$\det AB = \det \begin{bmatrix} UB \\ A_2 B \\ \vdots \\ A_n B \end{bmatrix} + \det \begin{bmatrix} VB \\ A_2 B \\ \vdots \\ A_n B \end{bmatrix}$$

Dividing both sides of this equality by $\det B$ shows that

$$D(A) = D\left(\begin{bmatrix} U \\ A_2 \\ \vdots \\ A_n \end{bmatrix}\right) + D\left(\begin{bmatrix} V \\ A_2 \\ \vdots \\ w, A_n \end{bmatrix}\right)$$

which proves additivity for D. We leave the row scalar property as an exercise for the reader. ∎

Proof of the uniqueness principle.   Suppose that D is some function which transforms $n \times n$ matrices into numbers and satisfies properties (a) and (b) from the statement of the uniqueness principle.

Let us first assume that A is in reduced echelon form. If A has rank n, then $A = I$ and (from property (a)) $D(A) = 1 = \det A$. If A has rank less than n, then the last row of A is $\mathbf{0}$. From the row scalar property

$$\det A = \det \begin{bmatrix} A_1 \\ A_2 \\ \vdots \\ A_{n-1} \\ \mathbf{0} \end{bmatrix} = \det \begin{bmatrix} A_1 \\ A_2 \\ \vdots \\ A_{n-1} \\ 00 \end{bmatrix} = 0 \det \begin{bmatrix} A_1 \\ A_2 \\ \vdots \\ A_{n-1} \\ \mathbf{0} \end{bmatrix} = 0$$

Precisely the same reasoning shows that $D(A) = 0$; hence $D(A) = \det A$.

Next, suppose that A can be brought into reduced echelon form using only one elementary row operation. Let A' be the reduced echelon form of A. From the previous discussion, $D(A') = \det A'$.

If two rows of A were interchanged to produce A', then $D(A) = -D(A')$ and $\det A = -\det A'$, showing that $D(A) = \det A$. If some row of A was multiplied by a non-zero scalar c to produce A', then $D(A) = c^{-1}D(A')$ and $\det A = c^{-1}\det A'$. Again the desired equality follows. Finally, suppose that a multiple of some row of A was added onto another row of A to produce A'. The argument used in the proof of Theorem 1 shows that $D(A) = D(A')$. Also, from Theorem 1, $\det A = \det A'$. Again the desired equality follows. Thus, regardless of how A' was produced $D(A) = \det A$.

Now, suppose that we have proven that $D(A) = \det A$ for any matrix that can be brought into reduced echelon form using n elementary row operations. Let A be a matrix that is reducible using $n + 1$ elementary row operations and let A' be the matrix produced by applying the first of these operation to A. Since A' can be reduced using only n elementary row operations, we know that $D(A') = \det A'$. The argument from the previous paragraph shows that then $D(A) = \det A$. Our theorem follows by mathematical induction. ∎

EXERCISES

(1) Use row reduction to compute the following determinants.

(a)
$$\begin{vmatrix} 1 & 0 & 1 \\ 2 & 1 & 1 \\ 3 & 2 & 1 \end{vmatrix}$$

(b)
$$\begin{vmatrix} -3 & 2 & 2 \\ 1 & 4 & 1 \\ 7 & 6 & -2 \end{vmatrix}$$

(c)
$$\begin{vmatrix} 2 & 0 & 2 & 0 \\ 1 & 1 & 1 & 1 \\ 0 & 0 & 3 & 2 \\ 1 & 0 & 0 & 5 \end{vmatrix}$$

(d)
$$\begin{vmatrix} 3 & 1 & 3 & 0 \\ 3 & 1 & 3 & 1 \\ 0 & 0 & 2 & 1 \\ 6 & 3 & 4 & 5 \end{vmatrix}$$

(e)
$$\begin{vmatrix} 1 & 2 & 3 & 4 \\ 5 & 6 & 7 & 8 \\ 9 & 10 & 11 & 12 \\ 13 & 14 & 15 & 16 \end{vmatrix}$$

(2) Compute the determinants in Exercise 1, using column reduction instead of row reduction.

(3) Compute the determinants in Exercise 1, Section 5.1, using row reduction.

(4) Compute the determinant in Example 1 of the text using column reduction.

(5) Suppose that
$$\begin{vmatrix} a & b & c \\ d & e & f \\ g & h & i \end{vmatrix} = 5$$

Find
$$\begin{vmatrix} 2a & 2b & 2c \\ 3d - a & 3e - b & 3f - c \\ 4g + 3a & 4h + 3b & 4i + 3c \end{vmatrix} \quad \text{and} \quad \begin{vmatrix} a + 2d & b + 2e & c + 2f \\ g & h & i \\ d & e & f \end{vmatrix}$$

(6) Use row reduction to prove that for all numbers x, y, and z
$$\begin{vmatrix} 1 & 1 & 1 \\ x & y & z \\ x^2 & y^2 & z^2 \end{vmatrix} = (y - x)(z - x)(z - y)$$

(7) Repeat Exercise 6 using column reduction instead of row reduction.

(8) Let
$$A = \begin{bmatrix} A_1 \\ A_2 \\ A_3 \end{bmatrix}$$

represent a 3×3 matrix and suppose that $A_3 = 2A_1 + A_2$. Use the row scalar and additive properties to prove that $\det A = 0$.

(9) Use the row scalar and additive properties to prove that any 3×3 matrix with dependent rows has zero determinant.

(10) Prove the row scalar property for the function D used in the proof of the product theorem.

(11) Suppose D is a function that transforms $n \times n$ matrices into numbers and satisfies the row scalar property. Let A be an $n \times n$ matrix whose first row is $\mathbf{0}$. Prove that $D(A) = 0$. Does the same proof apply to other rows?

(12) Let A be an $n \times n$ invertible matrix. Prove that $\det A^{-1} = 1/\det A$. (Think about the fact that $AA^{-1} = I$.)

(13) Two $n \times n$ matrices A and B are said to be similar if there is an invertible matrix Q such that $A = QBQ^{-1}$. Prove that similar matrices have the same determinant.

(14) Later we shall study $n \times n$ matrices with the property that $AA^t = I$. What are the possible values of the determinant of such a matrix?

(15) Let A be an $n \times n$ matrix. By

$$B = \begin{bmatrix} 1 & \mathbf{0} \\ \mathbf{0} & A \end{bmatrix}$$

we mean the $(n + 1) \times (n + 1)$ matrix B, which has $B_{11} = 1$, all other entries in the first row and column equal to 0, and A as an $n \times n$ submatrix in the lower right-hand corner. Thus, for example, if

$$A = \begin{bmatrix} 1 & 2 \\ 3 & 4 \end{bmatrix}$$

then

$$\begin{bmatrix} 1 & \mathbf{0} \\ \mathbf{0} & A \end{bmatrix} = \begin{bmatrix} 1 & 0 & 0 \\ 0 & 1 & 2 \\ 0 & 3 & 4 \end{bmatrix}$$

For any $n \times n$ matrix A let

$$D(A) = \det \begin{bmatrix} 1 & \mathbf{0} \\ \mathbf{0} & A \end{bmatrix}$$

(a) Compute (using row reduction techniques) $D(A)$ where A is as in Exercise 1a.
(b) Use the uniqueness principle to prove that $D(A) = \det A$ for all $n \times n$ matrices A.

(c) Prove that $D(A) = \det A$ for all $n \times n$ matrices A by expanding along the first row of B.

5.3   A FORMULA FOR INVERSES

In the first section of this chapter, we promised a formula for inverses. In this section, we deliver. To derive the formula, first let A be a 3×3 invertible matrix. We compute A^{-1} by solving $AX = Y$. This equation may be written

$$x_1 A_1 + x_2 A_2 + x_3 A_3 = Y$$

where A_i are the columns of A and $X = [x_1, x_2, x_3]^t$.

Since A is invertible, we know that a unique solution exists. To find it, consider the following computation:

$$
\begin{aligned}
\det \quad [Y, A_2, A_3] &= \det[x_1 A_1 + x_2 A_2 + x_3 A_3, A_2, A_3] \\
&= x_1 \det[A_1, A_2, A_3] + x_2 \det[A_2, A_2, A_3] + x_3 \det[A_3, A_2, A_3] \\
&= x_1 \det A
\end{aligned}
$$

(Note that a matrix with two equal columns will have 0 determinant.) We conclude that

$$x_1 = \frac{\det[Y, A_2, A_3]}{\det A}$$

In words, to compute x_1, we replace A_1 by Y, take the determinant, and divide by the determinant of A. A similar rule applies for computing the other variables.

$$x_2 = \frac{\det[A_1, Y, A_3]}{\det A}$$

$$x_3 = \frac{\det[A_1, A_2, Y]}{\det A}$$

These formulas are known collectively as Cramer's rule.

▶ EXAMPLE 1: Find x_1, x_2, and x_3 in terms of the y_i in the system below:

$$
\begin{aligned}
5x_1 + 2x_2 + x_3 &= y_1 \\
2x_1 + 2x_2 + 2x_3 &= y_2 \\
2x_1 + x_2 + x_3 &= y_3
\end{aligned}
$$

Use your answer to find A^{-1} where A is the coefficient matrix for the system.

Solution: According to Cramer's rule

$$x_1 = \frac{\begin{vmatrix} y_1 & 2 & 1 \\ y_2 & 2 & 2 \\ y_3 & 1 & 1 \end{vmatrix}}{\begin{vmatrix} 5 & 2 & 1 \\ 2 & 2 & 2 \\ 2 & 1 & 1 \end{vmatrix}} = \frac{-y_2 + 2y_3}{2}$$

$$x_2 = \frac{\begin{vmatrix} 5 & y_1 & 1 \\ 2 & y_2 & 2 \\ 2 & y_3 & 1 \end{vmatrix}}{\begin{vmatrix} 5 & 2 & 1 \\ 2 & 2 & 2 \\ 2 & 1 & 1 \end{vmatrix}} = \frac{2y_1 + 3y_2 - 8y_3}{2}$$

and

$$x_3 = \frac{\begin{vmatrix} 5 & 2 & y_1 \\ 2 & 2 & y_2 \\ 2 & 1 & y_3 \end{vmatrix}}{\begin{vmatrix} 5 & 2 & 1 \\ 2 & 2 & 2 \\ 2 & 1 & 1 \end{vmatrix}} = \frac{-2y_1 - y_2 + 6y_3}{2}$$

We write

$$\begin{bmatrix} x_1 \\ x_2 \\ x_3 \end{bmatrix} = \frac{1}{2}\begin{bmatrix} -y_2 + 2y_3 \\ 2y_1 + 3y_2 - 8y_3 \\ -2y_1 - y_2 + 6y_3 \end{bmatrix} = \frac{y_1}{2}\begin{bmatrix} 0 \\ 2 \\ -2 \end{bmatrix} + \frac{y_2}{2}\begin{bmatrix} -1 \\ 3 \\ -1 \end{bmatrix} + \frac{y_3}{2}\begin{bmatrix} 2 \\ -8 \\ 6 \end{bmatrix}$$

The matrix which expresses X in terms of Y is A^{-1}. Thus,

$$A^{-1} = \frac{1}{2}\begin{bmatrix} 0 & -1 & 2 \\ 2 & 3 & -8 \\ -2 & -1 & 6 \end{bmatrix}$$

◀

The general case of Cramer's rule is almost identical, as is the derivation:

Cramer's Rule.   *Let A be an n × n, invertible matrix. To solve the equation $AX = Y$ for X_i, replace the ith column of A by Y, take the determinant, and divide by det A.*

Proof: Since A is invertible, there is an X such that $AX = Y$. Thus, there are scalars x_i such that

$$x_1 A_1 + x_2 A_2 + \cdots + x_n A_n = Y$$

where the A_i are the columns of A. Then, from the column additivity property of the determinant,

$$\begin{aligned}
\det[Y, A_2, A_3, \ldots, A_n] &= x_1 \det[A_1, A_2, \ldots, A_n] \\
&\quad + x_2 \det[A_2, A_2, \ldots, A_n] \\
&\quad + \cdots + x_n \det[A_n, A_2, \ldots, A_n] \\
&= x_1 \det A
\end{aligned}$$

The last equality, which is true because the determinant of a matrix with two equal columns is zero, proves Cramer's rule for x_1. The proof for the other x_i is similar.   ∎

▶ EXAMPLE 2:   Use Cramer's rule to solve for z in the following system:

$$\begin{aligned}
4x + 5y + 3z + 3w &= 1 \\
2x + y + z + w &= 0 \\
2x + 3y + z + w &= 1 \\
5x + 7y + 3z + 4w &= 2
\end{aligned}$$

Solution: According to Cramer's rule, we replace the third column of the coefficient matrix by the constants on the right side of the equation, take the determinant, and divide by the determinant of the coefficient matrix. Thus:

$$z = \frac{\begin{vmatrix} 4 & 5 & 1 & 3 \\ 2 & 1 & 0 & 1 \\ 2 & 3 & 1 & 1 \\ 5 & 7 & 2 & 4 \end{vmatrix}}{\begin{vmatrix} 4 & 5 & 3 & 3 \\ 2 & 1 & 1 & 1 \\ 2 & 3 & 1 & 1 \\ 5 & 7 & 3 & 4 \end{vmatrix}} = -\frac{2}{4} = -\frac{1}{2}$$

◀

Example 1 demonstrates that we may use Cramer's rule to find inverses. To derive a general formula for inverses, let $A = [a_{ij}]$ be a 3×3 invertible matrix and $B = A^{-1}$. The equation $AB = I$ is equivalent with the three equations

$$AB_1 = I_1 \qquad AB_2 = I_2 \qquad AB_3 = I_3$$

where I_j is the jth column of the 3×3 identity matrix and the B_j are the columns of B. We apply Cramer's rule to the first equation, finding

$$b_{11} = \frac{\det[I_1, A_2, A_3]}{\det A}$$

$$b_{12} = \frac{\det[A_1, I_1, A_3]}{\det A}$$

$$b_{13} = \frac{\det[A_1, A_2, I_1]}{\det A}$$

Each of the terms on the right of these equalities is expressible in terms of 2×2 determinants. For example, the second equality states

$$b_{12} = \frac{\det[I_1, A_2, A_3]}{\det A} = \frac{\begin{vmatrix} 0 & a_{12} & a_{13} \\ 1 & a_{22} & a_{23} \\ 0 & a_{32} & a_{33} \end{vmatrix}}{\det A} = -\frac{\begin{vmatrix} a_{12} & a_{13} \\ a_{32} & a_{33} \end{vmatrix}}{\det A}$$

$$= -\frac{\det A_{21}}{\det A}$$

where A_{21} is A with its second row and first column deleted. (Note that we expanded along the first column.)

Similarly, we find that

$$b_{11} = \frac{\det A_{11}}{\det A}$$

$$b_{13} = \frac{\det A_{31}}{\det A}$$

The preceding calculations prove special instances of the following theorem.

Theorem 1.   *Let A be $n \times n$ and invertible. Then*

$$(A^{-1})_{ij} = \frac{(-1)^{i+j} \det(A_{ji})}{\det A} \tag{1}$$

where A_{ji} is the matrix obtained from A by deleting the jth row and ith columns.

Proof: Let $B = A^{-1}$. Then, as in Example 2, $AB = I$ is equivalent with the n systems $AB_j = I_j$ where B_j are the columns of B and I_j are the columns of the $n \times n$ identity matrix. Cramer's rule tells us that

$$b_{ij} = \frac{\det([A_1, \ldots, I_j, \ldots, A_n]}{\det A} A$$

where I_j replaces in the ith column of A. Formula (1) follows by expanding this determinant along this column. Note that the only non-zero coefficient in this column is the 1 in the jth position and from formula (4) in Section 5.1, its coefficient is $(-1)^{i+j} \det A_{ji}$. This proves our theorem. ∎

▶ EXAMPLE 3: Let A be the matrix below. Find $(A^{-1})_{23}$ and $(A^{-1})_{41}$.

$$A = \begin{bmatrix} 1 & 7 & 6 & -5 \\ 2 & 4 & 3 & 1 \\ 3 & 2 & 1 & 0 \\ 6 & -1 & 2 & 4 \end{bmatrix}$$

Solution: It is easily seen that $\det A = -264$. Hence, from Theorem 1,

$$(A^{-1})_{23} = \frac{(-1)^{2+3} \det A_{32}}{\det A}$$

$$= \frac{- \begin{vmatrix} 1 & 6 & -5 \\ 2 & 3 & 1 \\ 6 & 2 & 4 \end{vmatrix}}{-264} = \frac{68}{264}$$

Also

$$(-1)^{1+4} \det A_{14} = - \begin{vmatrix} 2 & 4 & 3 \\ 3 & 2 & 1 \\ 6 & -1 & 2 \end{vmatrix} = 35$$

yielding

$$(A^{-1})_{41} = -\frac{35}{264}$$

◀

EXERCISES

(1) Use Cramer's rule to solve the following system.

$$\begin{aligned} x + 2y + z + w &= 1 \\ x \quad\quad + 3z - w &= 0 \\ 2x + y + z + w &= 0 \\ x - y + z \quad\quad &= 1 \end{aligned}$$

(2) Use Cramer's rule to solve the following system for z in terms of p_1, p_2, and p_3. (Note that we have not asked for x or y.)

$$-x + 2y - 3z = p_1$$
$$3x + y - z = p_2$$
$$2x + 3y + 5z = p_3$$

(3) Use Cramer's rule to find functions $c_1(x)$ and $c_2(x)$ that satisfy the following equations for all x:

$$(e^x \cos x)c_1(x) + (\sin x)c_2(x) = 1 + x^2$$
$$(-e^x \sin x)c_1(x) + (\cos x)c_2(x) = x$$

(4) Use determinants to find the $(1, 2)$ entry for the inverse of each invertible matrix in Exercise 1, Section 5.2.

(5) Find the entry in the $(2, 4)$ position of the inverse of the coefficient matrix for the system in Exercise 1.

(6) Use Theorem 1 to compute A^{-1} where A is the coefficient matrix for the system in Example 1.

(7) In Example 1, we computed A^{-1} where A is the coefficient matrix of the given system. Use Theorem 1 to compute $(A^{-1})^{-1}$. What should you get?

(8) Find the inverse of the following matrix for all values of x for which the inverse exists.

$$\begin{bmatrix} x^2 & x^3 & x^7 \\ x & x^2 & x \\ x^5 & x & x^2 \end{bmatrix}$$

(9) As you may have noted, computing inverses often produces some nasty fractions. However, some matrices have inverses that involve no fractions. For example, from the formula for the inverse of a 2×2 matrix given at the beginning of Section 5.1,

$$\begin{bmatrix} 1 & 2 \\ 1 & 3 \end{bmatrix}^{-1} = \begin{bmatrix} 3 & -2 \\ -1 & 1 \end{bmatrix}$$

Find a 3×3 matrix A with all non-zero entries such that both A and A^{-1} have only integral entries. Note that we have not asked for A^{-1}. [*Hint:* Compute the determinant of the above matrix. How is this relevant?]

(10) Let A be an $n \times n$ matrix that has only integers as entries. State a necessary and sufficient condition on the determinant of such a matrix which guarantees that the

inverse will have only integral entries. Prove your condition. [*Hint:* Consider the property $AA^{-1} = I$.]

5 CHAPTER SUMMARY

This chapter was devoted to the ***determinant***, which is a number that we can compute from a given square matrix. One reason for the importance of the determinant is that it occurs as the denominator in the general formula for the inverse of an $n \times n$ matrix (formula (1) in Section 5.3). This suggests the important result that *an $n \times n$ matrix is invertible if and only if its determinant is non-zero* (Theorem 2, Section 5.2).

Our definition of the determinant was inductive, expressing the determinant of an $n \times n$ matrix in terms of determinants of $(n - 1) \times (n - 1)$ matrices. Explicitly, we defined the determinant in terms of a process called ***expansion along the first row*** (formula (2), Section 5.1). It turned out, however, that we can expand along any row we wish using the ***cofactor (Laplace)*** expansion (Theorem 1, Section 5.1). In fact, we can even expand along columns (formula (4), Section 5.1). This is a consequence of the fact that $\det A = \det A^t$ (Theorem 2, Section 5.1).

The ability to expand along any row of our choice has some important consequences, the most notable being the ***row additivity***, ***row scalar*** and ***row interchange*** properties from Section 5.1. In Section 5.2, these properties allowed us to understand exactly how the determinant of a matrix changes as we row-reduce the matrix, which in turn allowed us to use row reduction to compute determinants.

The ability to use reduction to compute determinants leads to some deep insights into properties of the determinant. For example, reduction was the basis for our proof that an $n \times n$ matrix is invertible if and only if its determinant is non-zero. Also, the ability to compute determinants via reduction told us that the determinant is the only transformation which assigns numbers to matrices and which satisfies the row additivity, row scalar and row interchange properties, together with the fact that the determinant of the identity matrix is one (the ***uniqueness theorem*** in Section 5.2). The fact that $\det AB = (\det A)(\det B)$ (the ***product theorem*** in Section 5.2) is, in turn, a consequence of the uniqueness theorem.

In Section 5.3 we showed that the determinant yields an important method for solving systems of equations (***Cramer's rule***). Typically, in solving a system where the coefficients are numbers, Cramer's rule will involve many more computations than row reduction. However, if the coefficient contain variables, or if they are, say, functions, then Cramer's rule can be useful. Cramer's rule also yielded a general formula for the inverse of a matrix.

DIAGONALIZATION AND MATRIX REPRESENTATIONS

6.1 EIGENVECTORS

One very important application of linear algebra is to the study of the long-term behavior of linear systems. (Such systems are called *linear dynamical systems.*) We begin with an example.

▶ EXAMPLE 1: Metropolis is served by two newspapers, the Planet and the Jupiter. The Jupiter, however, seems to be in trouble. Currently, the Jupiter has only a 38% market share. Furthermore, every year, 10% of its readership switches to the Planet while only 7% of the Planet's readers switch to the Jupiter. Assume that no one subscribes to both papers and that total newspaper readership remains constant. What is the long-term outlook for the Jupiter?

Solution: Currently, out of every hundred readers, 38 read the Jupiter and 62 read the Planet. Next year, the figures for the Jupiter and Planet will be, respectively,

$$0.9(38) + 0.07(62) = 38.54$$
$$0.1(38) + 0.93(62) = 61.46$$

This may be expressed as the matrix product

$$\begin{bmatrix} 0.9 & 0.07 \\ 0.1 & 0.93 \end{bmatrix} \begin{bmatrix} 38 \\ 62 \end{bmatrix} = \begin{bmatrix} 38.54 \\ 61.46 \end{bmatrix}$$

The vectors $V_1 = [38, 62]^t$ and $V_2 = [38.54, 61.46]^t$ are referred to as the (percentage) **state vectors** for the first and second years, respectively.

The preceding 2×2 matrix is called the **transition matrix** because multiplication of a state vector by it produces the state vector for the next year. Thus, for the third year (rounding to two decimal places), we have

$$V_3 = \begin{bmatrix} 0.9 & 0.07 \\ 0.1 & 0.93 \end{bmatrix} \begin{bmatrix} 38.54 \\ 61.46 \end{bmatrix} = \begin{bmatrix} 38.99 \\ 61.01 \end{bmatrix}$$

Repeatedly multiplying by the transition matrix, we obtain the state vectors through year 7:

$$V_4 = \begin{bmatrix} 39.36 \\ 60.64 \end{bmatrix} \quad V_5 = \begin{bmatrix} 39.67 \\ 60.33 \end{bmatrix} \quad V_6 = \begin{bmatrix} 39.93 \\ 60.07 \end{bmatrix} V_7 = \begin{bmatrix} 40.14 \\ 59.86 \end{bmatrix}$$

Not only is the Jupiter not in trouble; it is actually thriving. The market share grows year after year! What is happening is that even though the Jupiter is less popular than the Planet, there are enough disgruntled Planet subscribers to keep Jupiter growing.

It does seem, though, that the *rate* of growth is slowing; in the first year the circulation grows by 1.4%. In the seventh year it only grows by only 0.5%. We expect that eventually we will reach a state vector X for which there is no noticeable growth. This means that if P denotes the transition matrix, then, to within the accuracy of our calculations, $PX = X$.

If $X = [x, y]^t$, the system $PX = X$ is equivalent with

$$0.9x + 0.07y = x$$
$$0.1x + 0.93y = y$$

which is the same as

$$-0.1x + 0.07y = 0$$
$$0.1x - 0.07y = 0$$

It is striking that this system is dependent. This means that there really is a non-zero vector X such that $PX = X$. Explicitly, we find that $x = .7y$. Since x and y represent percentages of readers, we also know that $x + y = 100$. This yields $y = 100/1.7 \approx 58.82$ and $x = 70/1.7 \approx 41.18$. Thus, we guess that Jupiter will eventually wind up with 41.18% of the market and the Planet will get 58.82%. That is, $X = [41.18, 58.82]^t$.

We can, in fact, prove that our guess is correct. For this, let

$$Y = V_1 - X = [38, 62]^t - [41.18, 58.82]^t = [-3.18, 3.18]^t$$

Note that

$$PY = \begin{bmatrix} 0.9 & 0.07 \\ 0.1 & 0.93 \end{bmatrix} \begin{bmatrix} -3.18 \\ 3.18 \end{bmatrix} = \begin{bmatrix} -2.64 \\ 2.64 \end{bmatrix} = (0.83) \begin{bmatrix} -3.18 \\ 3.18 \end{bmatrix} = (0.83)Y$$

Now, let's start letting time pass. Since $V_1 = X + Y$, we see that after one year, we are in the state

$$V_2 = PV_1 = P(X + Y) = PX + PY = X + 0.83Y$$

In the next year, we are in the state

$$V_3 = P(X + 0.83Y) = PX + 0.83PY = X + (0.83)^2 Y$$

After n years, our state is

$$V_n = X + (0.83)^n Y \qquad (1)$$

As $n \to \infty$, $(0.83)^n \to 0$. Thus, our state vectors do in fact converge to $X = [41.18, 58.82]^t$. ◄

The analysis done in our newspaper example was quite remarkable. Initially, the only way we had of predicting the readership after a given number of years was to repeatedly multiply the state vector by the transition matrix. Once we had written our state V_1 as a linear combination of the vectors X and Y, however, we were able to predict the readership after any number of years [formula (1)]. This worked because each of the vectors X and Y has the property that multiplication by P transforms it into a multiple of itself. Specifically $PX = 1X$ and $PY = 038Y$. This allowed us to compute the effect of repeated multiplication by P *without doing any matrix multiplication.*

The same sort of analysis can be applied to a surprisingly large number of $n \times n$ matrices A. All one needs is sufficiently many non-zero column vectors X such that $AX = \lambda X$ for some scalar λ. Such vectors are called "eigenvectors."

Definition. *Let A be an* $n \times n$ *matrix. A non-zero vector X such that*

$$AX = \lambda X$$

for some scalar λ is called an eigenvector *for A. The scalar λ is called the* eigenvalue ▮

Example 2 demonstrates how to find eigenvectors and eigenvalues.

▶ EXAMPLE 2:   Compute $A^{10}B$ where

$$A = \begin{bmatrix} 3 & 1 \\ 1 & 3 \end{bmatrix} \qquad B = \begin{bmatrix} 1 \\ 2 \end{bmatrix}$$

Solution: We begin by looking for vectors X and scalars λ such that $AX = \lambda X$. If $X = [x, y]^t$, this equality is equivalent with

$$3x + y = \lambda x$$
$$x + 3y = \lambda y$$

which is the same as

$$(3 - \lambda)x + y = 0$$
$$x + (3 - \lambda)y = 0 \tag{2}$$

The coefficient matrix for this system is

$$C = \begin{bmatrix} 3 - \lambda & 1 \\ 1 & 3 - \lambda \end{bmatrix}$$

If C is invertible, then $X = C^{-1}0 = 0$. This would be a disappointing conclusion: we want non-zero X.

Thus, non-zero X can exist only when C is *not* invertible. This is true if and only if the determinant of C is zero—that is,

$$0 = \begin{vmatrix} 3 - \lambda & 1 \\ 1 & 3 - \lambda \end{vmatrix} = (3 - \lambda)^2 - 1 = \lambda^2 - 6\lambda + 8 = (\lambda - 2)(\lambda - 4)$$

Hence, $\lambda = 2$ or $\lambda = 4$.

For $\lambda = 2$, the system in formula (2) becomes

$$\begin{aligned} x + y &= 0 \\ x + y &= 0 \end{aligned}$$

The general solution to this system is $y[-1, 1]'$. In particular, $X = [-1, 1]'$ is an eigenvector corresponding to $\lambda = 2$. The reader should check that, indeed, $AX = 2X$.

For $\lambda = 4$, the system (2) becomes

$$\begin{aligned} -x + y &= 0 \\ -x + y &= 0 \end{aligned}$$

The general solution to this system is $y[1, 1]'$. In particular, $Y = [1, 1]'$ is an eigenvector corresponding to $\lambda = 4$. Again the reader should check that indeed, $AY = 4Y$.

Finally, to finish our example, we note that

$$B = [1, 2]' = -\frac{1}{2}[1, -1]' + \frac{3}{2}[1, 1]' = -\frac{1}{2}X + \frac{3}{2}Y \tag{3}$$

Since $AX = 2X$ and $AY = 4Y$, we see that

$$AB = -\frac{1}{2}AX + \frac{3}{2}AY = -\frac{1}{2}(2X) + \frac{3}{2}(4Y)$$

Multiplying both sides of this equation by A again produces

$$A^2 B = -\frac{1}{2}(2AX) + \frac{3}{2}(4AY) = -\frac{1}{2}(2^2 X) + \frac{3}{2}(4^2 Y)$$

Repeating the same argument ten times, we see that

$$A^{10} B = -\frac{1}{2}(2^{10} X) + \frac{3}{2}(4^{10} Y) = [1572352, 1573376]'$$

Thus, we have multiplied B by A 10 times without computing a single matrix product! We could just as easily multiply B by A as many times as desired.   ◄

A few general comments about the procedure just illustrated are in order. Any non-zero multiple of X is an eigenvector corresponding to $\lambda = 2$, and any non-zero multiple of Y is an eigenvector corresponding to $\lambda = 4$. We could, if we preferred, have used, say, $X' = [2, -2]^t$ and $Y' = [1/3, 1/3]^t$ in our calculation. This would have changed the coefficients in formula (3). The calculation would then proceed as before. The final result would, of course, be exactly the same. The only essential ingredient was having B expressible as a linear combination of eigenvectors.

This indeterminacy in the eigenvectors always occurs in eigenvector calculations. It occurs because, for any given λ and any $n \times n$ matrix A, the set of X such that $AX = \lambda X$ is a subspace of $\mathbb{R}^n$. To see this, suppose that $AX = \lambda X$ and $AY = \lambda Y$. Then, for any scalars a and b,

$$A(aX + bY) = aAX + bAY = a\lambda X + b\lambda Y = \lambda(aX + bY)$$

Thus, any non-zero linear combination of eigenvectors for λ is still an eigenvector for λ. The set of vectors X such that $AX = \lambda X$ is called the eigenspace corresponding to the eigenvalue λ. In Example 2, X and Y were used because they seemed to be convenient bases for their eigenspaces.

Notice that the matrix B from Example 2 may be written

$$\begin{bmatrix} 3 - \lambda & 1 \\ 1 & 3 - \lambda \end{bmatrix} = \begin{bmatrix} 3 & 1 \\ 1 & 3 \end{bmatrix} - \begin{bmatrix} \lambda & 0 \\ 0 & \lambda \end{bmatrix} = A - \lambda I$$

For a general $n \times n$ matrix A, the equation $AX = \lambda X$ may be written as $AX = \lambda IX$, which is equivalent with $(A - \lambda I)X = \mathbf{0}$. Just as in Example 2, if $A - \lambda I$ is invertible, then the only X that satisfies this equation is $X = \mathbf{0}$. Hence, for non-zero eigenvectors to exist, $A - \lambda I$ must be non-invertible. This is the same as saying that $p(\lambda) = 0$ where

$$p(\lambda) = \det(A - \lambda I)$$

The function $p(\lambda)$ is a polynomial of degree n. It is called the characteristic polynomial for A. The eigenvalues will be the roots of $p(\lambda)$.

Let us demonstrate these comments with another example.

▶ EXAMPLE 3:   Find all eigenvalues and a basis for the corresponding eigenspace for the matrix

$$A = \begin{bmatrix} 3 & 1 & 0 \\ 0 & 1 & 0 \\ 4 & 2 & 1 \end{bmatrix}$$

Use your answer to compute $A^{100}B$ where $B = [2, 2, 8]^t$.

Solution: We compute:

$$A - \lambda I = \begin{bmatrix} 3 & 1 & 0 \\ 0 & 1 & 0 \\ 4 & 2 & 1 \end{bmatrix} - \lambda \begin{bmatrix} 1 & 0 & 0 \\ 0 & 1 & 0 \\ 0 & 0 & 1 \end{bmatrix} = \begin{bmatrix} 3-\lambda & 1 & 0 \\ 0 & 1-\lambda & 0 \\ 4 & 2 & 1-\lambda \end{bmatrix}$$

Then

$$\det(A - \lambda I) = \begin{vmatrix} 3-\lambda & 1 & 0 \\ 0 & 1-\lambda & 0 \\ 4 & 2 & 1-\lambda \end{vmatrix} = -\lambda^3 + 5\lambda^2 - 7\lambda + 3 = -(\lambda - 1)^2(\lambda - 3)$$

(One could, for example, do a Laplace expansion along the first row.)

The roots are $\lambda = 1$ and $\lambda = 3$. These are the eigenvalues.

To find the eigenvectors, let us first consider the $\lambda = 1$ case. The equation $(A - I)X = \mathbf{0}$ says

$$\begin{bmatrix} 2 & 1 & 0 \\ 0 & 0 & 0 \\ 4 & 2 & 0 \end{bmatrix} \begin{bmatrix} x_1 \\ x_2 \\ x_3 \end{bmatrix} = \begin{bmatrix} 0 \\ 0 \\ 0 \end{bmatrix}$$

By inspection, this is a rank 1 system. Hence, we have a two-dimensional solution set. Both x_2 and x_3 are free and $x_1 = -x_2/2$. We find

$$X = [-\frac{x_2}{2}, x_2, x_3]^t = x_2[-0.5, 1, 0]^t + x_3[0, 0, 1]^t$$

A basis for the $\lambda = 2$ eigenspace is $X_1 = [-1, 2, 0]^t$ and $X_2 = [0, 0, 1]^t$.

Next consider $\lambda = 3$. Solving the equation $(A - 3I)X = \mathbf{0}$ yields the general solution $x_3[0.5, 0, 1]^t$. We choose $Y = [1, 0, 2]^t$ as our basis element.

Finally, to compute $A^{100}B$, we note that

$$B = [2, 2, 8]^t = X_1 + 2X_2 + 3Y$$

Hence

$$A^{100}B = A^{100}X_1 + 2A^{100}X_2 + 3A^{100}Y$$
$$= X_1 + 2X_2 + 3^{100}3Y = [-1 + 3^{101}, 2, 2(3^{101}) + 2]^t \qquad \blacktriangleleft$$

In Example 2, we could have computed $A^{100}B$ for any B in $\mathbb{R}^3$. All that is needed is to be able to express B as a linear combination of X_1, X_2, and Y. This is always possible since the set formed by these three vectors is easily seen to be independent. (Note that three independent vectors in $\mathbb{R}^3$ will span $\mathbb{R}^3$.)

Unfortunately, not all matrices are so nice:

▶ **EXAMPLE 4:** Attempt to compute $A^{100}B$ using the techniques just given where $B = [1, 1, 1]^t$ and

$$A = \begin{bmatrix} 3 & 1 & 1 \\ 0 & 3 & 1 \\ 0 & 0 & 5 \end{bmatrix}$$

Solution: Clearly (since A is triangular)

$$\det(A - \lambda I) = (3 - \lambda)^2(5 - \lambda)$$

Thus, we have two eigenvalues: $\lambda = 3$ and $\lambda = 5$. The equation $(A - 3I)X = \mathbf{0}$ says

$$\begin{bmatrix} 0 & 1 & 1 \\ 0 & 0 & 1 \\ 0 & 0 & 2 \end{bmatrix} \begin{bmatrix} x_1 \\ x_2 \\ x_3 \end{bmatrix} = \begin{bmatrix} 0 \\ 0 \\ 0 \end{bmatrix}$$

This is a rank 2 system having the general solution $X = x_1[1, 0, 0]^t$. This yields only one basis element: $X = [1, 0, 0]^t$.

Similarly $(A - 5I)X = \mathbf{0}$ yields the general solution $x_3[\frac{3}{4}, \frac{1}{2}, 1]^t$. A convenient basis element is $Y = [3, 2, 4]^t$.

So far, so good. When we attempt to express B as a linear combination of the eigenvectors, however, we run into difficulty. Clearly, there are no scalars a and b such that

$$[1, 1, 1]^t = a[1, 0, 0]^t + b[3, 2, 4]^t$$

Thus, the technique breaks down: we cannot compute $A^{100}B$ in this manner. ◀

The difference between Examples 3 and 4 is a matter of dimension. In Example 3 we found three independent eigenvectors. This meant that our eigenvectors formed a basis for $\mathbb{R}^3$, allowing us to express any given vector B in $\mathbb{R}^3$ as a linear combination of them. In Example 4, we found only two independent eigenvectors, which does not suffice to span $\mathbb{R}^3$. An $n \times n$ matrix that does not have n independent eigenvectors is called **deficient**. Thus, the matrix in Example 4 is deficient.

Notice that in Example 3, the characteristic polynomial factored as $-(\lambda - 1)^2$ $(\lambda - 3)$. The root $\lambda = 1$ is of multiplicity two. We say that λ is an eigenvalue of multiplicity 2. The corresponding eigenspace was also two-dimensional. It is a general theorem (which is beyond the scope of this text to prove) that if λ is an eigenvalue of multiplicity n, then the corresponding eigenspace can be at most n-dimensional.

In Example 4, the polynomial factored as $-(\lambda - 3)^2(\lambda - 5)$. As the preceding comments indicate, the root $\lambda = 5$ can produce only one independent eigenvector. To get three independent eigenvectors, we would need to get two from the $\lambda = 3$ root. Thus, the deficiency of this matrix is attributable to the fact that $\lambda = 3$ is an eigenvalue of multiplicity 2 which produces only one independent eigenvector. In general, *an n × n matrix will be deficient if any one of its eigenvalues produces fewer than k independent eigenvectors,* where k is the multiplicity of the eigenvalue.

The techniques described in this section also require modification if the characteristic polynomial does not have enough roots. Consider, for example, the matrix

$$A = \begin{bmatrix} 0 & 1 \\ -1 & 0 \end{bmatrix}$$

Then (as the reader may check)

$$\det(A - \lambda I) = \lambda^2 + 1$$

There are no real roots to $\lambda^2 + 1 = 0$, hence no eigenvectors, at least if we use only real numbers.

This is not a serious problem. Its solution, however, requires using complex matrices. These techniques are discussed in Section 6.3.

EXERCISES

(1) For the following matrices, determine which of the given vectors are eigenvectors and which are not. For those that are, give the eigenvalue.

(a)
$$\begin{bmatrix} 1 & 1 & 1 \\ 1 & 1 & 1 \\ 1 & 1 & 1 \end{bmatrix} \quad X = \begin{bmatrix} 1 \\ 1 \\ 1 \end{bmatrix} \quad Y = \begin{bmatrix} 1 \\ 1 \\ 2 \end{bmatrix}$$

(b)
$$\begin{bmatrix} 2 & 1 & -1 \\ 1 & 2 & 1 \\ 1 & 1 & 0 \end{bmatrix} \quad X = \begin{bmatrix} 1 \\ -1 \\ 1 \end{bmatrix} \quad Y = \begin{bmatrix} 1 \\ 2 \\ 1 \end{bmatrix}$$

(c)
$$\begin{bmatrix} 1 & 0 & 0 & 2 \\ 0 & -1 & 0 & 0 \\ -2 & 1 & -2 & 0 \\ 0 & 0 & 3 & 2 \end{bmatrix} \quad X = \begin{bmatrix} 1 \\ 3 \\ 1 \\ -1 \end{bmatrix} \quad Y = \begin{bmatrix} 0 \\ 0 \\ 0 \\ 0 \end{bmatrix}$$

(2) For the matrices A, verify that the given vectors are eigenvectors. Use this information to compute $A^{10}B$ where $B = [1, 1, 2]^t$.

(a)
$$A = \begin{bmatrix} 0 & 3 & -3 \\ 2 & 2 & -2 \\ -4 & -1 & 1 \end{bmatrix} \quad X = \begin{bmatrix} 1 \\ 1 \\ -1 \end{bmatrix} \quad Y = \begin{bmatrix} 1 \\ 0 \\ 1 \end{bmatrix} \quad Z = \begin{bmatrix} 0 \\ 1 \\ 1 \end{bmatrix}$$

(b)
$$A = \begin{bmatrix} 4 & -1 & 1 \\ 2 & 1 & 1 \\ 2 & -1 & 3 \end{bmatrix} \quad X = \begin{bmatrix} 1 \\ 0 \\ -2 \end{bmatrix} \quad Y = \begin{bmatrix} 0 \\ 1 \\ 1 \end{bmatrix} \quad Z = \begin{bmatrix} 1 \\ 1 \\ 1 \end{bmatrix}$$

(3) For the matrix A in Exercise 2b, find an eigenvector that has eigenvalue 2 and which has only *positive* entries. [*Note*: 0 is not positive!] Check that your answer really is an eigenvector.

(4) In the context of Example 1, choose a state vector (reader's choice) that initially has more Jupiter readers than Planet readers. (Make sure that the entries total 100%.) Use an argument similar to that of Example 1 to prove that readership levels eventually will approach the same vector X as in Example 1. Will the readership always approach X, regardless of the initial readership? Explain.

(5) For the following matrix, show that $\lambda = 2$ and $\lambda = 3$ are the only eigenvalues. Describe geometrically the corresponding eigenspaces.
$$\begin{bmatrix} 2 & 0 & 0 \\ 0 & 2 & 0 \\ 0 & 0 & 3 \end{bmatrix}$$

(6) True or false? "The sum of two eigenvectors is an eigenvector." Explain.

(7) The $\lambda = 0$ eigenspace was studied in Section 1.5. What name did it go by in that section?

(8) True or false? "It is possible for an invertible matrix to have $\lambda = 0$ as an eigenvalue." Explain.

(9) For the matrices in (a) and (b) of Exercise 1, find the characteristic polynomial and use this to find all eigenvalues. Finally, use your answer to compute a formula for $A^n B$ where $B = [3, 1, 1]^t$.

(10) The Fibonacci sequence F_n is defined by $F_1 = 1$, $F_2 = 1$ and $F_{n+1} = F_n + F_{n-1}$ for $n \geq 2$. Thus
$$F_3 = F_2 + F_1 = 1 + 1 = 2$$
$$F_4 = F_3 + F_2 = 2 + 1 = 3$$

(a) Compute F_n for $n = 5, 6$, and 7.
(b) Let A and X be as shown. Compute $A^n X$ for $n = 1, 2, 3, 4, 5$, and 6. What do you observe? Can you explain why this happens?
$$A = \begin{bmatrix} 0 & 1 \\ 1 & 1 \end{bmatrix} \quad X = \begin{bmatrix} 1 \\ 1 \end{bmatrix}$$

(c) Find the eigenvalues and a basis for the corresponding eigenspaces for A and express X as a linear combination of the basis elements.
(d) Use the expression in part (c) to compute $A^{10}X$. (Give an explicit decimal answer.) Use this answer to find F_{10}.
(e) Give a general formula for F_n.

(11) For the following matrices, find all real eigenvalues and a basis for each eigenspace. Which of these matrices are deficient?

(a)
$$\begin{bmatrix} -2 & -1 & 1 \\ -6 & -2 & 0 \\ 13 & 7 & -4 \end{bmatrix}$$

(b)
$$\begin{bmatrix} 1 & 2 & 0 \\ -3 & 2 & 3 \\ -1 & 2 & 2 \end{bmatrix}$$

(c)
$$\begin{bmatrix} 1 & 3 & 0 & 0 \\ 3 & 1 & 0 & 0 \\ 0 & 0 & -1 & 2 \\ 0 & 0 & -1 & -4 \end{bmatrix}$$

(12) Let A be an $n \times n$ matrix and let λ be an eigenvalue for A. Prove that λ^2 is an eigenvalue for A^2.

(13) Suppose that A in Exercise 12 is invertible. Prove that λ^{-1} is an eigenvalue for A^{-1}.

(14) Explain why the "characteristic polynomial" for an $n \times n$ matrix is a polynomial. Why does it have degree n? State a rule for determining the coefficient of λ^n in the characteristic polynomial.

(15) Suppose that A is a square matrix with characteristic polynomial $p(\lambda) = \lambda^2(\lambda + 5)^3(\lambda - 7)^5$.
(a) What is the size of A?
(b) Can A be invertible?
(c) What are the possible dimensions for the nullspace of A?
(d) What can you say about the dimension of the $\lambda = 7$ eigenspace?

(16) Let $A = \begin{bmatrix} a & b \\ c & d \end{bmatrix}$. Prove that the characteristic polynomial of A is $p(\lambda) = \lambda^2 - (a+d)\lambda + \det A$. Find two different 2×2 matrices with eigenvalues $\lambda = 2$ and $\lambda = 3$. [Hint: $p(\lambda) = (\lambda - 2)(\lambda - 3)$.]

(17) Using Exercise 16, state a condition on the entries of A which guarantee that A has no real eigenvalues. [*Hint:* The roots of $a\lambda^2 + b\lambda + c = 0$ are $\lambda = \frac{-b \pm \sqrt{b^2 - 4ac}}{2a}$.] Find an example of a 2×2 matrix with all its entries non-zero, which has no real eigenvalues.

ON LINE

In the "On Line" section for Section 5.1 we commented that virtually any computation for which you might use determinants, a computer would do otherwise. This includes finding eigenvalues. MATLAB uses a sophisticated computational technique which we shall not attempt to describe. This technique certainly does not involve finding the characteristic polynomial and finding its roots. In fact, it turns out that the algorithms for finding eigenvalues are so good that they often are used to find roots of polynomials! The exercises here explore this idea.

(1) Show that the characteristic polynomial for the following matrix is $p(\lambda) = \lambda^2 + a\lambda + b$. Use this to construct a matrix A having $p(\lambda) = \lambda^2 + 7\lambda + 1$ as its characteristic polynomial. Use the MATLAB command "eig(A)" to compute the eigenvalues of A, hence the roots of $p(\lambda)$. Check your calculation by using the quadratic formula to find the roots of $p(\lambda)$.

$$\begin{bmatrix} 0 & 1 \\ -b & -a \end{bmatrix}$$

(2) Compute the characteristic polynomial for the following matrix. Use this information and the MATLAB "eig" command to approximate the roots of $p(\lambda) = \lambda^3 + 8\lambda^2 + 17\lambda + 10$. Test your answer by substituting your roots into $p(\lambda)$.

$$\begin{bmatrix} 0 & 1 & 0 \\ 0 & 0 & 1 \\ -c & -b & -a \end{bmatrix}$$

(3) Let A be the matrix which you used to approximate the roots of $p(\lambda)$ in Exercise 2. The command "[X,D]=eig(A)" will produce a matrix X whose columns form a basis of eigenvectors for A. The corresponding eigenvalues appear as the diagonal entries of D. In MATLAB, let V be the column of X corresponding to $\lambda = -5$. Normalize V by entering "V=V/V(1,1)". What do you notice about the entries of V? Do the other eigenvectors of A exhibit a similar pattern? Use this observation to describe the eigenvectors for the matrix in the formula given in Exercise 2 in terms of the roots of the characteristic polynomial. Prove your answer.

(4) Find a 4×4 matrix A whose characteristic polynomial is $p(\lambda) = \lambda^4 + 3\lambda^2 - 5\lambda + 7$. Then use the MATLAB eig command to approximate the roots.

6.1.1 MARKOV PROCESSES

Example 1 in Section 6.1 can be described in terms of a concept from probability called a "Markov process." A probabilist would say that a news reader in Metropolis can be in one of two states: he or she is either a "Jupiter reader" or a "Planet reader." Let us refer to Jupiter reading as state 1 and Planet reading as state 2. Saying that 10% of the Jupiter readers change to the Planet every year means that the probability of a given reader changing from state 1 to state 2 in a given year is 0.1, which is exactly the (2,1) entry of the transition matrix. In general, the (i, j) entry is the probability of changing from state j to i. What makes this a Markov process is the independence of the transition probabilities of either the current or past readership.

Here is another example of a Markov process:

▶ EXAMPLE 1:   A car rental agency in Chicago has offices at two airports, O'Hare and Midway, as well as one downtown in the Loop. The table below shows the probabilities that a car rented at one particular location will be returned to some other location. Assuming that all cars are returned by the end of the day and that the agency does not move cars itself from office to office, what fractions of its total fleet, on the average, will be available at each of its offices?

Rentals	Returns		
	O'Hare	Midway	Loop
O'Hare	0.7	0.1	0.2
Midway	0.2	0.6	0.2
Loop	0.5	0.3	0.2

Solution: In this example we define the three states by saying that a given car can be at O'Hare or Midway or in the Loop. Let H, M, and L denote, respectively, the fraction of the total fleet initially in each place: O'Hare, Midway, and the Loop. Then at the end of a day we expect the fractions for each of the respective offices to be

$$H_1 = 0.7H + 0.2M + 0.5L$$
$$M_1 = 0.1H + 0.6M + 0.3L \tag{1}$$
$$L_1 = 0.2H + 0.2M + 0.2L$$

Thus, the transition matrix is

$$P = \begin{bmatrix} 0.7 & 0.2 & 0.5 \\ 0.1 & 0.6 & 0.3 \\ 0.2 & 0.2 & 0.2 \end{bmatrix}$$
◀

In a general (finite) Markov process, we are studying a system that is describable in terms of a finite number of states numbered 1 through n. We assume that each individual (a probabilist would say "sample") occupies one and only one of the states. It is also assumed that the system changes periodically and the probability p_{ij} of changing from state j to state i is known and is independent of how many individuals occupied any of the states at any given time. The matrix $P = [p_{ij}]$ is referred to as the **transition matrix**.

The p_{ij} are of course non-negative and less than 1, since they are probabilities. Notice that the entries in each column of the transition matrix from Example 1 total to 1. The same statement is true for all transition matrices because an individual who was in state j before the transition occurred must wind up in one of the n possible states after the transition. These comments motivate the following definition:

Definition. *An $n \times n$ matrix P is a probability matrix if the entries of P are all non-negative and the sum of the entries in each column is 1.* ◼

For a general Markov process, we usually use the fraction of the total population rather than percentages to measure the occupancy of any given state. Thus, our state vectors will be $n \times 1$ column vectors whose entries total to 1. Such vectors are called **probability vectors**.

If we begin in a state defined by a probability vector V_0, then after the first transition we will be in state $V_1 = PV_0$. After the next transition we will be in state $V_2 = PV_1 = P^2 V_0$. In general, the nth state will be $V_n = P^n V_0$. The sequence of vectors $V_0, V_1, \ldots, V_n, \ldots$ is a "Markov chain."

The following theorem is fundamental to the study of Markov processes.

Theorem 1. *Let P be a probability matrix with no entries equal to 0. Then there is a unique probability vector X such that $PX = X$. If V_0 is any probability vector, then $\lim_{n \to \infty} P^n V_0 = X$.* ◼

The vector X in Theorem 1 is called the **equilibrium vector**

EXERCISES

(1) The following exercises refer to Example 1 on page 309.

    (a) Suppose that on Monday morning, we have equal numbers of cars at each location. What will the distribution be on Tuesday? What will it be on Wednesday? Do not compute eigenvectors to answer this part.

    (b) Find the eigenvectors for the matrix P. To save you time, the characteristic polynomial for P is given: $p(\lambda) = -\lambda(\lambda - 1)(\lambda - 0.5)$. Use your answer to decide what the distribution will be in two weeks.

(c) What is the equilibrium distribution? Call this vector X. Assume that your computer can display 16 digits on its screen. How many days must pass before the distribution shows no apparent change from day to day?

(d) Show (without using Theorem 1) that you approach this same distribution X regardless of the initial distribution. [*Hint:* It is important to note that the initial distribution is a probability vector. You might also note that the totals of the entries in each of the eigenvectors other than X is zero.]

(e) Assuming that the agency does not itself move cars from office to office, what fractions of its total fleet, on the average, will be available at each of its offices?

(2) Find the equilibrium vector for the following matrices.

(a) $\begin{bmatrix} 0.3 & 0.5 \\ 0.7 & 0.5 \end{bmatrix}$ (b) $\begin{bmatrix} 0.1 & 0.3 & 0.2 \\ 0.4 & 0.7 & 0.1 \\ 0.5 & 0 & 0.7 \end{bmatrix}$

(c) $\begin{bmatrix} 0.6 & 0.1 \\ 0.4 & 0.9 \end{bmatrix}$ (d) $\begin{bmatrix} 0.2 & 0.8 & 0.2 \\ 0.4 & 0.1 & 0.2 \\ 0.4 & 0.1 & 0.6 \end{bmatrix}$

(3) Denote the matrix in part (a) of Exercise 2 by P. Let $V_0 = [1, 0]^t$. Compute values of $P^n V_0$ until the answer agrees with the computed equilibrium vector to within two decimal places. Repeat for the matrix in part (c) of Exercise 2.

(4) For parts (a) and (c) from Exercise 2, prove that $Y = [1, -1]^t$ is an eigenvector. What is the corresponding eigenvalue? Then reason as in Example 1, Section 6.1, to show that Theorem 1 is true for these matrices.

(5) Prove that for any 2×2 probability matrix P, the vector $Y = [1, -1]^t$ is an eigenvector. Then prove Theorem 1 in the 2×2 case.

(6) I tend to be rather moody at times. If I am in a good mood today, there is an 80% chance I'll still be in a good mood tomorrow. But if I'm grumpy today, there is only a 60% chance that my mood will be good tomorrow.

(a) Describe the transition matrix for this situation.

(b) If I am in a good mood today, what is the probability that I will be in a good mood three days from now?

(c) Over the long term, what percentage of the time am I in a good mood?

(7) In a certain market, there are three competing brands of cereal: brands X, Y, and Z. In any given month 40% of Brand X users will switch to either Brand Y or Brand Z, with equal numbers going to each. Similarly, 30% of Brand Y users and 20% of Brand Z users will switch, with the switchers dividing equally between the other two brands.

(a) What is the transition matrix for this situation?

(b) What is the expected market share for each brand?

(8) Invent a problem, similar in form to Example 1, but involving three radio stations. Give the transition matrix for your problem and find the equilibrium vector.

(9) Explain why the $n \times n$ identity matrix I is a probability matrix. Explain which of the conclusions of Theorem 1 fail for I. Which of the hypotheses of Theorem 1 does I not satisfy?

(10) Enter the matrix in part (b) of Exercise 2 into MATLAB and call it P. Give the command "[F,G]=eig(P)". MATLAB will display two matrices F and G. The columns of F are eigenvectors for P and the nonzero entries of G are the eigenvalues. To verify this, multiply each column of F by P and show that you get the same result if you multiply this column by the corresponding eigenvalue.

Use this information to prove (without using Theorem 1) that $\lim_{n \to \infty} P^n V_0 = X$ where V_0 is any given probability vector and X is the equilibrium state. Your reasoning should be similar to that used in Example 1, Section 6.1.

(11) Let P be the matrix from Exercise 2(a) and let Z be a 2×1 matrix. Prove that the sum of the entries in PZ is the same as the sum of the entries in Z. Note that this proves that multiplication by P transforms probability vectors into probability vectors.

(12) Repeat Exercise 11 using the transition matrix from Example 1, assuming now Z is 3×1.

(13) Let P be an $n \times n$ probability matrix and let Z be an $n \times 1$ matrix. Prove that the sum of the entries in PZ is the same as the sum of the entries in Z.

(14) Prove that the product of two probability matrices is a probability matrix. [*Hint:* One of the previous exercises might help.]

(15) Let

$$P = \begin{bmatrix} 0.1 & 0.5 & 0.2 \\ 0.3 & 0 & 0.4 \\ 0.6 & 0.5 & 0.4 \end{bmatrix}$$

(a) Which of the hypotheses of Theorem 1 does P not satisfy?

(b) Show that P^2 does satisfy all the hypotheses of Theorem 1.

(c) Use part (b), together with Theorem 1, to prove that for all probability vectors V, $\lim_{n \to \infty} P^n V = X$ where X is the unique probability vector such that $P^2 X = X$. Why does $PX = X$ for this X?

(16) Let P be an $n \times n$ probability matrix and let $X = [1, 1, \ldots, 1]^t$ be the $n \times 1$ matrix, all of whose entries are 1. What is $P^t X$? (Try some examples if you aren't sure.) How does it follow that $\det(P^t - I) = 0$? How does it follow that $\det(P - I) = 0$? How does this relate to proving Theorem 1?

(17) If X is a column vector, we let $M(X)$ denote the value of largest entry in X and $m(X)$ the value of the smallest. Let P be an $n \times n$ probability matrix and let X be an $n \times 1$ vector. Prove that each entry of $P^t X$ lies between $m(X)$ and $M(X)$. [*Hint:* First try proving this for the matrix in part (a) of Exercise 1 above.]
        Use this to prove

(a) All eigenvalues of P^t lie between -1 and 1.
(b) The eigenspace of P^t corresponding to the eigenvalue 1 is one-dimensional and is spanned by the vector described in Exercise 16.

(18) Use Exercise 17 to prove Theorem 1 in the case where P is $n \times n$ and has n distinct eigenvalues, and -1 is not an eigenvalue.

(19) Prove that in the context of Exercise 17, -1 is not an eigenvalue.

6.2   DIAGONALIZATION

Eigenvalues and eigenvectors yield considerable information about the "geometry" of linear transformations, as the next example shows.

▶ EXAMPLE 1:   Describe geometrically the effect of the transformation of $\mathbb{R}^2$ into $\mathbb{R}^2$ defined by multiplication by the matrix

$$A = \begin{bmatrix} 3 & 1 \\ 1 & 3 \end{bmatrix}$$

Solution: In Example 2 in Section 6.1, we found that A has eigenvalues 2 and 4 with $Q_1 = [-1, 1]^t$ and $Q_2 = [1, 1]^t$ as bases for the corresponding eigenspaces. Notice that the vectors Q_1 and Q_2 form two sides of a square in $\mathbb{R}^2$. Furthermore, multiplication by A transforms Q_1 into $2Q_1$ and Q_2 into $4Q_2$. Hence, multiplication by A transforms this square into the rectangle indicated in Figure 1.
        More generally, note that the Q_i form a basis for $\mathbb{R}^2$. For each X in $\mathbb{R}^2$, let X' denote its coordinates with respect to this basis. If $X' = [x', y']^t$, then

$$X = x'Q_1 + y'Q_2$$

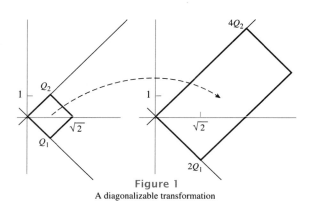

Figure 1
A diagonalizable transformation

Hence

$$AX = x'AQ_1 + y'AQ_2 = 2x'Q_1 + 4y'Q_2$$

These equations say that multiplication by A transforms the vector with coordinates $[x', y']^t$ into the vector with coordinates $[2x', 4y']^t$. Thus, we could describe the geometric effect of multiplication by A as follows: "Multiplication of X by A expands the x' coordinate of X by a factor of 2 and the y' coordinate of X by a factor of 4." ◀

We can express the conclusion of Example 1 algebraically. Notice that

$$\begin{bmatrix} 2x' \\ 4y' \end{bmatrix} = \begin{bmatrix} 2 & 0 \\ 0 & 4 \end{bmatrix} \begin{bmatrix} x' \\ y' \end{bmatrix}$$

The conclusion of Example 1 says that the coordinate vector for AX is DX' where D is the 2×2 matrix in this equality. In other words,

$$(AX)' = DX'$$

But $X' = Q^{-1}X$, and $AX = Q(AX)'$ where $Q = [Q_1, Q_2]$ is the point matrix for our basis. Thus

$$AX = Q(AX)' = QDX' = QDQ^{-1}X$$

Since this is true for all X in $\mathbb{R}^2$, it follows that

$$A = QDQ^{-1} \qquad (1)$$

Formula (1) allows us to extend the computations done in Example 2, Section 6.1. Specifically, we saw in Section 6.1 that we could compute $A^k B$ for any B in $\mathbb{R}^2$. It follows that we should be able to compute a formula for A^k itself. Formula (1) tells us how to do this. Notice that

$$A^2 = (QDQ^{-1})(QDQ^{-1}) = QD(Q^{-1}Q)DQ^{-1} = QD^2Q^{-1}$$

since $Q^{-1}Q = I$. Similarly,

$$A^3 = A^2A = QD^2Q^{-1}QDQ^{-1} = QD^3Q^{-1}$$

Repeating this argument k times shows that

$$A^k = QD^kQ^{-1} \tag{2}$$

But D^k is easily computed. We leave it as an exercise for the reader to show that

$$D^k = \begin{bmatrix} 2^k & 0 \\ 0 & 4^k \end{bmatrix} \tag{3}$$

Hence

$$\begin{bmatrix} 3 & 1 \\ 1 & 3 \end{bmatrix}^k = \begin{bmatrix} -1 & 1 \\ 1 & 1 \end{bmatrix} \begin{bmatrix} 2^k & 0 \\ 0 & 4^k \end{bmatrix} \begin{bmatrix} -1 & 1 \\ 1 & 1 \end{bmatrix}^{-1} = \frac{1}{2} \begin{bmatrix} 4^k + 2^k & 4^k - 2^k \\ 4^k - 2^k & 4^k + 2^k \end{bmatrix}$$

The same reasoning would apply to any $n \times n$ matrix A that has n independent eigenvectors $Q_1, Q_2, \ldots, Q_n$. These vectors form a basis for $\mathbb{R}^n$. For each X in $\mathbb{R}^n$, we let X' denote its coordinate vector with respect to this basis. Specifically, if $X' = [x_1', x_2', \ldots, x_n']^t$, then

$$X = x_1'Q_1 + x_2'Q_2 + \cdots + x_n'Q_n$$

Hence

$$AX = x_1'AQ_1 + x_2'AQ_2 + \cdots + x_n'AQ_n$$
$$= \lambda_1 x_1'Q_1 + \lambda_2 x_2'Q_2 + \cdots + \lambda_n x_n'Q_n$$

where λ_i is the eigenvalue corresponding to the eigenvector Q_i. Therefore, $(AX)' = [\lambda_1 x_1', \lambda_2 x_2', \ldots, \lambda_n x_n']^t = DX'$ where

$$D = \begin{bmatrix} \lambda_1 & 0 & \cdots & 0 \\ 0 & \lambda_2 & \cdots & 0 \\ \vdots & \vdots & \ddots & 0 \\ 0 & 0 & \cdots & \lambda_n \end{bmatrix} \tag{4}$$

Such matrices are called **diagonal** because their only non-zero entries lie on the main diagonal. The chain of equalities immediately preceding formula (1) is valid in our current context, where Q is the point matrix for our basis. We conclude that $A = QDQ^{-1}$. These comments lead to the following theorem, which is used to reduce questions about general matrices to questions about diagonal matrices.

Theorem 1.   *Suppose that A is an n × n matrix that has n independent eigenvectors $Q_1, Q_2, \ldots, Q_n$. Let λ_i be the corresponding eigenvalues. Then*

$$A = QDQ^{-1} \qquad (5)$$

where D is the diagonal matrix that has λ_i as its ith diagonal entry and $Q = [Q_1, Q_2, \ldots, Q_n]$. ∎

▶ **EXAMPLE 2:**   For the matrix A below, find a diagonal matrix D and an invertible matrix Q such that $A = QDQ^{-1}$. Use your answer to find A^{100}.

$$A = \begin{bmatrix} 3 & 1 & 0 \\ 0 & 1 & 0 \\ 4 & 2 & 1 \end{bmatrix}$$

Solution: This matrix was considered in Example 3 in Section 6.1. We found that the eigenvalues were $\lambda = 1$ and $\lambda = 3$. The $\lambda = 1$ eigenspace is two-dimensional. We choose $X_1 = [-1, 2, 0]^t$ and $X_2 = [0, 0, 1]^t$ as our basis. For $\lambda = 3$, we choose $Y = [1, 0, 2]^t$ as our basis element. Theorem 1 tells us that if we set

$$Q = \begin{bmatrix} -1 & 0 & 1 \\ 2 & 0 & 0 \\ 0 & 1 & 2 \end{bmatrix} \text{ and } D = \begin{bmatrix} 1 & 0 & 0 \\ 0 & 1 & 0 \\ 0 & 0 & 3 \end{bmatrix}$$

then it is *guaranteed* that $A = QDQ^{-1}$.

As before, we see that

$$A^{100} = QD^{100}Q^{-1}$$

Hence,

$$A^{100} = \begin{bmatrix} -1 & 0 & 1 \\ 2 & 0 & 0 \\ 0 & 1 & 2 \end{bmatrix} \begin{bmatrix} 1 & 0 & 0 \\ 0 & 1 & 0 \\ 0 & 0 & 3^{100} \end{bmatrix} \begin{bmatrix} 0 & \frac{1}{2} & 0 \\ -2 & -1 & 1 \\ 1 & \frac{1}{2} & 0 \end{bmatrix}$$

$$= \begin{bmatrix} 3^{100} & \frac{3^{100}-1}{2} & 0 \\ 0 & 1 & 0 \\ 2(3^{100}-1) & 3^{100}-1 & 1 \end{bmatrix}$$

◀

In general, we say that an $n \times n$ matrix is **diagonalizable** if there is an $n \times n$ diagonal matrix D and an invertible matrix Q such that formula (5) holds. In this case, formula (5) is referred to as a **diagonalization** of A. Theorem 1 says that we can use eigenvectors to produce diagonalizations. It is interesting that this is, in essence, the only way of producing diagonalizations. To explain this, notice that formula (5) is equivalent with the identity

$$AQ = QD$$

Let Q_i be the columns of Q so that $Q = [Q_1, Q_2, \ldots, Q_n]$. Then

$$QD = Q \begin{bmatrix} \lambda_1 & 0 & \cdots & 0 \\ 0 & \lambda_2 & \cdots & 0 \\ \vdots & \vdots & \ddots & 0 \\ 0 & 0 & \cdots & \lambda_n \end{bmatrix} = [\lambda_1 Q_1, \lambda_2 Q_2, \ldots, \lambda_n Q_n]$$

On the other hand

$$AQ = [AQ_1, AQ_2, \ldots, AQ_n]$$

These two expressions are equal only if for all i, $AQ_i = \lambda_i Q_i$. Thus the Q_i are all eigenvectors. Furthermore, since Q is invertible, the Q_i must form an independent set. These comments, together with Theorem 1 prove the following theorem:

Theorem 2. *An $n \times n$ matrix A is diagonalizable if and only if A has a set of n independent eigenvectors.* ∎

Recall that in Section 6.1 we said that an $n \times n$ matrix A is deficient if it does not have n independent eigenvectors. Theorem 2 says that the terms "deficient" and "diagonalizable" are opposites—an $n \times n$ matrix A is diagonalizable if and only if it is not deficient.

Before ending this section, we should make one final comment concerning diagonalizing bases. Notice that in all of the examples we never bothered showing that the eigenvectors were independent. This is because eigenvalues corresponding to different eigenvalues are *always* independent due to the following theorem.

Theorem 3. *Let A be an $n \times n$ matrix and let Q_1, Q_2, $\ldots$, Q_k be eigenvectors corresponding to the eigenvalues l_i. Suppose that the l_i are all different. Then the Q_i form an independent set of vectors.*

Proof: Let suppose that the Q_i form a dependent set. We shall prove that this is nonsense. If the Q_i are dependent, then there is some smallest number l such that $S = \{Q_1, Q_2, \ldots, Q_l\}$ is dependent. Of course, $l > 1$ since $Q_1 \neq \mathbf{0}$.

Consider the dependency equation

$$c_1 Q_1 + c_2 Q_2 + \cdots + c_l Q_l = \mathbf{0}. \tag{6}$$

We multiply both sides of this equation by A, producing

$$\mathbf{0} = c_1 A Q_1 + c_2 A Q_2 + \cdots + c_l A Q_l = l_1 c_1 Q_1 + l_2 c_2 Q_2 + \cdots + l_l c_l Q_l. \tag{7}$$

We now multiply both sides of formula 6 above by l_l and subtract it from formula 7. Notice that Q_l will drop out. We get:

$$\mathbf{0} = (l_1 - l_l)c_1 Q_1 + (l_2 - l_l)c_2 Q_2 + \cdots + (l_{l-1} - l_l)c_{l-1} Q_{l-1}.$$

From the choice of l, $\{Q_1, \ldots, Q_{l-1}\}$ is independent. Furthermore $l_i - l_l \neq 0$ for $i < l$. It follows that $c_i = 0$ for $1 \leq i < l$. Then formula 6 shows that $c_l = 0$ as well. This contradicts the fact that S is dependent, proving our theorem. ∎

EXERCISES

(1) Show that for all real numbers a, b, c, and d

$$\begin{bmatrix} a & 0 \\ 0 & b \end{bmatrix} \begin{bmatrix} c & 0 \\ 0 & d \end{bmatrix} = \begin{bmatrix} ac & 0 \\ 0 & bd \end{bmatrix}$$

How does formula (3) in the text follow from this?

(2) Generalize Exercise 1 to the product of two 3×3 diagonal matrices.

(3) For the matrices in Exercise 2, Section 6.1, find a diagonal matrix D and an invertible matrix Q such that $A = QDQ^{-1}$. Use this information to find a formula for A^n. (You may leave Q^{-1} unevaluated.)

(4) For the matrices in parts (a) and (b) in Exercise 1, Section 6.1, find the characteristic polynomial and use this to find all eigenvalues. Finally, find a diagonal matrix D and a matrix Q such that $A = QDQ^{-1}$ and a formula for A^n. (You may leave Q^{-1} unevaluated.)

(5) For the following matrices, find (if possible) a matrix Q and a diagonal matrix D such that $A = QDQ^{-1}$. [*Note*: You are not asked to find Q^{-1}.]

(a)
$$\begin{bmatrix} -2 & -1 & 1 \\ -6 & -2 & 0 \\ 13 & 7 & -4 \end{bmatrix}$$

(b)
$$\begin{bmatrix} 1 & 2 & 0 \\ -3 & 2 & 3 \\ -1 & 2 & 2 \end{bmatrix}$$

(c)
$$\begin{bmatrix} 1 & 3 & 0 & 0 \\ 3 & 1 & 0 & 0 \\ 0 & 0 & -1 & 2 \\ 0 & 0 & -1 & -4 \end{bmatrix}$$

(6) For the following matrices, find (if possible) an invertible matrix Q and a diagonal matrix D such that $A = QDQ^{-1}$. [*Note:* You are not asked to find Q^{-1}.]

(a)
$$\begin{bmatrix} -7 & 3 \\ -18 & 8 \end{bmatrix}$$

(b)
$$\begin{bmatrix} 7 & -1 \\ 9 & 1 \end{bmatrix}$$

(c)
$$\begin{bmatrix} 1 & 2 & -2 \\ 0 & 2 & 1 \\ 0 & 0 & 3 \end{bmatrix}$$

(d)
$$\begin{bmatrix} 1 & 0 & 0 \\ 3 & 2 & -1 \\ 3 & 0 & 1 \end{bmatrix}$$

(7) Let A, D, and Q be as in Example 2 from this section. Prove (by direct computation) that $AQ = QD$.

(8) Find values of a, b, and c, all nonzero, such that the matrix A below *is* diagonalizable.

$$A = \begin{bmatrix} 2 & a & b \\ 0 & -5 & c \\ 0 & 0 & 2 \end{bmatrix}$$

(9) If the characteristic polynomial of a 4×4 matrix A has 4 distinct roots, must A be diagonalizable? Explain.

(10) In Examples 1 and 2 in the text, find matrices B such that $B^2 = A$. Check your answer by direct computation.

(11) Suppose that A is diagonalizable and A has only ± 1 as eigenvalues. Show that $A^2 = I$.

(12) Suppose that A is diagonalizable and A has only 0 and 1 as eigenvalues. Show that $A^2 = A$.

(13) Suppose that A is diagonalizable and A has only 2 and 4 as eigenvalues. Show that $A^2 - 6A + 8I = 0$.

(14) Suppose that A is diagonalizable with eigenvalues λ_i. Suppose also that

$$q(\lambda) = a_n \lambda^n + a_{n-1} \lambda^{n-1} + \cdots + a_0$$

is some polynomial such that $q(\lambda_i) = 0$ for all i. Prove that

$$a_n A^n + a_{n-1} A^{n-1} + \cdots + a_0 I = \mathbf{0}$$

Remark: One example of a polynomial that is zero on all of the eigenvalues is the characteristic polynomial $p(\lambda) = \det(A - \lambda I)$. This problem proves that if we substitute a diagonalizable matrix into its characteristic polynomial, we get the zero matrix. This result is actually true even if the matrix is not diagonalizable. It is known as the Cayley–Hamilton theorem.

(15) Prove formula (5) from the text. Your proof should be similar to the discussion following formula (1).

(16) Suppose that A and B are $n \times n$ matrices such that $A = QBQ^{-1}$ for some invertible matrix Q. Prove that A and B have the same characteristic polynomials. Suppose that X is an eigenvector for B. Show that QX is an eigenvector for A.

(17) Suppose that A is an $n \times n$ matrix with n distinct eigenvalues λ_i. Why is A diagonalizable? Show that $\det A$ is just the product of the λ_i. [*Hint:* Use formula (5).]

ON LINE

In this exercise, you will create your own eigenvalue problem. You should first declare "format short" in MATLAB. Then, enter a 3×3, rank 3, matrix P of your own choice into MATLAB. (You can use the MATLAB rank command to check the rank of your matrix.) To avoid trivialities, make each of the entries of P be non-zero. Next, enter a 3×3 *diagonal* matrix D. Choose D so that its diagonal entries are 2, 2, and 3 (in that order). Finally, let

```
A=P*D*inv(P)
```

The general theory makes two predictions:

(a) Each column of P is an eigenvector for A.
(b) The eigenvalues of A are 2 and 3.

(1) Verify prediction (a) above by multiplying each column of P by A and checking to see that they are indeed eigenvectors corresponding to the stated eigenvalues.

(2) Verify prediction (b) above with the MATLAB command "eig(A)".

(3) MATLAB can also compute eigenvectors. Enter the command "[Q,E]=eig(A)". This will produce a matrix whose first three columns are eigenvectors for A and whose last three columns form the corresponding diagonal form of A. As in Exercise 1, check that the columns of Q really are eigenvectors of A. The column of Q that corresponds to the eigenvalue 3 should be a multiple of the third column of P. (Why?) Check that this really is the case. How should the other two columns of Q relate to the columns of P? Check that this really is true.

(4) In the text, we defined the characteristic polynomial of A to be $p(x) = \det(A - xI)$. MATLAB uses $q(x) = \det(xI - A)$. The relationship between these two is simple: if A is $n \times n$, then $q(x) = (-1)^n p(x)$. (Explain!) The advantage of MATLAB's version is that the program's characteristic polynomials always have x^n as their highest order term. There is only one degree 3 polynomial with roots 2, 2, and 3 that has x^3 as its highest degree term. What is this polynomial? [*Hint:* Write it as a product of linear factors and then expand.] You can check your work with the MATLAB command "poly(A)," which produces a vector whose components are the coefficients of the characteristic polynomial.

6.3   COMPLEX EIGENVECTORS

Before beginning our discussion of complex eigenvectors, let us take a moment to review the complex numbers. The symbol i represents $\sqrt{-1}$. Thus, $i^2 = -1$. A complex number is an expression of the form $a + bi$ where a and b are real numbers. Thus, $2 + 3i$ is a complex number. We say that two complex numbers $a + bi$ and $c + di$ are equal if and only if $a = c$ and $b = d$.

We add complex numbers according to the formula

$$(a + bi) + (c + di) = a + c + (b + d)i \tag{1}$$

The formula for multiplication is based on the fact that $i^2 = -1$. Hence

$$\begin{aligned}(a + bi)(c + di) &= a(c + di) + bi(c + di) \\ &= ac + adi + bci + bdi^2 = ac - bd + (ad + bc)i\end{aligned} \tag{2}$$

Thus, for example

$$(2 + 3i) + (7 - 6i) = 9 - 3i$$
$$(2 + 3i)(7 - 6i) = 2 \cdot 7 - 3 \cdot (-6) + (3 \cdot 7 + 2 \cdot (-6))i = 32 + 9i$$

Students often feel uneasy about using complex numbers because they have a belief that they don't really exist—they are not "real." This is wrong. We think of $2 + 3i$ as just another way of representing the point $[2, 3]'$ in $\mathbb{R}^2$. Thus the set of complex numbers is just $\mathbb{R}^2$. Formula (1) says that addition of complex numbers is simply vector addition (Figure 1).

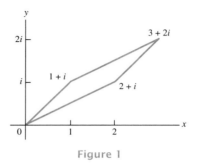

Figure 1

The set of real numbers is the set of numbers of the form $x + 0i$. This is the same as the set of points $[x, 0]'$ which is just the x axis. From this point of view, formula (2) is just a special way of multiplying points in $\mathbb{R}^2$. Note that if we accept formula (2) as a definition, then we can prove that $i^2 = -1$ since

$$i^2 = (0 + 1i)(0 + 1i) = 0 \cdot 0 - 1 \cdot 1 + (0 \cdot 1 + 1 \cdot 0)i = -1$$

Thus, $\sqrt{-1}$ does exist; it is the point $[0, 1]'$ in $\mathbb{R}^2$.

Actually, the multiplication of complex numbers has an interesting geometric interpretation. Any point $[x, y]'$ in $\mathbb{R}^2$ may be expressed in polar coordinates. This amounts to writing

$$x = r \cos \theta \text{ and } y = r \sin \theta \tag{3}$$

where r and θ are as in Figure 2. From Figure 2,

$$r = \sqrt{x^2 + y^2} \text{ and } \theta = \tan^{-1} \frac{y}{x}$$

Notice that r is just the length of the vector $[x, y]'$. From formula (3) we may write

$$x + iy = r \cos \theta + (r \sin \theta)i = r(\cos \theta + i \sin \theta)$$

The angle θ is called the argument of the complex number.

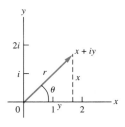

Figure 2

Now, let $u + vi$ be another complex number written in polar form as

$$u + vi = s(\cos \psi + i \sin \psi)$$

Then, from formula (2),

$$(x + yi)(u + vi) = rs(\cos\theta \cos \psi - \sin\theta \sin \psi + (\sin\theta \cos \psi + \cos\theta \sin \psi)i)$$
$$= rs(\cos(\theta + \psi) + i \sin(\theta + \psi)) \qquad (4)$$

The last equality came from the angle addition formulas for the sine and cosine functions. The term on the right side defines a point in polar form with argument $\theta + \psi$ and length rs. Thus, we multiply two complex numbers by adding their arguments and multiplying their length.

In particular, we square a complex number by squaring its length and doubling its argument. In fact, if we apply formula (4) repeatedly, we obtain the following formula, which is known as DeMoivre's theorem:

$$(r(\cos\theta + i \sin\theta))^n = r^n(\cos n\theta + i \sin n\theta) \qquad (5)$$

▶ EXAMPLE 1:   Compute $(1 + \sqrt{3}i)^{20}$.

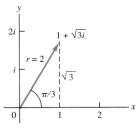

Figure 3

Solution: As Figure 3 indicates, the argument of $1 + \sqrt{3}i$ is $\tan^{-1}\sqrt{3} = \pi/3$. Also $r = \sqrt{1 + 3} = 2$. Hence,

$$1 + \sqrt{3}i = 2(\cos\frac{\pi}{3} + i\sin\frac{\pi}{3})$$

Thus, from formula (5),

$$(1 + \sqrt{3}i)^{20} = 2^{20}(\cos\frac{20\pi}{3} + i\sin\frac{20\pi}{3}) = -2^{19} + 2^{19}\sqrt{3}i$$

◀

Now, let's turn to the main topic of this section. As usual, we begin with an example.

▶ EXAMPLE 2:   Compute A^{20} where A is the matrix given.

$$A = \begin{bmatrix} 1 & -3 \\ 1 & 1 \end{bmatrix}$$

Solution: This problem looks very much like some problems that we solved in Section 6.2. The technique was to diagonalize A. Specifically, we found a diagonal matrix D and an invertible matrix Q such that $A = QDQ^{-1}$. Then formula (2) in Section 6.2 shows that $A^{20} = QD^{20}Q^{-1}$.

To find Q, we need to find the eigenvalues of A. We compute

$$\begin{vmatrix} 1-\lambda & -3 \\ 1 & 1-\lambda \end{vmatrix} = \lambda^2 - 2\lambda + 4$$

Using the quadratic formula, we find

$$\lambda = \frac{2 \pm \sqrt{4-16}}{2} = 1 \pm \sqrt{3}i$$

Now we are concerned—the roots are complex.

Let us, however, proceed with our computation and see what happens. To find the eigenvectors, we must solve the equation $(A - \lambda I)X = 0$. Assume first that $\lambda = 1 + \sqrt{3}i$. Then our equation is

$$\begin{bmatrix} -\sqrt{3}i & -3 \\ 1 & -\sqrt{3}i \end{bmatrix}\begin{bmatrix} x \\ y \end{bmatrix} = \begin{bmatrix} 0 \\ 0 \end{bmatrix}$$

This corresponds to the system

$$(-\sqrt{3}i)x - 3y = 0$$
$$x - (\sqrt{3}i)y = 0$$

Notice that if we multiply the second equation by $-\sqrt{3}i$ we obtain

$$0 = -\sqrt{3}i(x - (\sqrt{3}i)y) = (-\sqrt{3}i)x + 3i^2y = (-\sqrt{3}i)x - 3y$$

This is the first equation. Thus our system is equivalent with the system consisting of only the second equation. The dependency of this system should not surprise us since a system of two equations in two unknowns cannot have any non-zero solutions unless it is dependent.

We solve this system by letting y be arbitrary and setting $x = (\sqrt{3}i)y$. In vector form, this becomes

$$[x, y]^t = [\sqrt{3}i, 1]^t y$$

A basis for the solution set is the single complex vector

$$Q_1 = [\sqrt{3}i, 1]^t$$

As a check on our work, we compute

$$\begin{bmatrix} 1 & -3 \\ 1 & 1 \end{bmatrix} \begin{bmatrix} \sqrt{3}i \\ 1 \end{bmatrix} = \begin{bmatrix} \sqrt{3}i - 3 \\ \sqrt{3}i + 1 \end{bmatrix} = (1 + \sqrt{3}i) \begin{bmatrix} \sqrt{3}i \\ 1 \end{bmatrix}$$

Thus, Q_1 really is an eigenvector with eigenvalue $1 + \sqrt{3}i$!

Similarly, if we use $\lambda = 1 - \sqrt{3}i$, we find the eigenvector

$$Q_2 = [-\sqrt{3}i, 1]^t$$

Hence, our diagonalizing matrix is

$$Q = \begin{bmatrix} \sqrt{3}i & -\sqrt{3}i \\ 1 & 1 \end{bmatrix}$$

Does this matrix really diagonalize A? To check it, we recall that the inverse of a general 2×2 matrix is given by the formula

$$\begin{bmatrix} a & b \\ c & d \end{bmatrix}^{-1} = \frac{1}{ad - cb} \begin{bmatrix} d & -b \\ -c & a \end{bmatrix}$$

Applying this to Q (and hoping that it works for complex numbers), we find

$$Q^{-1} = \frac{1}{2\sqrt{3}i} \begin{bmatrix} 1 & \sqrt{3}i \\ -1 & \sqrt{3}i \end{bmatrix} = \frac{-\sqrt{3}i}{6} \begin{bmatrix} 1 & \sqrt{3}i \\ -1 & \sqrt{3}i \end{bmatrix}$$

Now, it is easily computed that indeed, $Q^{-1}Q = I$ and

$$Q^{-1}AQ = \begin{bmatrix} 1 + \sqrt{3}i & 0 \\ 0 & 1 - \sqrt{3}i \end{bmatrix}$$

Let's call this latter matrix D. From Example 1 (and a similar computation for $1 - \sqrt{3}i$), we see that

$$D^{20} = \begin{bmatrix} (1 + \sqrt{3}i)^{20} & 0 \\ 0 & (1 - \sqrt{3}i)^{20} \end{bmatrix} = 2^{19} \begin{bmatrix} -1 + \sqrt{3}i & 0 \\ 0 & -1 - \sqrt{3}i \end{bmatrix}$$

Hence, doing the required matrix multiplication, we find that

$$A^{20} = QD^{20}Q^{-1} = 2^{19} \begin{bmatrix} -1 & 3 \\ -1 & 1 \end{bmatrix}$$
◀

Having done this work and having arrived at an answer, let us step back to ask whether we can be confident that our answer is really correct. The heart of our computation was formula (2) from Section 6.2. This in turn was based on the associative law of matrix multiplication. Specifically, we used arguments such as

$$(QDQ^{-1})(QDQ^{-1}) = QD(Q^{-1}Q)DQ^{-1} = QDIDQ^{-1} = QD^2Q^{-1}$$

What is new here is that the matrices have complex entries. It is clear what we mean by a complex matrix—it is just a rectangular array of complex numbers. Clearly, we may define multiplication of complex matrices in exactly the same way that we define multiplication of real matrices. For the most part, the algebraic properties of complex matrices will be identical with the properties of real matrices. This is because the basic algebraic properties of the complex numbers are, for the most part, identical with those of the real numbers. Specifically, all of the following properties hold regardless of whether we interpret a, b, and c as elements of $\mathbb{C}$ or $\mathbb{R}$.

(a) $a + b = b + a$.

(b) $a + (b + c) = (a + b) + c$.

(c) $a + 0 = 0 + a = a$.

(d) For every element a there is an element $-a$ such that $a + (-a) = 0$.

(e) $a(bc) = (ab)c$.

(f) $a1 = 1a = a$.

(g) For every element $a \neq 0$, there is an element a^{-1} such that $aa^{-1} = 1$.

(h) $ab = ba$.

(i) $a(b + c) = ab + ac$.

The above properties are referred to as the field properties. All the basic properties of matrix algebra (such as the associative law for matrix multiplication) may be proved directly from the field properties. Hence, they will all be true for complex matrices. Thus, our calculation really is correct.

Let us ask a more philosophical question: Why were we forced to use the complex numbers to solve a problem that involved only real numbers? The answer is, of course, that the characteristic polynomial for our matrix had complex roots. This demonstrates why the complex numbers were invented in the first place. To find the roots of a second-degree polynomial, one often needs to be able to take the square root of negative numbers. It is one of the most surprising and useful facts of mathematics that once we are able to form square roots of negative numbers, we can (in principle) factor any polynomial. This result is called the fundamental theorem of algebra.

Fundamental Theorem of Algebra.   *Every polynomial $p(x)$ with complex coefficients has a factorization of the form*

$$p(x) = a(x - r_1)^{n_1}(x - r_2)^{n_2} \ldots (x - r_k)^{n_k}$$

where the n_i are natural numbers and a and r_i are complex numbers.   ■

A few comments are necessary. First, any real number, such as 2, is also a complex number: $2 = 2 + 0i$. Hence the theorem does not deny the possibility that all the roots are real. Similarly, the theorem is valid for polynomials with real coefficients as well.

The set of all $n \times 1$ complex matrices is denoted $\mathbb{C}^n$. The concepts of complex eigenvectors and complex eigenvalue are defined just as in the case of real numbers:

Definition.   *Let A be a complex $n \times n$ matrix. A non-zero element X in $\mathbb{C}^n$ such that*

$$AX = \lambda X$$

for some scalar λ is called a complex eigenvector for A. The scalar λ is called the corresponding complex eigenvalue.   ■

The complex eigenvalues are the complex roots of the characteristic polynomial $p(\lambda) = \det(A - \lambda I)$. One consequence of the fundamental theorem of algebra is that the characteristic polynomial of any $n \times n$ matrix A will always have at least one complex root. It follows, just as in the real case, that A must therefore have at least one eigenvalue. Thus, we conclude:

Theorem 1.   *Every $n \times n$ matrix A has at least one complex eigenvalue.*   ■

Again, we stress that the complex eigenvalue could be real.

Notice that the eigenvalues for the matrix A from Example 2 were $1 + \sqrt{3}i$ and $1 - \sqrt{3}i$. Recall that for a complex number $z = a + bi$, the **complex conjugate** is

$$\bar{z} = a - ib$$

Thus the two eigenvalues are conjugate to each other. The corresponding eigenvectors are also conjugate:

$$\overline{\begin{bmatrix} \sqrt{3}i \\ 1 \end{bmatrix}} = \begin{bmatrix} -\sqrt{3}i \\ 1 \end{bmatrix}$$

(If $A = [a_{ij}]$ is a matrix with complex entries, then we define $\bar{A} = [\bar{a}_{ij}]$.)

The eigenvalues and eigenvectors for real matrices always occur in conjugate pairs. To prove this, let A be a real matrix and let λ be an eigenvalue for A. Then there is a complex vector X such that

$$\lambda X = AX$$

For all complex numbers z and w, $\overline{zw} = \bar{z}\,\bar{w}$. This extends to matrices as

$$\bar{A}\,\bar{B} = \overline{AB} \tag{6}$$

Hence

$$\bar{\lambda}\,\bar{X} = \overline{\lambda X} = \overline{AX} = \bar{A}\,\bar{X} = A\bar{X}$$

The last equality follows because A is real. This shows that $\bar{X}$ is an eigenvector for A corresponding to the eigenvalue $\bar{\lambda}$, as claimed.

EXERCISES

(1) Compute AB and BA for the matrices A and B.

$$A = \begin{bmatrix} 1+i & 2i \\ 2 & 3i \end{bmatrix} \qquad B = \begin{bmatrix} -i & 3 \\ 2+i & 4i \end{bmatrix}$$

(2) Compute A^{20} for the matrix A shown. [*Note*: The argument of the complex number in question is not one of the standard angles from trigonometry. You will need to use a calculator to approximate the answer.]

$$A = \begin{bmatrix} 1 & -4 \\ 1 & 1 \end{bmatrix}$$

(3) Find all eigenvalues for the following matrix A.

$$A = \begin{bmatrix} 2 & 3 & 6 \\ 6 & 2 & -3 \\ 3 & -6 & 2 \end{bmatrix}$$

It might help to know that -7 is one eigenvalue.

(4) Consider the transformation $T : \mathbb{C} \to \mathbb{C}$ given by $T(z) = (2 + 3i)z$. We commented that we may interpret the complex numbers as being $\mathbb{R}^2$. Thus, we may think of T as transforming $\mathbb{R}^2$ into $\mathbb{R}^2$. Explicitly, for example, T transforms the point $[1, 2]'$ into $[-4, 7]'$ since $(2 + 3i)(1 + 2i) = -4 + 7i$.
Show that as a transformation of $\mathbb{R}^2$ into $\mathbb{R}^2$, T is linear. Find a real 2×2 matrix that represents T. Find all eigenvalues of this matrix.

(5) Repeat Exercise 4 with $a + bi$ in place of $2 + 3i$.

(6) Compute the eigenvalues and eigenvectors for the rotation matrix

$$R(\theta) = \begin{bmatrix} \cos\theta & \sin\theta \\ -\sin\theta & \cos\theta \end{bmatrix}$$

(7) Prove that for all complex numbers z and w, $\overline{zw} = \overline{z}\,\overline{w}$. Use this to prove formula (6) in the text. [*Hint:* Use formula (3) in Section 3.2.]

(8) If A is a complex matrix, then we let $A^* = \overline{A}^t$.
  (a) Let A and B be as in Exercise 1. Show by direct computation that $(AB)^* = B^*A^*$.
  (b) Prove that the property stated in part (a) is true for any complex matrices that can be multiplied.
  (c) If A is an $n \times n$ matrix with complex entries, we say that A is invertible if there is an $n \times n$ matrix B such that $AB = BA = I$. Prove that if A is invertible, then so is A^*, and $(A^{-1})^* = (A^*)^{-1}$.

(9) Let A be a complex $n \times n$ matrix and let X be a complex eigenvector with eigenvalue λ. Show that $\overline{X}$ is an eigenvector for $\overline{A}$. What is the corresponding eigenvalue?

(10) Let $p(x) = a_n x^n + a_{n-1} x^{n-1} + \cdots + a_0$ where the a_i are real numbers. Show that if α is a complex root of $p(x)$, then so is $\overline{\alpha}$. Use this fact to prove that if λ is an eigenvalue for a real $n \times n$ matrix A, then so is $\overline{\lambda}$.

(11) Prove that an $n \times n$ matrix can have at most n different eigenvalues. (There is a proof based on the linear independence of the eigenvectors. There is also a proof based on the characteristic polynomial.)

(12) Prove that a real 3×3 matrix must have at least one real eigenvalue.

(13) Let A be a real, $n \times n$ orthogonal matrix. Let λ be a complex eigenvalue for A. Show that $|\lambda| = 1$. [*Hint:* Let $AX = \lambda X$. Consider $AX \cdot A\overline{X}$.]

> Remark: If $\lambda = a + bi$, then $|\lambda| = 1$ means that λ represents a point on the unit circle in $\mathbb{R}^2$. Such points may be expressed in the form $\cos\theta + (\sin\theta)i$ for some angle θ. Suppose that A is 3×3. What do you suppose is the geometric significance of θ?

(14) Let A be an $n \times n$ matrix with complex entries. Let X be an eigenvector for A with eigenvalue λ, and let Y be an eigenvector for A^t with eigenvalue β. Show that if $\lambda \neq \beta$, then $X \cdot Y = 0$. [*Hint:* Begin your proof with $\lambda X \cdot Y = AX \cdot Y = (AX)^t Y = \ldots$.]

(15) Let A be an $n \times n$ matrix with complex entries. We say that A is **Hermitian symmetric** (or just Hermitian) if $A^* = A$, where the notation is as in Exercise 8. Give an example of a 3×3 Hermitian matrix. Keep all entries non-zero and use as few real entries as possible.

(16) Let A be an $n \times n$ Hermitian matrix.

(a) Show that all eigenvalues of A are real. [*Hint:* Exercises 9 and 14 might help.]
(b) Let X and Y be eigenvectors for A corresponding to different eigenvalues α and β. Show that $X \cdot \overline{Y} = 0$.

(17) Recall that $\mathbb{C}^n$ is the set of all $n \times 1$ column matrices with complex entries. Let X and Y be elements of $\mathbb{C}^n$. Define

$$< X, Y > = Y^* X$$

where Y^* is as defined in Exercise 8. This is called the Hermitian scalar product on $\mathbb{C}^n$.

(a) Compute $< [1 + i, 2i, 3]^t, [2 - i, 1 + i, -5i]^t >$.
(b) Prove that for all X, Y, and Z in $\mathbb{C}^n$

$$< X + Y, Z > = < X, Z > + < Y, Z >$$

(c) Prove that $< X, Y > = \overline{< Y, X >}$.
(d) Prove that for all complex numbers α

$$< \alpha X, Y > = \alpha < X, Y > = < X, \overline{\alpha} Y >$$

(e) $< X, X > \geq 0$ with equality if and only if $X = \mathbf{0}$.

> Remark: If $X \in \mathbb{C}^n$, we define $||X|| = \sqrt{< X, X >}$. We consider this as representing the "length" of the complex vector X.

(18) Prove that for all X and Y in $\mathbb{C}^n$,

$$< X, Y >= \frac{1}{4}(||X + Y||^2 - ||X - Y||^2 + i(||X + iY||^2 - ||X - iY||^2)$$

This identity is called the polarization identity. It plays the same role for complex vectors as the law of cosines does for real vectors.

(19) Recall that an orthogonal matrix is a real $n \times n$ matrix such that for all $X \in \mathbb{R}^n$, $|AX| = |X|$. The corresponding concept for complex matrices is called "unitary." Explicitly, we say that an $n \times n$ matrix U with complex entries is unitary if $||UX|| = ||X||$ for all $X \in \mathbb{C}^n$. Let U be unitary. Use Exercise 18 to prove that for all X and Y in $\mathbb{C}^n$,

$$< UX, UY > = < X, Y >$$

(20) Show that all real orthogonal matrices are unitary.

(21) What can you say about the columns of a unitary matrix? (See Section 4.4.) Prove your answer.

(22) Let U be a unitary matrix. Prove that $U^*U = I$.

ON LINE

(1) Try the MATLAB command "[B,D]=eig(A)" where A is as in Example 2. This demonstrates that MATLAB "knows about" complex eigenvalues.

(2) Let A be an $n \times n$ matrix with complex entries. We say that A is Hermitian symmetric if $A^* = A$. (See Exercise 8.) Give an example of a 3×3 Hermitian matrix. Keep all entries non-zero and use as few real entries as possible. Enter your matrix into MATLAB. (Complex numbers such as $2 + 3i$ are entered into MATLAB as "2+3*i".) Then get MATLAB to find the eigenvalues. You should discover (remarkably) that they are all real!

(3) Change one of the entries of the matrix A from Exercise 2 to make it non-Hermitian. (This can be done simply by reassigning the value of, say, A(2,1).) Now compute the eigenvalues. Are they still real?

(4) You may be under the impression that "A'" is MATLAB for the transpose of A. Try it on a few complex matrices and see what you get. Read the MATLAB help entry for "transpose" for an explanation.

(5) Enter into MATLAB a 3×3, real, symmetric matrix A. Make as many of the entries of A as possible distinct. In MATLAB, let "B=eye(3)+i*A". ["eye(3)" is

MATLAB's notation for the 3×3 identity matrix.] Let "C=eye(3)+inv(B)/2" and compute $C * C'$ and $C' * C$. Can you prove that what you observe is always true? [*Hint:* Begin with the equality $B + B^* = 2I$ and multiply by B^{-1} on the left and $(B^*)^{-1}$ on the right.]

6.4   MATRIX OF A LINEAR TRANSFORMATION

We begin by reviewing the discussion of Example 1 in Section 6.2. In this example, we studied the transformation of $\mathbb{R}^2$ into $\mathbb{R}^2$ defined by multiplication by the matrix

$$A = \begin{bmatrix} 3 & 1 \\ 1 & 3 \end{bmatrix}$$

This matrix has two independent eigenvectors $Q_1 = [-1, 1]^t$ and $Q_2 = [1, 1]^t$ corresponding to the eigenvalues 2 and 4.

If we use the Q_i as a basis for a coordinate system on $\mathbb{R}^2$, then an element X in $\mathbb{R}^2$ has coordinate vector $X' = [x', y']^t$ if and only if

$$X = x'Q_1 + y'Q_2$$

It follows that

$$AX = x'AQ_1 + y'AQ_2 = 2x'Q_1 + 4y'Q_2$$

Hence, multiplication by A transforms the vector with coordinates $[x', y']^t$ into the vector with coordinates

$$\begin{bmatrix} 2x' \\ 4y' \end{bmatrix} = \begin{bmatrix} 2 & 0 \\ 0 & 4 \end{bmatrix} \begin{bmatrix} x' \\ y' \end{bmatrix}$$

Let D be the 2×2 matrix in this equation. This matrix describes our transformation in terms of coordinates: multiplication by A transforms the point with coordinates X' into the point with coordinates DX'. From this point of view, the eigenvectors help us to find coordinates in which our transformation is describable by a particularly simple matrix.

The idea of using a special choice of coordinates to study a given linear transformation has applications far beyond the diagonalization of matrices. In the general case, we will be given an n-dimensional vector space $\mathcal{V}$, an m-dimensional vector space

$\mathcal{W}$, and a linear transformation T of $\mathcal{V}$ into $\mathcal{W}$. We will also be given an ordered basis $B = \{Q_1, Q_2, \ldots, Q_n\}$ for $\mathcal{V}$ and an ordered basis $B' = \{P_1, P_2, \ldots, P_m\}$ for $\mathcal{W}$. For each X in $\mathcal{V}$ let X' be its coordinate vector computed with respect to the basis B and for each Y in $\mathcal{W}$ let Y'' be its coordinate vector computed with respect to B'. We say that an $m \times n$ matrix M represents T if for all X in $\mathcal{V}$

$$(T(X))'' = MX' \tag{1}$$

Thus, M describes the transformation T in the sense that it tells us how to compute the coordinate vector for $T(X)$ from the coordinate vector for X. This matrix is called **the matrix of T with respect to the given bases.** In most applications, $\mathcal{V} = \mathcal{W}$ and we use the same basis for the domain and target space of T. In this case, we refer to M as **the matrix of T with respect to the given basis.**

There is another way of expressing formula (1) which explains why M exists. Recall that an element X of $\mathcal{V}$ has coordinate vector $X' = [x_1', x_2', \ldots, x_n']^t$ if and only if

$$X = x_1'Q_1 + x_2'Q_2 + \cdots + x_n'Q_n$$

Let S_Q be the transformation of $\mathbb{R}^n$ into $\mathcal{V}$ which transforms each X' in $\mathbb{R}^n$ to X where X is given by the preceding formula. (Note that if $\mathcal{V} = \mathbb{R}^n$, then S_Q is multiplication by the point matrix for the basis B.) The transformation which transforms each X in $\mathcal{V}$ to its coordinate vector X' is the inverse transformation S_Q^{-1}. The argument used in the proof of the linearity properties for matrix multiplication in Section 1.5 shows that S_Q is a linear transformation, and the argument in part (b) of Theorem 2 in Section 4.2 proves that S_Q^{-1} is also linear.[1]

Let S_P denote the analogous transformation of $\mathbb{R}^m$ into $\mathcal{W}$ defined using the basis B'. Then for all X in $\mathcal{V}$,

$$X = S_Q(X') \text{ and } T(X)'' = S_P^{-1}(T(X))$$

Thus,

$$(T(X))'' = S_P^{-1}(T(S_Q(X'))) = S_P^{-1} \circ T \circ S_Q(X')$$

The composition of linear transformations is linear. (See Exercise 27, Section 3.2.) Hence, $S_P^{-1} \circ T \circ S_Q$ is a linear transformation. It follows from the matrix representation theorem in Section 3.1 that this transformation is described by matrix multiplication, proving the existence of M.

---

[1] Doing a separate argument for the linearity of S_Q^{-1} is actually unnecessary since it is a theorem that the inverse of a linear transformation is always linear. We have not, however, proved this theorem.

If $V = \mathbb{R}^n$ and $Q = \mathbb{R}^m$, then S_Q and S_P are, respectively, multiplication by the matrices $Q = [Q_1, Q_2, \ldots, Q_n]$ and $P = [P_1, P_2, \ldots, P_m]$. If additionally T is multiplication by an $m \times n$ matrix A, then the preceding equality tells us that

$$M = P^{-1}AQ \qquad (2)$$

These comments prove the following theorem.

Theorem 1.   Let $T : V \to W$ be a linear transformation between two vector spaces. Let $B = \{Q_1, Q_2, \ldots, Q_n\}$ be an ordered basis for V and $B' = \{P_1, P_2, \ldots, P_m\}$ be an ordered basis for W. For each X in V let X' be its coordinate vector with respect to the basis B and for each Y in W let Y'' be its coordinate vector with respect to B'. Then there is a unique $m \times n$ matrix M such that formula (1) holds for all X in V. If $V = \mathbb{R}^n$ and $Q = \mathbb{R}^m$, then formula (2) holds as well where Q and P are, respectively, the point matrices for the bases B and B'.   ∎

One of the most important uses of Theorem 1 is to study *non-diagonalizable* matrices. Recall that an $n \times n$ matrix is said to be **lower triangular** if all its entries above the main diagonal are zero. For example, the following matrix M is lower triangular.

$$M = \begin{bmatrix} 1 & 0 & 0 \\ 0 & 3 & 0 \\ 4 & 5 & 6 \end{bmatrix}$$

An $n \times n$ matrix A is said to be **triangularizable** if there is a lower triangular matrix M and an invertible matrix Q such that

$$A = QMQ^{-1}$$

in which case, this equation is referred to as a triangularization of A. Note that since all diagonal matrices are also triangular, all diagonalizable matrices are also triangularizable. The preceding equality is equivalent with $Q^{-1}AQ = M$. Thus, Theorem 1 says that A is triangularizable if there is a basis for $\mathbb{R}^n$ for which A is representable by a triangular matrix M.

We know that not all matrices are diagonalizable. It is one of the most fundamental results of linear algebra that all $n \times n$ matrices are at least triangularizable, provided we allow the use of complex numbers. Furthermore, it turns out triangularizations are often sufficient to solve many of the same types of problems which we solved with diagonalizations in Section 6.2. For example, in several of the exercises you will use triangularizations to compute powers of matrices.

The proof of the statement that every matrix can be triangularized is beyond the scope of this text. In fact, we will provide systematic techniques for producing triangularizations only in the simplest of cases. One such simple case is demonstrated by the next example.

▶ EXAMPLE 1:   Find an invertible matrix Q and a triangular matrix M such that $A = QMQ^{-1}$ where

$$A = \begin{bmatrix} 8 & 6 & -14 \\ 2 & 0 & -2 \\ 5 & 3 & -8 \end{bmatrix}$$

Solution: We compute that

$$A^2 = \begin{bmatrix} 6 & 6 & -12 \\ 6 & 6 & -12 \\ 6 & 6 & -12 \end{bmatrix} \text{ and } A^3 = AA^2 = \begin{bmatrix} 0 & 0 & 0 \\ 0 & 0 & 0 \\ 0 & 0 & 0 \end{bmatrix}$$

This shows that A is what is referred to as a "nilpotent" matrix. In general, an $n \times n$ matrix B is said to be nilpotent if there is a natural number k such that $B^k = \mathbf{0}$. The smallest such k is called the degree of nilpotency of N. For nilpotent matrices, we may use what are called "chain bases" to produce triangularizations. Such bases are produced by repeatedly multiplying some fixed vector by the matrix in question. Thus, for our matrix A, we let Q_1 be any convenient vector in $\mathbb{R}^3$. We then let

$$Q_2 = AQ_1, \qquad Q_3 = AQ_2 = A^2Q_1 \tag{3}$$

Notice that

$$AQ_3 = A^3Q_1 = \mathbf{0}$$

It follows from the nilpotency of A that as long as $Q_3 \neq \mathbf{0}$, the Q_i will be independent. (See Exercise 8 on page 341.) Since $\mathbb{R}^3$ is three-dimensional, the Q_i will then form a basis for $\mathbb{R}^3$. For example, we might let $Q_1 = [1, 0, 0]^t$, in which case $Q_2 = [8, 2, 5]^t$ and $Q_3 = [6, 6, 6]^t$.

To find the matrix of the transformation defined by A with respect to this basis, we note that

$$A(x'Q_1 + y'Q_2 + z'Q_3) = x'AQ_1 + y'AQ_2 + z'AQ_3 = x'Q_2 + y'Q_3$$

Hence multiplication by A transforms the vector with coordinates $[x', y', z']^t$ into the vector with coordinates $[0, x', y']^t$. This transformation is described by multiplication by the triangular matrix

$$M = \begin{bmatrix} 0 & 0 & 0 \\ 1 & 0 & 0 \\ 0 & 1 & 0 \end{bmatrix}$$

According to Theorem 1, $A = QMQ^{-1}$ where

$$Q = \begin{bmatrix} 1 & 8 & 6 \\ 0 & 2 & 6 \\ 0 & 5 & 6 \end{bmatrix}$$

◄

Remark: In Example 1, the computation of M was simple because in formula (3), each AQ_k was written as a linear combination of the Q_j. Expressing the image of each element of the domain basis in terms of the image basis is one common way of computing matrix representations.

At first sight, it might seem that nilpotent matrices are too special to occur frequently. Actually, they play a fundamental role in the general theory. It turns out, for example, that *if an $n \times n$ matrix A has only one eigenvalue λ, then $A - \lambda I$ is nilpotent.* Example 2 demonstrates the use of this result in triangularizing a 2×2 nondiagonalizable matrix.

▶ **EXAMPLE 2:** Find a triangularization for the matrix A. Describe geometrically the effect of the transformation of $\mathbb{R}^2$ into $\mathbb{R}^2$ defined by multiplication by A.

$$A = \begin{bmatrix} 1.5 & -0.5 \\ 0.5 & 0.5 \end{bmatrix}$$

Solution: It is easily seen that the only eigenvalue for A is $\lambda = 1$. It follows from the preceding comments that

$$B = A - I = \begin{bmatrix} 0.5 & -0.5 \\ 0.5 & -0.5 \end{bmatrix}$$

is a nilpotent matrix. Indeed, the reader may check that $B^2 = \mathbf{0}$.

To triangularize A, we construct a chain basis for B. Thus, let Q_1 be any vector in $\mathbb{R}^2$ such that $Q_2 = BQ_1 \neq \mathbf{0}$. Then $BQ_2 = B^2Q_1 = \mathbf{0}$. For example, we might choose $Q_1 = [1, 0]'$ in which case $Q_2 = [0.5, 0.5]'$.

Since $A = B + I$, we see that

$$AQ_1 = BQ_1 + Q_1 = Q_2 + Q_1$$
$$AQ_2 = BQ_2 + Q_2 = Q_2$$

To find the matrix for the transformation described by A, we compute

$$A(x'Q_1 + y'Q_2) = x'AQ_1 + y'AQ_2 = x'(Q_2 + Q_1) + y'Q_2$$
$$= x'Q_1 + (x' + y')Q_2$$

Thus, if X has coordinate vector $[x', y']^t$, then AX has coordinate vector $[x', x' + y']^t$. This transformation is described by multiplication by

$$M = \begin{bmatrix} 1 & 0 \\ 1 & 1 \end{bmatrix}$$

From Theorem 1, $A = QMQ^{-1}$ where

$$Q = \begin{bmatrix} 1 & 0.5 \\ 0 & 0.5 \end{bmatrix}$$

The transformation described by M was studied in Example 1, Section 3.1. In natural coordinates, it transforms rectangles with sides on the coordinate axes into parallelograms by shearing them parallel to the y axis (Figure 4, Section 3.1). This means that multiplication by A will transform parallelograms with sides on the x' and y' axes into parallelograms by shearing them parallel to the y' axis (Figure 1).

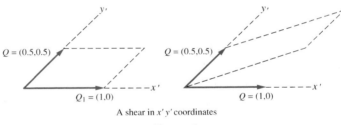

A shear in $x'y'$ coordinates

Figure 1

◀

Remark: The technique demonstrated in Example 2, together with Theorem 1, allows us to triangularize any 2×2 matrix A that has only real eigenvalues: if A has two distinct eigenvalues, then A is diagonalizable and we use Theorem 1 from Section 6.2. If A has only one real eigenvalue λ, then $B = A - \lambda I$ is nilpotent. If $B = \mathbf{0}$, then $A = \lambda I$ so that A is already diagonal. If $B \neq \mathbf{0}$, then any chain basis for B will triangularize A.

Sometimes formula (2) is the easiest method to compute the matrix representation of a given transformation, as the next example shows.

▶ **EXAMPLE 3:** In Example 1 in Section 4.4 we showed that the following matrix A

is orthogonal. Show that the transformation defined by A is a rotation by some angle θ about a line Λ through the origin in $\mathbb{R}^3$. What is θ?

$$A = \frac{1}{2} \begin{bmatrix} 1 & \sqrt{2} & 1 \\ -\sqrt{2} & 0 & \sqrt{2} \\ 1 & -\sqrt{2} & 1 \end{bmatrix}$$

Solution: Figure 2 illustrates a rotation about a line through the origin. The transformation acts by rotating points by a fixed angle about the line of rotation. Notice that the points on the line of rotation are transformed into themselves. Thus, if A does indeed describe a rotation, then the line of rotation should be the set of all X such that $AX = X$, which is equivalent with $(A - I)X = \mathbf{0}$. This equation is described by the augmented matrix

$$\begin{bmatrix} -\frac{1}{2} & \frac{\sqrt{2}}{2} & \frac{1}{2} & 0 \\ -\frac{\sqrt{2}}{2} & -\frac{1}{2} & \frac{\sqrt{2}}{2} & 0 \\ \frac{1}{2} & -\frac{\sqrt{2}}{2} & -\frac{1}{2} & 0 \end{bmatrix}$$

Reducing, we find that the solution set is spanned by the vector $Y = [1, 0, 1]^t$. Thus, if A describes a rotation, then the axis of rotation is the line spanned by Y.

   In studying a rotation, it seems natural to choose an orthogonal coordinate system for $\mathbb{R}^3$ with one axis along the line of rotation, in which case the other two axes will lie in the plane $\mathcal{W}$ through the origin perpendicular to this line (Figure 2). The plane perpendicular to Y is the subspace defined by the equation

$$0 = Y \cdot X = x + z$$

Using the technique of Example 3 in Section 4.2, we find that the vectors $Q_1 = [\frac{1}{\sqrt{2}}, 0, \frac{-1}{\sqrt{2}}]^t$ and $Q_2 = [0, 1, 0]^t$ form an orthonormal basis for $\mathcal{W}$.

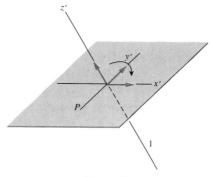

Figure 2

Let $Q_3 = Y/|Y| = [\frac{1}{\sqrt{2}}, 0, \frac{1}{\sqrt{2}}]^t$. Then the vectors Q_1, Q_2, and Q_3 form an orthonormal basis for $\mathbb{R}^3$. Thus the point matrix

$$Q = \begin{bmatrix} \frac{1}{\sqrt{2}} & 0 & \frac{1}{\sqrt{2}} \\ 0 & 1 & 0 \\ \frac{-1}{\sqrt{2}} & 0 & \frac{1}{\sqrt{2}} \end{bmatrix}$$

is orthogonal. From formula (2), $M = Q^{-1}AQ = Q^t A Q$. Performing the matrix multiplication, we find that

$$M = \begin{bmatrix} 0 & 1 & 0 \\ -1 & 0 & 0 \\ 0 & 0 & 1 \end{bmatrix}$$

According to formula (2) in Section 4.4, multiplication by

$$R_\theta^z = \begin{bmatrix} \cos\theta & -\sin\theta & 0 \\ \sin\theta & \cos\theta & 0 \\ 0 & 0 & 1 \end{bmatrix}$$

rotates by θ radians around the z axis. This matrix equals M if $\theta = -\frac{\pi}{2}$ and the z-axis in our new coordinates is the line spanned by Y. Thus, multiplication by A rotates points by 90 degrees about the line spanned by Y. ◀

Remark: It is tempting to say that the rotation in Example 3 is clockwise since the angle of rotation is negative. However, what we mean by a "clockwise rotation" depends on which direction along the line of rotation we happen to be looking. All we can really say about the direction of rotation is that it is from the y' axis to the x' axis.

One important use of Theorem 1 is to study transformations between vector spaces other than $\mathbb{R}^n$. Recall that P_n represents the space of all functions $p(x)$ of the form

$$p(x) = a_0 + a_1 x + \cdots + a_n x^n$$

The functions $1, x, x^2, \ldots, x^n$ form a basis for P_n, which is called the standard basis. The coordinate vector for $p(x)$ with respect to this basis is

$$[a_0, a_1, \ldots, a_n]^t$$

▶ **EXAMPLE 4:** Let $T : P_2 \to P_2$ be the linear transformation defined by

$$T(y) = (x+1)\frac{dy}{dx} - y.$$

Find bases for the image and nullspace of T.

Solution: Note that

$$T(a + bx + cx^2) = (x + 1)\frac{d}{dx}(a + bx + cx^2) - (a + bx + cx^2)$$
$$= (x + 1)(b + 2cx) - (a + bx + cx^2) = b - a + 2cx + cx^2$$

Hence T transforms the polynomial with coordinates $[a, b, c]'$ into the polynomial with coordinates

$$\begin{bmatrix} -a + b \\ 2c \\ c \end{bmatrix} = \begin{bmatrix} -1 & 1 & 0 \\ 0 & 0 & 2 \\ 0 & 0 & 1 \end{bmatrix} \begin{bmatrix} a \\ b \\ c \end{bmatrix}$$

Let M be the 3×3 matrix above. This is the matrix of T with respect to the given basis.

The nullspace of T is the set of X in P_2 such that $T(X) = \mathbf{0}$. From formula (2), this is true if and only if $MX' = \mathbf{0}$. Solving the corresponding system, we find that

$$X' = c[1, 1, 0]' = [c, c, 0]'$$

Thus, the nullspace of T is the set of polynomials of the form $c + cx$. The basis is the polynomial $1 + x$.

An element Y is in the image of T if and only if there is an X such that $T(X) = Y$. From formula (1), this is true if and only if $Y'' = MX'$—i.e. Y'' is in the image of the transformation defined by M. The columns $M_2 = [1, 0, 0]'$ and $M_3 = [0, 2, 1]'$ form a basis for this image. These correspond to the polynomials 1 and $2x + x^2$. These polynomials form the basis for the image of T.   ◄

EXERCISES

(1) Suppose that A is a 3×3 matrix and that Q_1, Q_2, and Q_3 form a basis for $\mathbb{R}^3$. Suppose also that

$$AQ_1 = 3Q_1 + 2Q_2 + Q_3$$
$$AQ_2 = Q_1 + Q_3$$
$$AQ_3 = 2Q_2 + Q_3$$

What is the matrix of the transformation defined by A with respect to the given basis? [*Hint:* Simplify $A(x'Q_1 + y'Q_2 + z'Q_3)$.]

(2) Suppose that A is a 3×3 matrix and that Q_1, Q_2, and Q_3 form a basis for $\mathbb{R}^3$. Suppose also that

$$AQ_1 = 2Q_2 + 5Q_3$$
$$AQ_2 = 4Q_3$$
$$AQ_3 = \mathbf{0}$$

Find a matrix M such that $A = QMQ^{-1}$ where $Q = [Q_1, Q_2, Q_3]$. Use your result to prove that A is nilpotent.

(3) For the matrices in parts (a) and (b):

      (i) for each i, express AQ_i as a linear combination of the Q_i,

      (ii) find the matrix M of the linear transformation defined by A with respect to the ordered basis formed by the Q_i,

      (iii) verify that $AQ = QM$ where Q is the point matrix for the given basis.

(a)
$$A = \begin{bmatrix} -1 & 16 \\ -1 & 7 \end{bmatrix} \qquad Q_1 = \begin{bmatrix} 4 \\ 1 \end{bmatrix} \qquad Q_2 = \begin{bmatrix} 3 \\ 1 \end{bmatrix}$$

(b)
$$A = \begin{bmatrix} 7 & -2 & 0 \\ 8 & -1 & 0 \\ 3 & -3 & 6 \end{bmatrix} \qquad Q_1 = \begin{bmatrix} 1 \\ -1 \\ 1 \end{bmatrix} \qquad Q_2 = \begin{bmatrix} 1 \\ 2 \\ 1 \end{bmatrix} \qquad Q_3 = \begin{bmatrix} 0 \\ 0 \\ 1 \end{bmatrix}$$

(4) Use the method of Example 2 to find a lower triangular matrix M and an invertible matrix Q such that $A = QMQ^{-1}$ where

(a) $$A = \begin{bmatrix} 1 & 2 \\ -2 & 5 \end{bmatrix}$$
      (b) $$A = \begin{bmatrix} -11 & 27 \\ -3 & 7 \end{bmatrix}$$

(5) Find an invertible matrix Q such that $A = QMQ^{-1}$ where
$$A = \begin{bmatrix} 3 & 2 & -4 \\ 0 & 3 & 1 \\ 0 & 0 & 3 \end{bmatrix} \text{ and } M = \begin{bmatrix} 3 & 0 & 0 \\ 1 & 3 & 0 \\ 0 & 1 & 3 \end{bmatrix}$$
[*Hint:* Let $B = A - 3I$ and use a chain basis for B.]

(6) This exercise shows that triangularizations may be used to compute powers of matrices in much the same way as diagonalizations.

    (a) For the matrix M in Example 2 from the text, find a formula for M^k. (This may require some experimentation.)

    (b) Use part (a), together with formula (2), to find a formula for A^k where A is as in Example 2.

(7) In Example 4, Section 6.1, we attempted to compute $A^{100}B$ where $B = [1, 1, 1]^t$ and
$$A = \begin{bmatrix} 3 & 1 & 1 \\ 0 & 3 & 1 \\ 0 & 0 & 5 \end{bmatrix}$$

We were unsuccessful because B is not a linear combination of eigenvectors of A. We can now solve this problem. For this let $Q_2 = [1, 0, 0]^t$ and $Q_3 = [3, 2, 4]^t$. In Section 6.1, we saw that Q_2 and Q_3 are eigenvectors for A corresponding to the eigenvalues 3 and 5, respectively.

(a) Let $B = A - 3I$. Find a specific vector Q_1 such that $BQ_1 = Q_2$.
(b) Show that the matrix of the transformation defined by A with respect to the ordered basis formed by Q_1, Q_2, and Q_3 is

$$M = \begin{bmatrix} 3 & 0 & 0 \\ 1 & 3 & 0 \\ 0 & 0 & 5 \end{bmatrix}$$

(c) Find a formula for M^k. (This may require some experimentation.) Use your answer to find a formula for A^k and $A^k B$.

(8) Let A be a 3×3 matrix. Let Q_1 be a vector in $\mathbb{R}^3$ and set $Q_2 = AQ_1$ and $Q_3 = AQ_2$. Suppose that $Q_3 \neq 0$ but $AQ_3 = 0$.

(a) Show that the Q_i form an independent set in $\mathbb{R}^3$. [*Hint:* Multiply both sides of the dependency equation by A and A^2.]
(b) Find the matrix M of the transformation described by A with respect to the given basis.
(c) Prove that A is nilpotent of degree 3. [*Hint:* $A = QMQ^{-1}$.]

(9) Let A be an $n \times n$ matrix. Let Q_1 be a vector in $\mathbb{R}^n$ and set $Q_2 = AQ_1$, $Q_3 = AQ_2$ and, in general, $Q_{k+1} = AQ_k$. Suppose that $Q_n \neq 0$ but $AQ_n = 0$.

(a) Show that the Q_i form an independent set in $\mathbb{R}^3$. [*Hint:* Multiply both sides of the dependency equation by A^{n-1}.]
(b) Find the matrix M of the transformation described by A with respect to the given basis. Explain why it is lower triangular.

(10) Suppose that A and B are $n \times n$ matrices such that $A = QBQ^{-1}$ for some $n \times n$ invertible matrix Q. Show that B is the matrix of the linear transformation defined by A where we use the columns of Q as a basis of both the domain and target spaces for A. [*Hint:* It suffices to show that for all X in $\mathbb{R}^n$, the coordinate vector for AX is BX' where X' is the coordinate vector for X.]

(11) In the text, it was commented that we usually use the same basis for the domain and target space when these spaces are equal. One common exception to this rule involves the identity matrix.

(a) Compute the matrix M of the transformation defined by the 2×2 identity matrix I where we use $Q_1 = [1, 1]^t$ and $Q_2 = [1, -1]^t$ as an ordered basis for the domain and where we use $P_1 = [1, 0]^t$ and $P_2 = [1, 1]^t$ as an ordered basis for the range. [*Hint:* Use formula (2).]

(b) Let $X = [3, 2]^t$ and let X' and X'' be, respectively, the coordinate vectors for X with respect to the Q_i and P_i bases. Compute X' and X'' and show that $M X' = X''$ where M is as in part (a). Use formula (1) to explain why this is true.

(12) Let V be a vector space and let I be the transformation of V into itself defined by stipulating that $I(X) = X$ for all X in V. (This transformation is called the "identity transformation.") Suppose that we are given two bases for V: B_1 and B_2. Let M be the matrix for I where we consider B_1 as the domain basis and B_2 as the target space basis. Prove that for all X in V, $M X' = X''$ where X' is the coordinate vector for X with respect to B_1 and X'' is the coordinate vector for X with respect to X''. [*Hint:* See Theorem 1.]

> Remark: The matrix M is called the **change of basis matrix**.

(13) Let P_2 denote the vector space of all polynomials of degree less than or equal to 2. Let $D : P_2 \to P_2$ be the differentiation transformation $D(y) = y'$.

 (a) Compute the matrix M of D with respect to the standard basis of P_2.
 (b) Find a basis for the nullspace of M. (The nullspace is one-dimensional.) Show that if you interpret the basis vector as the coordinate vector for an element $p(x)$ of P_2, then $D(p(x)) = \mathbf{0}$. Explain, on the basis of Theorem 1, why this should be true.
 (c) Find a basis for the image of M. Use your answer to find a basis for the image of D. Find explicit polynomials which D transforms onto your basis elements.

(14) Repeat Exercise 13 with P_3 in place of P_2.

(15) The finite difference operator is the operator Δ on functions defined by

$$\Delta(f)(x) = f(x + 1) - f(x).$$

Thus, for example,

$$\Delta(x^2) = (x + 1)^2 - x^2 = 2x + 1.$$

Consider Δ as a transformation of P_2 into itself. Find the matrix M which describes Δ relative to the standard basis of P_3. Use your answer to find bases for the nullspace and image of Δ. How does the dimension of these spaces relate to the rank of M?

(16) Repeat Exercise 15 with P_3 in place of P_2.

(17) Find the matrix of the operator Δ from Exercise 15 with respect to the following basis of P_2.

$$p_0(x) = 1, \quad p_1(x) = x - 1, \quad p_2(x) = x(x - 1)$$

[*Hint:* Show that $\Delta(p_2) = 2p_1$ and $\Delta(p_1) = p_0$.)] Compare your answer with that from Exercise 13. Use your answer to find bases for the nullspace and image of Δ.

(18) Repeat Exercise 17 with P_3 in place of P_2 using the following basis

$$p_0(x) = 1, \quad p_1(x) = x - 1, \quad p_2(x) = x(x - 1), \quad p_3(x) = x(x - 1)(x - 2)$$

[*Hint:* Show that $\Delta(p_3) = 3p_2$, $\Delta(p_2) = 2p_1$ and $\Delta(p_1) = p_0$.]

6.5 QUADRATIC FORMS: ORTHOGONAL DIAGONALIZATION

In analytic geometry, one often studies curves defined by equations of the form

$$ax^2 + bxy + cy^2 = d \tag{1}$$

where a, b, c, and d are constants. If $b = 0$, then the equation takes the form

$$ax^2 + cy^2 = d$$

In this case, the curve is typically either an ellipse or a hyperbola in standard form. For example

$$x^2 + \frac{y^2}{9} = 1$$

is an ellipse while

$$x^2 - \frac{y^2}{9} = 1$$

is a hyperbola.

In the general ($b \neq 0$) case, the equation still is (usually) an ellipse or a hyperbola, rotated by some fixed angle in the xy-plane. In this section we describe how linear algebra can be used to find rotated coordinates that realize the curve via an equation in standard form.

To make the connection with linear algebra, we first express our equation in what we call "symmetric form":

$$ax^2 + \frac{b}{2}xy + \frac{b}{2}yx + cy^2 = d$$

We factor x from the first two terms and y from the second two:

$$x\left(ax + \frac{b}{2}y\right) + y\left(\frac{b}{2}x + cy\right) = d$$

This may be expressed as a matrix product as

$$d = \begin{bmatrix} x & y \end{bmatrix} \begin{bmatrix} ax + \frac{b}{2}y \\ \frac{b}{2}x + cy \end{bmatrix} = \begin{bmatrix} x & y \end{bmatrix} \begin{bmatrix} a & \frac{b}{2} \\ \frac{b}{2} & c \end{bmatrix} \begin{bmatrix} x \\ y \end{bmatrix}$$

Thus, our curve is described by the matrix equality

$$d = X^t A X$$

where $X = [x, y]^t$ and

$$A = \begin{bmatrix} a & \frac{b}{2} \\ \frac{b}{2} & c \end{bmatrix} \tag{2}$$

▶ **EXAMPLE 1:**   Plot the graph of the following equation

$$3x^2 + 2xy + 3y^2 = 4$$

Solution: As before, we write this equation as

$$4 = 3x^2 + xy + yx + 3y^2 = x(3x + y) + y(x + 3y) = \begin{bmatrix} x & y \end{bmatrix} \begin{bmatrix} 3 & 1 \\ 1 & 3 \end{bmatrix} \begin{bmatrix} x \\ y \end{bmatrix}$$

Let A be the 2×2 matrix in the equation.

Now, suppose that we rotate the natural coordinates by some fixed angle. As Figure 1 indicates, the new coordinates will be defined by an orthonormal basis $\{P_1, P_2\}$. (Recall that orthonormal bases consist of mutually perpendicular vectors of length one.) Let $P = [P_1, P_2]$ be the point matrix for this basis. Note that saying that the basis is orthonormal is the same as saying that P is an orthogonal matrix.

The formula for our curve in the rotated coordinates is easily described. Let $X = [x, y]^t$ have coordinate vector $X' = [x', y']^t$ so that $X = PX'$. Our curve, then, is described by

$$4 = 3x^2 + 2xy + 3y^2 = X^t A X = (PX')^t A P X' = (X')^t P^t A P X'.$$

Thus, in the new coordinates, the curve is described by the matrix $P^t A P$.

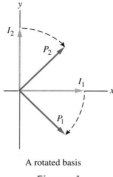

A rotated basis

Figure 1

Now comes the main point. The curve in formula (1) is in standard form if and only if $b = 0$. From formula (2), this is the same as saying that the matrix that represents it is diagonal. Hence, our curve will be in standard form in the rotated coordinates if and only if $P^t A P$ is diagonal. Since $P^t = P^{-1}$, this is the same as

$$P^{-1}AP = D$$

where D is a diagonal matrix. But this equation is equivalent with $A = PDP^{-1}$, which is our usual diagonalization equation. Hence, we see that we can solve our problem by finding an orthonormal, diagonalizing basis for A. Furthermore, the matrix that represents the curve will be D, the diagonal form of A.

Diagonalizing A is simple. In fact, this particular matrix was considered in Example 2, Section 6.1. It has eigenvectors $Q_1 = [1, -1]^t$ and $Q_2 = [1, 1]^t$ corresponding to the eigenvalues 2 and 4, respectively. The corresponding diagonal matrix is

$$D = \begin{bmatrix} 2 & 0 \\ 0 & 4 \end{bmatrix}$$

It is easily computed that $Q_1 \cdot Q_2 = 0$; hence these vectors form an orthogonal basis for $\mathbb{R}^2$. This basis is not normal, since $|Q_i| = \sqrt{2}$. Thus, we normalize, setting

$$P_1 = [1, -1]^t/\sqrt{2} \text{ and } P_2 = [1, 1]^t/\sqrt{2}$$

From the preceding discussion, it is clear that if we use the P_i to define new coordinates, our curve will be described by the formula

$$4 = (X')^t DX' = \begin{bmatrix} x' & y' \end{bmatrix} \begin{bmatrix} 2 & 0 \\ 0 & 4 \end{bmatrix} \begin{bmatrix} x' \\ y' \end{bmatrix} = 2(x')^2 + 4(y')^2$$

which describes an ellipse with intercepts $x' = \pm\sqrt{2}$ and $y' = \pm 1$. Its graph is indicated in Figure 2:

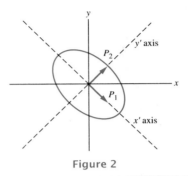

Figure 2 ◀

In Example 1, we seem to have been very lucky in that the eigenvectors $[1, -1]^t$ and $[1, 1]^t$ turned out to be perpendicular to each other. If this had not been the case, we would not have been able to find an orthonormal diagonalizing basis for A; hence we would not have been able to solve our problem. This actually was not luck at all. Notice that the matrix A of formula (2) satisfies

$$A = A^t$$

Matrices that satisfy this property are called **symmetric**. The following simple theorem explains our "stroke of luck."

Theorem 1.   *Eigenvectors corresponding to distinct eigenvalues of a symmetric matrix A are orthogonal.*

Proof: Let X and Y be two eigenvectors for A that correspond to eigenvalues λ and β where $\lambda \neq \beta$. Note that

$$AX \cdot Y = (AX)^t Y = X^t A^t Y = X^t(AY) = X \cdot AY \tag{3}$$

It follows that

$$(\lambda X) \cdot Y = (X \cdot \beta Y)$$

and thus

$$0 = (\lambda - \beta)(X \cdot Y)$$

Since $\lambda \neq \beta$, this implies $X \cdot Y = 0$, proving the perpendicularity.   ■

Comets and planets orbits are approximated by quadratic curves relative to appropriately chosen coordinates in the plane of motion.

This theorem tells us that if $A = A^t$ and A has n distinct eigenvalues, then any diagonalizing basis for A will consist of mutually perpendicular vectors, hence will form an orthogonal basis. By normalizing each basis element, we may construct an orthonormal basis, hence an orthogonal matrix P such that $P^t A P$ is diagonal. Actually, as long as $A = A^t$, we can find such a matrix even if A has fewer than n eigenvalues. The following is one of the most important theorems of linear algebra. It has applications far beyond the sphere of quadratic curves. We will prove it at the end of this section.

Spectral Theorem.   Let A be an n × n, real symmetric matrix. Then A has an orthonormal diagonalizing basis. Hence, there is an orthogonal matrix P such that $P^t A P$ is diagonal.   ■

This theorem tells us that the analysis we did in Example 1 generalizes to higher dimensions with almost no modification. In general, if $X = [x_1, x_2, \ldots, x_n]^t \in \mathbb{R}^n$, and if $A = [a_{ij}]$ is an $n \times n$ matrix, then the solution set to

$$\sum a_{ij} x_i x_j = d$$

is called a **quadratic variety**. In matrix language, we may write this equation as

$$X^t A X = d$$

The function which transforms X into $X^t A X$ is referred to as a quadratic form. The matrix A is referred to as the matrix of the form.

Given any quadratic form, we may always find a symmetric matrix A which represents it. The quadratic form is said to be in standard form if its matrix is diagonal. Hence, the corresponding variety will be represented by an equation of the form

$$a_{11}x_1^2 + a_{22}x_2^2 + \cdots + a_{nn}x_n^2 = d$$

An almost immediate consequence of the spectral theorem is the following result:

Principal Axis Theorem.   Let a quadratic variety be defined by the equation

$$d = X^t A X$$

where $X = [x_1, x_2, \ldots, x_n]^t$, A is an $n \times n$ symmetric matrix, and d is a scalar. Then the corresponding quadratic form is in standard form relative to the coordinates defined by any orthonormal diagonalizing basis for A. Furthermore, in these coordinates, the variety is given by

$$\lambda_1 (x_1')^2 + \lambda_2 (x_2')^2 + \cdots + \lambda_n (x_n')^2 = d \qquad (4)$$

where the λ_i are the eigenvalues of A (listed in an order consistent with the ordering of the basis).

Proof: Let the variety be defined by

$$X^t A X = d$$

where A is symmetric. From the spectral theorem, there is an orthogonal matrix P such that $D = P^t A P$ is diagonal. The columns of P form an orthonormal basis for $\mathbb{R}^n$. If $X \in \mathbb{R}^n$ has coordinates X' with respect to this basis, then $X = PX'$. Hence

$$X^t A X = (PX')^t A P X' = (X')^t P^t A P X' = (X')^t D X'$$

This is equivalent with formula (4).   ∎

▶ EXAMPLE 2:   What is the surface described by the following equation?

$$2x_1^2 + 2x_2^2 + 2x_3^2 + 6\sqrt{2}x_1x_2 - 6\sqrt{2}x_2x_3 = 8$$

Solution: We first write the equation in symmetric form:

$$2x_1^2 + 2x_2^2 + 2x_3^2 + 3\sqrt{2}x_1x_2 + 3\sqrt{2}x_2x_1 - 3\sqrt{2}x_2x_3 - 3\sqrt{2}x_3x_2 = 8$$

This may be expressed as

$$8 = \begin{bmatrix} x_1 & x_2 & x_3 \end{bmatrix} \begin{bmatrix} 2 & 3\sqrt{2} & 0 \\ 3\sqrt{2} & 2 & -3\sqrt{2} \\ 0 & -3\sqrt{2} & 2 \end{bmatrix} \begin{bmatrix} x_1 \\ x_2 \\ x_3 \end{bmatrix}$$

Let A be the 3×3 matrix in the equation. We find that

$$\det(A - \lambda I) = (2 - \lambda)(\lambda^2 - 4\lambda - 32) = (2 - \lambda)(\lambda + 4)(\lambda - 8)$$

Thus, our eigenvalues are 2, -4, and 8. From the spectral theorem, our variety is described by

$$2(x_1')^2 - 4(x_2')^2 + 8(x_3')^2 = 8$$

Our equation describes a hyperboloid of one sheet opening along the y'-axis. Roughly, it would appear as shown in Figure 3.

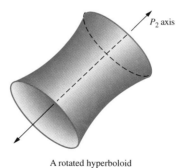

A rotated hyperboloid

Figure 3

Notice that we did not need to find the eigenvectors in order to identify the surface. Had we been asked for coordinates which produce standard form, then we would have had to compute an orthonormal diagonalizing basis. In this particular case, this would not have been difficult. Since A has distinct eigenvalues, any diagonalizing basis for A will automatically form an orthogonal basis (from Theorem 1). Thus, we need only to find a diagonalizing basis and then normalize it. If A had not had distinct eigenvalues, then we probably would have needed to use the Gram–Schmidt process from Section 4.2 to produce an orthogonal diagonalizing basis, as the next example demonstrates.

▶ EXAMPLE 3: Find a diagonal matrix D and an orthogonal matrix P such that $A = PDP^t$ where A is as follows:

$$A = \begin{bmatrix} 2 & 2 & 6 & -2 \\ 2 & -1 & 3 & -1 \\ 6 & 3 & 7 & -3 \\ -2 & -1 & -3 & -1 \end{bmatrix}$$

Solution: Using a computer, we discover that

$$\det(A - \lambda I) = -104 - 148\lambda - 66\lambda^2 - 7\lambda^3 + \lambda^4$$

which factors as $(\lambda - 13)(\lambda + 2)^3$. Thus, the eigenvalues are 13 and -2. We find bases for the eigenspaces as usual, finding that a basis for the -2 eigenspace is formed by the vectors

$$X_1 = [-1, 2, 0, 0]^t \quad X_2 = [-3, 0, 2, 0]^t \quad X_3 = [1, 0, 0, 2]^t$$

while the 13 eigenspace is spanned by $Y = [-2, -1, -3, 1]^t$. Direct computation shows that the vector Y is perpendicular to the vectors X_1, X_2, and X_3. This is only to be expected in light of Theorem 1.

The vectors X_1, X_2, and X_3 are not perpendicular to each other. We can, however, use the Gram–Schmidt process to produce an orthogonal basis for their span. Actually, these three vectors were considered in Example 3 in Section 4.2, where they were denoted A_1, A_2, and A_3. There, we found that the following vectors form an orthogonal basis for their span:

$$Q_1 = [-1, 2, 0, 0]^t \quad Q_2 = \frac{2}{5}[-6, -3, 5, 0]^t \quad Q_3 = [2, 1, 3, 14]^t / 7$$

The Q_i are, of course, perpendicular to Y since, again, eigenvectors corresponding to different eigenvalues are perpendicular. This means that the set $\{Q_1, Q_2, Q_3, Y\}$ forms an orthogonal, diagonalizing basis for A. We obtain an orthonormal basis by normalizing each element of this basis. The basis elements then become the columns of P. We find that

$$P = \frac{1}{\sqrt{5}} \begin{bmatrix} -1 & \frac{-6}{\sqrt{14}} & \frac{2}{\sqrt{42}} & \frac{-2}{\sqrt{3}} \\ 2 & \frac{-3}{\sqrt{14}} & \frac{1}{\sqrt{42}} & \frac{-1}{\sqrt{3}} \\ 0 & \frac{5}{\sqrt{14}} & \frac{3}{\sqrt{42}} & \frac{-3}{\sqrt{3}} \\ 0 & 0 & \frac{14}{\sqrt{42}} & \frac{1}{\sqrt{3}} \end{bmatrix}$$

The diagonal matrix in question is

$$D = \begin{bmatrix} -2 & 0 & 0 & 0 \\ 0 & -2 & 0 & 0 \\ 0 & 0 & -2 & 0 \\ 0 & 0 & 0 & 13 \end{bmatrix}$$

◀

We end this section by proving two fundamental theorems, the spectral theorem and the theorem that states that every eigenvalue of a symmetric matrix is real (Theorem 2). We prove the spectral theorem first, despite the fact that our proof uses Theorem 2. We do so because the proof of Theorem 2 requires some results from Section 6.3. A student who has not studied Section 6.3 can still follow the proof of the spectral theorem, as long as he/she is willing to assume Theorem 2.

Proof of the Spectral Theorem.   We shall do a proof by contradiction. This means that we will imagine that the spectral theorem might fail for some matrices. We will then show that this assumption leads to nonsense, proving that the theorem must be true for all matrices.

Thus, assume that the spectral theorem is false for some matrices. Let A be an $n \times n$ matrix for which it fails. Assume that A is chosen so that its size is as small as possible. Then the theorem must be true for all smaller matrices. Notice that $n > 1$ since the theorem is clearly true for 1×1 matrices.

From Theorem 2, A has at least one real eigenvector, X_0, which we normalize to have length 1. Let λ_0 be the corresponding eigenvalue. Let $\mathcal{W}$ be the set of vectors W such that $X_0 \cdot W = 0$. Since $X_0 \cdot W = (X_0)^t W$, $\mathcal{W}$ is the nullspace of the $1 \times n$ matrix X_0^t. Since this matrix has rank 1, $\mathcal{W}$ is an $n - 1$ dimensional subspace of $\mathbb{R}^n$. Since $n > 1$, $\mathcal{W}$ is non-zero.

In Section 4.2, we said that any k-dimensional subspace of $\mathbb{R}^n$ can be thought of as being equal to $\mathbb{R}^k$. Thus, we can think of $\mathcal{W}$ as being $\mathbb{R}^{n-1}$. More precisely, we showed in Theorem 2, Section 4.2, that if we use an orthonormal basis to define coordinates for $\mathcal{W}$, then the transformation which transforms each X in $\mathcal{W}$ to its coordinate vector X' in $\mathbb{R}^{n-1}$ is a one-to-one, onto, linear transformation. Furthermore, for all X and Y in $\mathcal{W}$,

$$X \cdot Y = X' \cdot Y' \tag{5}$$

We shall prove that multiplication by A transforms $\mathcal{W}$ into itself. This, then, defines a linear transformation of $\mathbb{R}^{n-1}$ into itself. From the matrix representation theorem in Section 3.1 (or better, from Theorem 1 in Section 6.4) there is an $(n-1) \times (n-1)$ matrix M such that for all X in $\mathcal{W}$,

$$(AX)' = MX' \tag{6}$$

We will show that M is a *symmetric* matrix. Then, since the size of M is smaller than that of A, the spectral theorem is true for M. Thus, there is an orthonormal basis $\{P_1', P_2', \ldots, P_{n-1}'\}$ for $\mathbb{R}^{n-1}$ consisting of eigenvectors for M. Let P_i denote the corresponding vectors in $\mathcal{W}$. Then, from formula (5), the P_i form an orthonormal basis for $\mathcal{W}$ and from formula (6)

$$(AP_i)' = MP_i' = \lambda_i P_i' = (\lambda_i P_i)'$$

where λ_i are scalars. It follows that $AP_i = \lambda_i P_i$, showing that P_i are eigenvectors for A.

Note that X_0 is perpendicular to the P_i since the P_i all belong to $\mathcal{W}$. Thus the set $\{X_0, P_1, P_2, \ldots, P_{n-1}\}$ is an orthonormal diagonalizing basis for A. Hence, the spectral theorem is true for A! But we assumed that it was false for A. This contradiction means that there are no matrices for which it is false.

To finish our proof, we must prove (i) that multiplication by A transforms $\mathcal{W}$ into itself and (ii) that M is symmetric. To prove (i), let $W \in \mathcal{W}$. Then, from the reasoning in formula (3),

$$X_0 \cdot AW = AX_0 \cdot W = (\lambda_0 X_0) \cdot W = \lambda_0 (X_0 \cdot W) = 0$$

Since AW is perpendicular to X_0, it belongs to $\mathcal{W}$.

For (ii), we note that from the reasoning in formula (3), for all X and Y in $\mathbb{R}^n$,

$$AX \cdot Y = X \cdot AY$$

It follows from formulas (5) and (6) that for all X' and Y' in $\mathbb{R}^{n-1}$,

$$M(X') \cdot Y' = X' \cdot M(Y') \tag{7}$$

We leave it as an exercise for the reader to prove that this implies that M must be symmetric. This finishes our proof. ∎

Theorem 2. *Let A be a real, symmetric, $n \times n$ matrix. Then every eigenvalue of A is real. In particular, A has at least one real eigenvector.*

Proof: Suppose that X is a complex eigenvector for A corresponding to the eigenvalue λ. Then, $\overline{X}$ is also an eigenvector for A corresponding to $\overline{\lambda}$. Assume that λ is not real so that $\lambda \neq \overline{\lambda}$.

In Theorem 1, we proved that for a symmetric matrix, if X and Y are eigenvectors corresponding to different eigenvalues, then $X \cdot Y = 0$. An inspection of the proof shows that the proof is valid for complex eigenvectors as well. It follows that $X \cdot \overline{X} = 0$. If $X = [z_1, z_2, \ldots, z_n]^t$, then,

$$0 = X \cdot \overline{X} = z_1 \overline{z}_1 + z_2 \overline{z}_2 + \cdots + z_n \overline{z}_n \tag{8}$$

If $z_k = x_k + iy_k$, then (as the reader may verify)

$$z_k \overline{z}_k = x_k^2 + y_k^2 \geq 0$$

Thus, the only way for equation (8) to hold is for all x_k and y_k to be zero, proving that $X = \mathbf{0}$. But, eigenvectors must be non-zero. Thus, λ must be real.

Finally, to prove the last statement in the theorem, we know that A has at least one eigenvalue λ which, from the above reasoning, must be real. Then the matrix $A - \lambda I$ has determinant zero which means that there is a non-zero, real vector X such that $(A - \lambda I)X = \mathbf{0}$. Thus, A has a real eigenvector. ∎

EXERCISES

(1) In Example 1, we found a matrix A such that the quadratic curve was described by $X^t A X = d$. Show that the matrix

$$A' = \begin{bmatrix} 3 & 0 \\ 2 & 3 \end{bmatrix}$$

also describes the same curve. What are all matrices that describe this curve? Show that the matrix A from Example 1 is the only *symmetric* matrix that describes this curve.

(2) Express each of the varieties below in standard form. (See Example 2.) Give an orthonormal basis which defines the coordinates.

(a) $2x^2 + 4xy - y^2 = 1$

(b) $x_1^2 + x_2^2 + x_3^2 + \sqrt{2}x_1 x_2 + \sqrt{2}x_2 x_3 = 4$

(c) $18x^2 + 5y^2 + 12z^2 - 12xy - 12yz = 7$

(3) In Example 1, suppose that the eigenvalues had been 2 and -4 instead of 2 and 4. What (geometrically) would the curve have represented? What if they were -2 and -4? State a general theorem in terms of the sign of the eigenvalues concerning what kind of figure a given 2×2 matrix will represent. Try to cover all cases.

(4) Express each curve (a)–(c) in standard form. (You need not find the corresponding orthonormal basis.) What, geometrically, does each represent?

(a) $x^2 + xy + 2y^2 = 1$

(b) $x^2 + 4xy + 2y^2 = 1$
(c) $x^2 + 4xy + 4y^2 = 1$

(5) Find a, b, and c such that the curve defined by the equation below describes an ellipse that has its major axis of length 6 on the line spanned by the vector $[3, 4]$ and its minor axis of length 4. [*Hint:* If you can find Q and D, then you can let $A = QDQ^{-1}$.]

$$ax^2 + bxy + cy^2 = 1$$

(6) In Exercise 5, find a, b, and c such that the curve defined by the given equation represents a hyperbola with one axis on the stated line. What is the formula for your hyperbola in rotated coordinates?

(7) Let a, b, and c be numbers with $a + c > 0$. Prove that the formula $ax^2 + 2bxy + cy^2 = 1$ represents an ellipse if and only if $b^2 < ac$. What does the curve represent if $b^2 = ac$?

(8) Let A be as shown. Find an orthogonal matrix P and a diagonal matrix D such that $P'AP = D$. To save you time, we have provided the eigenvalues (there are only two).

$$A = \begin{bmatrix} 3 & 1 & 1 & 2 \\ 1 & 3 & 1 & 2 \\ 1 & 1 & 3 & 2 \\ 2 & 2 & 2 & 6 \end{bmatrix} \quad \lambda_1 = 9, \quad \lambda_2 = 2$$

(9) A symmetric matrix A is said to be positive definite if $X'AX > 0$ for all $X \neq 0$. Show that the matrix A from Example 1 is positive definite.

(10) Prove that a symmetric matrix A is positive definite if and only if all its eigenvalues are positive.

(11) Find a 2×2 matrix that has only positive entries and is not positive definite.

(12) Show that the matrix A in Example 2 is not positive definite by explicitly exhibiting a vector X such that $X'AX < 0$.

(13) Let B be an $m \times n$ matrix. Let $A = B'B$. Prove that A is symmetric. Prove that $X'AX \geq 0$ for all X. This condition is called "positive semidefinite." What does this condition tell you about the eigenvalues of A?

(14) Let A be an $n \times n$ matrix. Suppose that there is an *orthogonal* matrix P and a diagonal matrix D such that $A = PDP^{-1}$. Prove that A must be symmetric.

Thus, symmetric matrices are the only matrices that may be diagonalized using an orthogonal diagonalizing basis.

(15) Let A be an $n \times n$ matrix. Let X be an eigenvector for A with eigenvalue λ and let Y be an eigenvector for A^t with eigenvalue β. Show that if $\lambda \neq \beta$, then $X \cdot Y = 0$.

(16) Let B be an $n \times n$ matrix such that for all X and Y in $\mathbb{R}^n$,

$$BX \cdot Y = X \cdot BY$$

Prove that B must be symmetric. [*Hint:* What does the above equation say if $X = I_j$ and $Y = I_k$ where $\{I_1, I_2, \ldots, I_n\}$ is the standard basis for $\mathbb{R}^n$?]

 ## ON LINE

The purpose of this exercise set is to check graphically a few of the general principles described above. We begin by graphing the quadratic variety

$$2y^2 + xy + x^2 = 1$$

To graph it, we first must express y as a function of x. We do so by writing the formula for the variety as

$$2y^2 + xy + (x^2 - 1) = 0$$

We then solve for y using the quadratic equation:

$$y = \frac{-x \pm \sqrt{x^2 - 8(x^2 - 1)}}{4} = \frac{-x \pm \sqrt{8 - 7x^2}}{4}$$

Notice that the square root is real for $-\sqrt{8/7} \leq x \leq \sqrt{8/7}$.
We plot our curve in two pieces with:

```
axis equal;  grid on;  hold on;
fplot('(-x+sqrt(8-7*x^ 2))/4',[-1.08,1.08]);
fplot('(-x-sqrt(8-7*x^ 2))/4',[-1.08,1.08]);
```

We see that the graph appears to be a rotated ellipse. Do the following exercises:

(1) According to the general theory, the major and minor axes of the ellipse should lie along the coordinate axes determined by the eigenvectors for the matrix A that describes the ellipse. To verify this in MATLAB, enter the matrix A for the ellipse and then set "[Q,D]=eig(A)". The matrix Q is then a matrix whose columns are eigenvectors of A. They are even normalized to have length 1. We plot the corresponding basis vectors (in red) with the commands:

```
V(:,1)=Q(:,1); V(:,3)=Q(:,2)
plot(V(1,:),V(2,:), 'r')
```

Get this plot printed.

(2) On the printed plot from Exercise 1, draw in the axes determined by the eigenvectors. On these axes, put tick marks every quarter unit, noting that each eigenvector is one unit long. Use a ruler to guarantee the accuracy of your tick marks. According to the general theory, the ellipse should cross the new axes at the points $(\pm\lambda_1, 0)$ and $(0, \pm\lambda_2)$ where λ_1 and λ_2 are eigenvalues of A. Verify this by reading the appropriate coordinates off of your graph.

(3) Find a formula for the ellipse centered at $[0, 0]^t$ whose major axis is 4 units long and lies along the line determined by the vector $[3, 4]^t$ and whose minor axis is 2 units long.

[*Hint:* If you can figure out what Q and D should be, then you can set $A = QDQ^t$.]

Get MATLAB to graph the ellipse as well as the unit eigenvectors for the corresponding symmetric matrix. Draw in the axes determined by the eigenvectors and put tick marks on them as before to demonstrate that your answer is correct.

6 CHAPTER SUMMARY

This chapter was devoted to the study of *eigenvalues* and *eigenvectors*, which we began in Section 6.1. An eigenvector for an $n \times n$ matrix A is a non-zero column vector X such that $AX = \lambda X$ for some scalar λ (the eigenvalue). The computation of eigenvalues is done in two steps. First we find the roots of the polynomial $p(\lambda) = \det(A - \lambda I)$ (the *characteristic polynomial*). These are the eigenvalues. Next, for each eigenvalue λ, we find all solutions to $(A - \lambda I)X = \mathbf{0}$. These are the eigenvectors. Usually, we are most interested in finding a basis for this solution space.

Our first application of these concepts was to computing formulas for $A^m B$ where B was a given column vector and A is an $n \times n$ matrix. The idea was that if we could find scalars c_i such that

$$B = c_1 X_1 + \cdots + c_k X_k$$

where X_i are eigenvectors, then

$$A^m B = c_1 \lambda_1^m X_1 + \cdots + c_k \lambda_k^m X_k$$

where the λ_i are the eigenvalues corresponding to the X_i. (See Example 2, Section 6.1.) This can be done for all B in $\mathbb{R}^n$ if the X_i form a basis for $\mathbb{R}^n$, in which case we say the A

is *diagonalizable* (over $\mathbb{R}$). Matrices which are not diagonalizable over $\mathbb{R}$ are *deficient* over $\mathbb{R}$.

The term diagonalizable actually arose in Section 6.2 where we noted that if A has a basis of eigenvectors, then the linear transformation defined by A has a particularly simple description in terms of the coordinates defined by this basis. Specifically, we saw that in these coordinates, A is defined by multiplication by a *diagonal* matrix D. (A matrix is diagonal if and only if all of the entries off of the main diagonal are zero.) This observation allowed us to prove the important result (Theorem 1, Section 6.2) that if A is diagonalizable, then $A = QDQ^{-1}$ where Q is the point matrix for the basis of eigenvectors and D is the diagonal matrix that has the corresponding eigenvalues as its diagonal entries. This formula allowed us to prove that $A^n = QD^nQ^{-1}$ (formula 2 in Section 6.2), which allowed us to compute powers of matrices.

These ideas were extended in Section 6.4 where we saw that many other linear transformations could be effectively studied using coordinates defined by suitably chosen bases (Theorem 1, Section 3.4). Specifically, we saw that we could *triangularize* many matrices which were not diagonalizable and that triangularizations could be used to compute matrix powers of non-diagonalizable matrices. In fact, there is a fundamental theorem that states that any matrix can be triangularized, provided we use the complex numbers. We also saw that matrices could be used to study linear transformations between vector spaces other than $\mathbb{R}^n$.

Initially, we worked only with real numbers. However, in Section 6.3 we were forced to work with complex numbers because the characteristic polynomial could have complex roots. The *fundamental theorem of algebra* (which says that every polynomial factors completely over the complex numbers) allowed us to prove that every matrix has at least one complex eigenvector. Moreover, the use of complex eigenvalues allowed us to compute powers some of matrices which did not diagonalize the real numbers.

In Section 6.5, we applied our analysis of eigenvectors and eigenvalues to the study of curves defined by equations of the form $ax^2 + bxy + cy^2 = d$ (*quadratic curves*). We saw that such curves could be written as

$$X^t M X = d$$

where $X = [x, y]^t$ and

$$M = \begin{bmatrix} a & \frac{b}{2} \\ \frac{b}{2} & c \end{bmatrix}$$

It turned out that such curves have a particularly simple form if we use coordinates defined by the eigenvectors of the matrix. Specifically, M has an orthonormal basis of eigenvectors and if the corresponding coordinates are denoted x' and y', then the curve is described by the equation $\lambda_1(x')^2 + \lambda_2(y')^2 = d$ where λ_i are the eigenvalues of M. The generalization of these statements to $n \times n$ *symmetric* matrices is the *spectral theorem*. (A matrix A is symmetric if $A = A^t$.) The general spectral theorem allows us to study *quadratic* forms in much the same manner done in two dimensions.

ANSWERS AND HINTS TO ODD-NUMBERED EXERCISES

SECTION 1.1

(1) (a) not equivalent (b) equivalent (c) not equivalent (d) not equivalent

(3) (a) not perpendicular (b) perpendicular (c) perpendicular (d) perpendicular

(5) The cone consisting of all vectors which make an angle of $30°$ or less with respect to the vector A.

(7) Any pair of points C and D such that $D - C = B - A$ e.g. $C = (1, 1, 1)$, $D = (1, 1, 7)$. (Now you find another pair.)

(9) Let (x, y) be a point on the line. Then $x = 2 + t$, $y = 1 + t$ so $y - 1 = t = x - 2$ or $y = x - 1$. Conversely suppose that (x, y) satisfies $y = x - 1$. Let $t = x - 2$. Then $x = 2 + t$, $y = 1 + t$, showing that (x, y) is a point on the line.

(11) All of $\mathbb{R}^2$.

(13) Find two points A and B on the line by assigning values to x. Then proceed as in Example 3 in the text. One correct answer is $(x, y) = (0, 2) + t(1, 3)$. (There are many correct answers.) For the proof note that $x = t$, $y = 2 + 3t = 2 + 3x$. Hence (x, y) satisfies the original equation. Conversely, if (x, y) satisfies the original equation, then $(x, y) = (0, 2) + t(1, 3)$ where $t = x$, proving that (x, y) is a point on the line.

(15) $(2, -2, 1)t + (1, -1, 1)$ works. So does $(2, -2, 1)t + (3, -3, 2)$

(17) $(3, 2, 6)$, but why? [*Hint*: Rename the parameter in the second line.]

(19) The point $(2, 4, 4)t + (4, 4, 8)$ on the first line is on the second line if there is a number s such that $(2, 4, 4)t + (4, 4, 8) = (1, 2, 2)s + (2, 0, 4)$. This is true if $s = 2t + 2$. (Check this.) Thus every point on the first line is on the second. Now you should prove that every point on the second is also on the first.

(21) (a) $A = (-1, 1, 0)$, $B = (-1, -1, -1)$, $C = (2, -1, 0)$ work (b) Compute $(2, 3, -6) \cdot (A - B)$ and $(2, 3, -6) \cdot (C - B)$. You should get 0. (c) $(x, y, z) - B = (x + 1, y + 1, z + 1)$ which is perpendicular to $(2, 3, 6)$ if and only if $0 = (x + 1, y + 1, z + 1) \cdot (2, 3, -6) = 2x + 3y - 6z - 1$, so the set of solutions is the plane through B perpendicular to $(2, 3, -6)$.

SECTION 1.2

(1) (a)
$$\begin{bmatrix} -1 & -4 & -7 & -10 \\ 1 & -2 & -5 & -8 \\ 3 & 0 & -3 & -6 \end{bmatrix}$$

third row: $[3, 0, -3, -6]$,

second column
$$\begin{bmatrix} -4 \\ -2 \\ 0 \end{bmatrix}$$

359

(b) $\begin{bmatrix} 1 & 8 \\ 4 & 32 \\ 9 & 72 \end{bmatrix}$ third row: [9, 72],

second column $\begin{bmatrix} 8 \\ 32 \\ 72 \end{bmatrix}$

(c) $\begin{bmatrix} \frac{1}{2} & -\frac{1}{2} \\ -\frac{1}{2} & -\frac{1}{2} \\ -1 & 0 \end{bmatrix}$

third row: [−1, 0], second column $\begin{bmatrix} -\frac{1}{2} \\ -\frac{1}{2} \\ 0 \end{bmatrix}$

(3) $C = A + B$.

(5) (a) $[1, 1, 4] = [1, 1, 2] + 2[0, 0, 1]$ (b) The second matrix is a linear combination of the other three. (The coefficient of the first matrix is zero.) (d) Compute the sum of the third and fourth matrices.

(7) Each vector has a non-zero entry in a position where the other two vectors have zeros.

(9) (a) Try a negative coefficient on X and a positive coefficient on Y (b) Let $[x, y, z] = aX + bY = [-a - b, a + 3b, -a + 2b]$ and substitute into $5x + 3y - 2z$. You should get 0. (c) Any point $[x, y, z]$ which does not solve the equation $5x + 3y - 2z = 0$ will work, e.g., $[1, 1, 1]$.

(11) For the first part, use various values of a, b and c in $aX + bY + cZ$. For the second part, what can you say about the (2, 1) entry of the general element of the span?

(13) Let V and W be elements of the span. Then $V = aX + bY$ and $W = cX + dY$. Simplify, $xV + yW$ until you obtain an expression of the form $eX + fY$ where e and f are scalars. This proves that $xV + yW$ is an element of the span of X and Y.

(15) $3\begin{bmatrix} 4 \\ 1 \\ 1 \end{bmatrix} - \begin{bmatrix} 6 \\ 2 \\ 1 \end{bmatrix} = \begin{bmatrix} 6 \\ 1 \\ 2 \end{bmatrix}$

(17) You could simply construct a matrix whose fourth row is the sum of the first three. Or you could make the second row a multiple of the first.

(19) (a) $\sinh x = \frac{e^x - e^{-x}}{2} = \frac{1}{4}(2e^x) - \frac{1}{6}(3e^{-x})$.
(b) Compute $\sinh x - \cosh x$.
(c) Look up the double angle formula for the cosine function.
(d) See the hint for (c). [*Note*: $\sin^2 x = 1 - \cos^2 x$]
(e) Use the angle addition formulas for $\sin x$ and $\cos x$.
(f) $(x + 3)^2 = x^2 + 6x + 9$
(h) Use the properties that $\ln \frac{a}{b} = \ln a - \ln b$ and $\ln a^b = b \ln a$.

(21) (a) Let $B = \begin{bmatrix} x & y \\ z & w \end{bmatrix}$. Then
$A + B = \begin{bmatrix} x + a & y + b \\ z + c & w + d \end{bmatrix} = \begin{bmatrix} x & y \\ z & w \end{bmatrix}$
Hence $x + a = x$, $y + b = y$, $z + c = c$, and $w + d = d$ which imply that $x = y = z = w = 0$. Hence $B = \mathbf{0}$. (b) Solved similarly to (a).

(23) In order, we used properties (c), (e), (d), (f), (g), (j).

(25) Here are the first four steps, but you still must put in the reasons.

$$-(aX) + (aX + (bY + cZ)) = -(aX) + \mathbf{0}$$
$$(-(aX) + aX) + (bY + cZ) = -(aX)$$
$$\mathbf{0} + (bY + cZ) = -1(aX)$$
$$bY + cZ = (-a)X$$

Next, multiply both sides by $(-a)^{-1}$ and simplify some more.

SECTION 1.3

(1) X is a solution. Y is not.

(3) $b = 0$, all a.

(5) (Roman numerals are equation numbers)
(a) Solution: $[-\frac{59}{9}, \frac{20}{9}, \frac{8}{9}]^t$, spanning: $\mathbf{0}$, translation: $[-\frac{59}{9}, \frac{20}{9}, \frac{8}{9}]^t$.

(b) Solution: the line $[1, 0, 0]^t + t[-\frac{17}{2}, \frac{5}{2}, 1]^t$,
spanning: $[-\frac{17}{2}, \frac{5}{2}, 1]^t$, translation:
$[1, 0, 0]^t$, $2I + II = III$ (c) Inconsistent:
2I+II contradicts III.
(g) Solution: the plane $[\frac{5}{4}, -\frac{1}{4}, 0, 0]^t +$
$r[-\frac{3}{4}, -\frac{1}{4}, 1, 0]^t + s[-1, 0, 0, 1]^t$, transla-
tion: $[\frac{5}{4}, -\frac{1}{4}, 0, 0]^t$, spanning: $[-\frac{3}{4}, -\frac{1}{4}, 1, 0]^t$
and $[-1, 0, 0, 1]^t$, IV=I+2II, III=4I-II.
(h) inconsistent

(7) *Hint:* For a line, you want only one free
variable. How many free variables do you
want for a plane?

(9) A point (x, y) solves the system if and only
if it lies on both lines. Since the lines are
parallel, there is no solution to the system.

(11) False.

SECTION 1.4

(1) (a) $\begin{bmatrix} 1 & 0 & -\frac{1}{2} & 0 & 5 \\ 0 & 1 & 1 & 0 & -1 \\ 0 & 0 & 0 & 1 & 2 \end{bmatrix}$

(c) $\begin{bmatrix} 1 & 0 & 0 & 0 \\ 0 & 1 & 0 & 0 \\ 0 & 0 & 1 & 0 \\ 0 & 0 & 0 & 1 \end{bmatrix}$

(3) (a) neither (b) echelon (c) neither

(5) (a) $\begin{bmatrix} 1 & 0 & 1 & 0 & 3 \\ 0 & 1 & -1 & 0 & 1 \\ 0 & 0 & 0 & 1 & 0 \end{bmatrix}$

(c) $\begin{bmatrix} 1 & 0 & \frac{10}{3} \\ 0 & 1 & \frac{1}{3} \end{bmatrix}$

(e) $\begin{bmatrix} 1 & 0 \\ 0 & 1 \end{bmatrix}$

(7) (a) In the span (b) in the span (c) not in the
span.

(9) This isn't hard. Just pick a vector at
random. The chances are that it won't
be in the span. To prove it, reason as in
Exercise 7.

(11) *Hint:* If the reduced form has two non-zero
rows, what must it be? If it has only one

non-zero row, then the first non-zero entry
in this row can be either the first or the
second entry. Finally, it might have no
non-zero entries.

(13) *Hint:* What is the simplest solution to
the system below that you can think of?
How does this prove that the system is
consistent?

$$x + 2y + 3z + 4w + 5t + 6u + 7v = 0$$
$$x - 3y + 7z - 3w + 3t + 9u - 6v = 0$$

(15) (a) No. (b) *Hint:* Can the system have any
free variables? (c) No. (d) No. (e) No.

(17) $x[1, 0, 1, -2]^t + y[0, 1, -1, 1]^t$

SECTION 1.5

(1) (a) $[0, 5, -11]^t$ (b) $[7, 10, 7, 5]^t$
(c) $[x_1 + 2x_2 + 3x_3, 4x_1 + 5x_2 + 6x_3]^t$

(3) For (c), for example, you might choose

$$x + 2y + 3z = 17$$
$$4x + 5y + 6z = -4$$

(5) The nullspace is spanned by : (a)
$\{[-1, 1, 1, 0, 0]^t, [-3, -1, 0, 0, 1]^t\}$
(c)$\{[-10, -1, 3]^t\}$ (e) $\{[0, 0]\}$

(9) See the hint for Exercise 13, Section 1.4.

(11) Yes, I am correct.

(13) True.

(15) (a) *Hint:* Let the equation be $ax + by +$
$cz = d$. Substitute various vectors that are
known to solve the system to determine
$a, b, c,$ and d. Question: If we multiply
the both sides of preceding equation by,
say, 2 we get a valid equation. What does
this tell you about the uniqueness of the
coefficients? (b) This is easy, once you
have done (a).

(17) No, the two answers are not consistent.

(19) (a) If W belongs to span $\{X, Y, Z\}$
then $W = aX + bY + cZ = aX + bY +$
$c(2X + 3Y) = (a + 2c)X + (b + 3c)Y$
which belongs to span $\{X, Y\}$. Conversely, if W belongs to span $\{X, Y\}$,

then $W = aX + bY = aX + bY + 0Z$ which belongs to span $\{X, Y, Z\}$. Thus, the two sets have the same elements and are therefore equal.

(23) $\mathcal{W}$ is the first quadrant in $\mathbb{R}^2$. $\mathcal{W}$ is not a subspace.

(25) Only lines through the origin. (Why?)

(29) *Hint:* Does $\left(\frac{2}{3}\right)X + \left(\frac{1}{3}\right)Y$ work? Does $2X + 3Y$ work?

SECTION 2.1

(1) (a) independent (c) independent (e) dependent: The third matrix is -3 times the first plus 4 times the second. (g) independent

(3) Dependent.

(5) *Hint:* Set up the dependency equation as a system of linear equations and look at the first, third and fifth equations.

(7) In each case, the first, third and fifth columns work.

(9) *Hint:* Consider formula (M1) on p. 61.

SECTION 2.2

(1) Let $B = [2, 7]^t$. Then $B = \frac{20}{7}A_1 - \frac{3}{7}A_2$. (Other answers are possible.)

(3) *Hint:* What, geometrically, is the span of $[1, 2]^t$?

(5) Yes.

(7) The dimension is 2.

(9) The dimension is 6.

(11) The dimension is 6.

(13) (a) The first two vectors form a basis for the span which has dimension 2. (b) The first three vectors form a basis for the span which has dimension 3. (c) These vectors are independent and form a basis for the span which has dimension 3.

(15) See Exercise 16 for a hint.

(17) (b) Yes, it is always possible, but why?

(19) $\{A, B, C\}$ is dependent, but why? $\{B, C\}$ need not be a basis of $\mathcal{W}$. Why?

(31) The dimension of P_n is $n + 1$.

SECTION 2.2.1

(1) (a) $\cosh x = \frac{1}{2}e^x + \frac{1}{2}e^{-x}$. (b) Look up the double angle formula for the cosine function (e) Note that $\ln ab = \ln a + \ln b$. Also, $\ln 3 = (\ln 3)1$. (g) Note that $x^2 = ((x-1)+1)^2 = (x-1)^2 + 2(x-1) + 1$.

(5) Compute y_1' and y_1''. Then check that $y_1'' + y_1' + 13y_1 = \mathbf{0}$. The proof for y_2 is similar. The proof of independence is similar to Example 1. The general solution is all linear combinations of y_1 and y_2. (Why?)

(7) The sum of any two solutions will solve the equation $y'' + 3y' + 2y = 2x^2$, not $y'' + 3y' + 2y = x^2$. (But you must show this!)

(9) The general solution is $\frac{1}{10}\cos x + \frac{3}{10}\sin x + C_1 e^{-x} + C_2 e^{-2x}$.

(11) The set of all polynomials is a subspace. The set of polynomials with integer coefficients is not.

(13) The general solution is all polynomials of degree three or less.

(15) The dimension of P_n is $n + 1$. The space of all polynomial functions is infinite dimensional.

SECTION 2.3

(1) X is in the row space but Y is not.

(3) It is not in the row space.

(7) Basis $X_1 = [1, 0, \frac{1}{11}, \frac{2}{11}]^t$, $X_2 = [0, 1, \frac{3}{11}, \frac{6}{11}]^t$.

(9) $X_1 = [1, 0, -\frac{5}{3}]^t$, $X_2 = [0, 1, \frac{7}{3}]^t$

(11) (a) Look for relations between the rows. (b) What is the dimension of the column space? (c) What is the dimension of the nullspace? (d) Note that $[1, 1, 2, 3]^t$ is the first column of A. (e) Think about (c) and (d) and the translation theorem.

(13) $m - n + d$ Why?

SECTION **3.1**

(1)

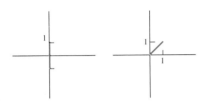

(3) $\frac{u^2}{4} + \frac{v^2}{9} = \frac{(2x)^2}{4} + \frac{(3y)^2}{9} = x^2 + y^2 = 1$

(5) Either $\begin{bmatrix} 2 & 4 \\ 4 & 2 \end{bmatrix}$ or $\begin{bmatrix} 4 & 2 \\ 2 & 4 \end{bmatrix}$. But how did you find it?

(7)

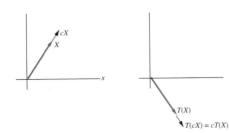

$T(X + Y) = T(X) + Y(Y)$

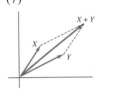

$T(cX) = cT(X)$

(11) The transformation rotates θ radian counterclockwise about the x-axis.

(13) No

(15) (a) $[1, -5]^t$ The matrix which describes T is $\begin{bmatrix} 0 & -2 & 3 \\ 3 & 0 & -5 \end{bmatrix}$

(25) *Hint:* Suppose that $X_1 = aX_2 + bX_3$. What can you say about the relationship of $T(X_1)$ with $T(X_2)$ and $T(X_3)$? What if X_2 is a linear combination of the others? How about X_3? Question: Can you find a more efficient proof using the test for independence and Exercise 26?

(27) You need to show that the nullspace is closed under linear combinations. Suppose that X and Y belong to the nullspace. What does this say about $T(X)$ and $T(Y)$? What can you say about $T(aX + bY)$ where a and b are scalars?

SECTION **3.2**

(1) You should compute AB, $(AB)C$, BC, and $A(BC)$. The final answer is $A(BC) = (AB)C = \begin{bmatrix} 4 & 7 & -1 \\ 12 & 20 & -4 \end{bmatrix}$.

(3) $(AB)^t = \begin{bmatrix} 2 & 4 \\ 1 & 4 \end{bmatrix}$. You should compute the rest yourself.

(5) $(ABC)^t = C^t B^t A^t$.

(11) (a) The quadralateral with vertices $[0, 0]^t$, $[\sqrt{2}, \frac{\sqrt{2}}{2}]^t$, $[\sqrt{2}, \sqrt{2}]^t$, $[0, \frac{\sqrt{2}}{2}]^t$.

(b) $\begin{bmatrix} \sqrt{2} & 0 \\ \frac{\sqrt{2}}{2} & \frac{\sqrt{2}}{2} \end{bmatrix}$

(13) (a) The ellipse $\frac{u^2}{16} + \frac{v^2}{81} = 1$. (b) $C = \begin{bmatrix} 4 & 0 \\ 0 & 9 \end{bmatrix}$.

(15) (a)

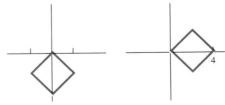

(b) $M = \begin{bmatrix} \frac{\sqrt{2}}{2} & -\frac{\sqrt{2}}{2} \\ -\frac{\sqrt{2}}{2} & -\frac{\sqrt{2}}{2} \end{bmatrix}$,

$Y = \begin{bmatrix} \frac{\sqrt{2}}{2} & \frac{\sqrt{2}}{2} \\ \frac{\sqrt{2}}{2} & -\frac{\sqrt{2}}{2} \end{bmatrix}$

(19) *Hint:* Try making one of the matrices equal to I.

(21) *Hint:* Almost any randomly chosen A and B will work. For the condition, expand $(A + B)(A - B)$ carefully using the distributive laws and set it equal to $A^2 - B^2$.

(25) One example is $\begin{bmatrix} -1 & 0 & 0 \\ 0 & 1 & 0 \\ 0 & 0 & 1 \end{bmatrix}$.

SECTION 3.3

(1) (a) Rank: 3, image: $\{[1, 2, 2]^t, [2, 3, 10]^t, [0, 2, 4]^t\}$, nullspace: $\{[2, 0, -\frac{7}{2}, 1]^t\}$ (b) Rank: 3, image: $\{[-1, 4, 3]^t, [4, 4, 0]^t, [-2, 2, -3]^t\}$, nullspace: no basis (the nullspace is $\{\mathbf{0}\}$) (c) Rank: 2, image: $\{[2, 1, 5]^t, [1, 3, 0]^t\}$, nullspace: $\{[-1, -3, 5]^t\}$ (d) Rank: 1, image: $\{[1, 2, 3, -2]^t\}$, nullspace: $\{[-2, 1, 0]^t, [-2, 0, 1]^t\}$

(3) No

(9) No.

(11) The fourth and sixth rows.

SECTION 3.4

(1) $\begin{bmatrix} 2 & -1 & 0 \\ 0 & 2 & -1 \\ -1 & -1 & 1 \end{bmatrix}$

(3) (a) $[-\frac{16}{5}, 3, \frac{7}{5}]^t$, (b) $[2, 1, 0]^t$, (c) not invertible, (d) $[\frac{3}{4}, \frac{5}{4}, \frac{1}{2}]^t$, (e) not invertible (f) $[-\frac{46}{3}, -\frac{4}{3}, -5, 11]^t$, (g) $[0, 0, 0, 1]^t$, (h) $[\frac{1}{a}, \frac{2}{b}, \frac{3}{c}, \frac{4}{d}]^t$

(11) $\frac{1}{ad-bc} \begin{bmatrix} d & -b \\ -c & a \end{bmatrix}$

(13) $\begin{bmatrix} 3 & 4 & 7 \\ 2 & 2 & 4 \\ 6 & 8 & 14 \end{bmatrix}$

(21) (a) *Hint:* $(QDQ^{-1})^2 = (QDQ^{-1})(QDQ^{-1})$

(31) *Hint:* Simplify $(I - N)(I + N + N^2)$

SECTION 3.5

(1) (a) $[-8, 3, -7, 6]^t$
(b) $[5, -7 - x_4, 3 - x_4, x_4]^t$

(3) You must interchange rows 2 and 3 of A for it to work. The **LU** decomposition for the resulting matrix is

$\begin{bmatrix} 1 & 0 & 0 \\ 2 & 1 & 0 \\ 2 & 0 & 1 \end{bmatrix} \begin{bmatrix} 1 & 1 & 3 \\ 0 & -1 & -2 \\ 0 & 0 & -1 \end{bmatrix}$

(5) *Hint:* Write $U = DU'$ where U' is unipotent and D is diagonal.

(9) You should still get A. Why?

SECTION 4.1

(1) (a) Let $X = [1, 1]^t$ and $Y = [1, -1]^t$. Then $[x, y]^t = x'X + y'Y = [x' + y', x' - y']^t$ so $1 = xy = (x' + y')(x' - y') = (x')^2 - (y')^2$, which is of the desired form.

(3) (a) The third basis (b) $X' = [\frac{13}{14}, -\frac{2}{3}, -\frac{5}{42}]^t$
(d) $\begin{bmatrix} \frac{1}{14} & \frac{3}{14} & \frac{1}{7} \\ -\frac{1}{3} & \frac{1}{3} & -\frac{1}{3} \\ \frac{5}{42} & \frac{1}{42} & -\frac{2}{21} \end{bmatrix}$

(5) $X' = [\frac{4}{7}, \frac{16}{35}, 0, -\frac{3}{5}]^t$

(7) *Hint:* Let $Q_4 = X$. Then $0 = Q_1 \cdot X = Q_2 \cdot X = Q_3 \cdot X$. This is a system of three equations for X. How can you be sure that it has non-zero solutions? What is the rank of the augmented matrix? What is the dimension of the nullspace of the coefficient matrix?

(13) (a) $[-3, 1, 0]^t$ (b) $[1, -3, 1]^t$
(c) $[32, -15, 2]^t$

SECTION **4.2**

(1) Projection: $[\frac{47}{42}, \frac{85}{42}, \frac{40}{21}]^t$

(3) Projection: $\frac{1}{5}[8, 4, 4, -12]^t$

(5) (a) $\{[0, 1, 1]^t, [1, 0, 0]^t\}$

(b) $\{[1, 2, 1, 1]^t, \ [-2, 1, 1, -1]^t, \frac{1}{7}[-6, -2, 0, 10]\}$

(7) (c) $\frac{(x+2y+z-w)}{7}[1, 2, 1, -1]^t + \frac{(-y+z-w)}{3}[0, -1, 1, -1]^t$

(9) (a) $\{[-3, 1, 0, 0]^t, [-1, -3, 30, 20]^t\}$

(b) $\frac{1}{65500}[26223, 78669, 6431, 6504]^t$

(11) Basis: $\{[1, 3, 1, -1]^t, [5, 15, -19, 31]^t\}$

SECTION **4.3**

(1) To prove (i)

$$((k+l)f)(x) = (k+l)f(x)$$
$$= kf(x) + lf(x)$$
$$= (kf + lf)(x)$$

(3) To prove the first equality in (c):

$$(cf, g) = \int_{-1}^{1} cf(x)g(x)dx$$
$$= c\int_{-1}^{1} f(x)g(x)dx$$
$$= c(f, g)$$

(5) $f_o(x) = \sum_{k=1}^{n}[\frac{12}{k^3\pi^3} - \frac{2}{k\pi}](-1)^k \sin k\pi x$

(7) (a) *Hint:* $(\cos a)(\cos b) = \frac{1}{2}(\cos(a + b) + \cos(a - b))$ (b) $(p_n, p_n) = 1$ if $n \neq 0$ and $(p_0, p_0) = 2$ (c) $f_o(x) = \frac{1}{3} + \sum_{k=1}^{n}[\frac{4}{k^2\pi^2}](-1)^k \cos k\pi x$

(9) (a) *Hint:* $(\sin a)(\cos b) = \frac{1}{2}(\sin(a + b) + \sin(a - b))$

(b)

$$f_o(x) = \frac{1}{4} + \sum_{k=1}^{n} \frac{((-1)^k - 1)}{k^2\pi^2} \cos k\pi x$$
$$+ \sum_{k=1}^{n} \frac{(-1)^{k+1}}{k\pi} \sin k\pi x$$

(11) $c_k = \frac{(-1)^{k+1}(.002)}{k\pi}$

(13) (a) 18 (b) *Hint:* A degree two polynomial $p(x)$ which is zero at three different values of x must be the zero polynomial.
(d) $c_1 = 0, c_2 = -1, c_3 = 2$.

(17) $n + 1$

(23) *Hint:* $||v + w||^2 = (v + w, v + w)$

(25) *Hint:* Try it first with only two v_i so $w = c_1 v_1 + c_2 v_2$.

SECTION **4.4**

(1) $|X| = |AX| = \sqrt{60}$

(3) Compute $(R_\theta^x)^t R_\theta^x$, $(R_\theta^y)^t R_\theta^y$, and $(R_\theta^z)^t R_\theta^z$ to see that you do get I.

(7) Note that $(AB)^t = B^t A^t$.

(9) *Another hint:* A matrix is orthogonal if and only if its transpose equals its inverse.

SECTION **4.5**

(1) $X = [30.9306, 2.0463]^t$, $B_0 = [31.9537, 33.1815, 34.0, 35.2278, 35.637]^t$

(3) (b) $[a, b, c]^t = [30.9622, 1.98957, .019944]^t$.

(5) $P = 226.512e^{.00959t}$. In 2005, $P = 287.665$.

(7) $B_0 = \frac{1}{14}[3, 4, 2, 3, 2]^t$

(9) *Hint:* Try to compute A^{-1}.

(11) *Hint:* Since each A_i is a linear combination of the C_i, there is an invertible matrix B such that $A = CB$ where A has the A_i as columns and B has the B_i as columns. (Why?)

(15) (a) In C, the third row is the sum of the first two. (b) In the augmented matrix, the third row is the sum of the first two.

(17) (a) $S^\perp$ is the line through the origin perpendicular to the original line, $(S^\perp)^\perp$ is the original line. (b) $S^\perp$ is the line through the origin which is perpendicular

to the original vectors, $(S^{\perp})^{\perp}$ is the plane spanned by the vectors. (Why?)

SECTION 5.1

(1) (a) 14 (b) 0 (c) 0 (d) 7(ac−db) (e) −16 (f) 0 (g) 0 (h) 16 (i) 34 (j) −6 (k) 144 (l) −2

(5)

$$\alpha = \begin{vmatrix} -13 & 7 & 9 & 5 \\ 16 & -37 & 99 & 64 \\ -42 & 78 & 55 & -3 \\ 47 & 29 & -14 & -8 \end{vmatrix}$$

$$- 4\begin{vmatrix} 24 & -7 & 9 & 5 \\ 11 & -37 & 99 & 64 \\ 31 & 78 & 55 & -3 \\ 62 & 29 & -14 & -8 \end{vmatrix}$$

$$+ 2\begin{vmatrix} 24 & -13 & 9 & 5 \\ 11 & 16 & 99 & 64 \\ 31 & -42 & 55 & -3 \\ 62 & 47 & -14 & -8 \end{vmatrix}$$

$$- 2\begin{vmatrix} 24 & -13 & 7 & 5 \\ 11 & 16 & -37 & 64 \\ 31 & -42 & 78 & -3 \\ 62 & 47 & 29 & -8 \end{vmatrix}$$

$$- 3\begin{vmatrix} 24 & -13 & 7 & 9 \\ 11 & 16 & -37 & 99 \\ 31 & -42 & 78 & 55 \\ 62 & 47 & 29 & -14 \end{vmatrix}$$

SECTION 5.2

(1) (a) 0 (b) 16 (c) 34 (d) −6 (e) 0
(3) See the answers for Exercise 1, Section 5.1.
(5) 120 and −5
(9) *Hint:* Do Exercise 8 first.
(11) See the argument on p. 274.
(13) *Hint:* Use the product theorem.

SECTION 5.3

(1) $x = -\frac{9}{4}, y = -\frac{5}{4}, z = 2, w = \frac{15}{4}$

(3) $c_1(x) = e^{-x}(1 + x^2)\cos x - e^{-x}x\sin x,$
    $c_2(x) = x\cos x + (1 + x^2)\sin x$
(5) −1
(7) You get A. Why?

SECTION 6.1

(1) (a) X is an eigenvector with eigenvalue 3, Y is not an eigenvector (b) both X and Y are eigenvectors (c) X is an eigenvector, Y is not. (By definition, eigenvectors must be non-zero.)
(3) *Hint:* non-zero linear combinations of eigenvectors corresponding to the same eigenvalue are eigenvectors.
(5) One eigenspace is the z axis. What is the other?
(9) (a) $p(\lambda) = -\lambda^2(\lambda - 3)$, eigenvalues and corresponding basis: $\lambda = 0$: $[1, 0, 1]^t$ and $[1, -1, 0]^t$, $\lambda = 3$, $[1, 1, 1]^t$, $A^n B = 5[3^{n-1}, 3^{n-1}, 3^{n-1}]^t$ for $n > 1$.
(11) (a) $\lambda = -8$, basis: $[-1, -1, 5]^t$. This matrix is deficient over the real numbers (but not over the complex numbers.) (b) is deficient and (c) is not.
(15) *Hint:* See Exercise 7 and the comments on p. 292.

SECTION 6.1.1

(1) (a) Tuesday: $[0.4667, 0.3333, 0.2]^t$, Wednesday: $[0.4933, 0.3067, 0.2]^t$ (b) In two weeks the state vector is $[0.519993, 0.280007, 0.2]^t$ (c) The equilibrium distribution is $[0.52, 0.28, 0.2]^t$. After 50 days there is no noticeable change. (We used logarithms to find our answer.) (e) The entries of the equilibrium state vector give the expected fractions.
(3) (a) takes four products, (b) takes one product
(5) *Hint:* Since the columns of P total to 1,

$$P = \begin{bmatrix} a & b \\ 1 - a & 1 - b \end{bmatrix}$$

(7) (a) $\begin{bmatrix} .6 & .15 & .10 \\ .2 & .7 & .10 \\ .2 & .15 & .8 \end{bmatrix}$

(15) *Hint:* Consider $\lim_{n\to\infty} P^{2n} V$ and $\lim_{n\to\infty} P^{2n+1} V = \lim_{n\to\infty} P^{2n} P V$. To prove that $PX = X$, note that $\lim_{n\to\infty} P^n V = X$.

(19) *Hint:* Assume that $PX = X$. Then for each i,

$$|x_i| = |p_{1i}x_1 + \ldots + p_{ni}x_n|$$
$$\leq p_{1i}|x_1| + \ldots + p_{ni}|x_n|$$

Apply this with the entry x_i of X with the largest absolute value.

SECTION 6.2

(3) (a)

$$Q = \begin{bmatrix} 1 & 1 & 0 \\ 1 & 0 & 1 \\ -1 & 1 & 1 \end{bmatrix}$$

$$D = \begin{bmatrix} 6 & 0 & 0 \\ 0 & -3 & 0 \\ 0 & 0 & 0 \end{bmatrix}$$

$$A^n = Q \begin{bmatrix} 6^n & 0 & 0 \\ 0 & (-3)^n & 0 \\ 0 & 0 & 0 \end{bmatrix} Q^{-1}$$

(5) (a) Not diagonalizable over $\mathbb{R}$. (b) Not diagonalizable.
(c)

$$Q = \begin{bmatrix} 1 & 1 & 0 & 0 \\ 1 & -1 & 0 & 0 \\ 0 & 0 & 2 & -1 \\ 0 & 0 & -1 & 1 \end{bmatrix}$$

$$D = \begin{bmatrix} 4 & 0 & 0 & 0 \\ 0 & -2 & 0 & 0 \\ 0 & 0 & -2 & 0 \\ 0 & 0 & 0 & -3 \end{bmatrix}$$

(9) Yes

(11) *Hint:* Write $A = QDQ^{-1}$ where D is diagonal. What can you say about D?

(13) Same hint as 11.

SECTION 6.3

(1)

$$AB = \begin{bmatrix} -1 + 3i & -5 + 3i \\ -3 + 4i & -6 \end{bmatrix},$$

$$BA = \begin{bmatrix} 7 - i & 2 + 9i \\ 1 + 11i & -14 + 4i \end{bmatrix}$$

(3) $-7, 6.5 \pm \frac{3\sqrt{3}}{2}i$

(5) $M = \begin{bmatrix} a & -b \\ b & a \end{bmatrix}$. Eigenvalues $a \pm bi$

(17) (a) $3 + 20i$

(21) They satisfy $\|U_i\| = 1$ and $< U_i, U_j > = 0$ if $i \neq j$. For the proof, see the argument for Proposition 1 in Section 4.4. In this argument, what plays the role of the preservation of angles theorem?

SECTION 6.4

(1) $\begin{bmatrix} 3 & 1 & 0 \\ 2 & 0 & 2 \\ 1 & 1 & 1 \end{bmatrix}$

(3) (a) $M = \begin{bmatrix} 3 & 1 \\ 0 & 3 \end{bmatrix}$

(7) (c) *Hint:* $M^k = 3^k N^k$ where

$$N = \begin{bmatrix} 1 & 0 & 0 \\ \frac{1}{3} & 1 & 0 \\ 0 & 0 & \frac{5}{3} \end{bmatrix}.$$ You should be able

to find a formula for N^k. The final answer for A^k is

$$\begin{bmatrix} 3^k & k3^k & \frac{3(5^k) - 2k3^{k-1} - 3^{k+1}}{4} \\ 0 & 3^k & \frac{5^k - 3^k}{2} \\ 0 & 0 & 5^k \end{bmatrix}$$

(9) *Hint:* Do Exercise 8 first.

(11) (a) $M = \begin{bmatrix} 0 & 2 \\ 1 & -1 \end{bmatrix}$

(b) $X' = [\frac{5}{2}, \frac{1}{2}]^t$, $X'' = [1, 2]^t$.

(13) (a) $M = \begin{bmatrix} 0 & 1 & 0 \\ 0 & 0 & 2 \\ 0 & 0 & 0 \end{bmatrix}$, nullspace is

spanned by $[1, 0, 0]^t$ which corresponds to the constant function $p_0(x) = 1$. The image is spanned by $[0, 2, 0]^t$ and $[1, 0, 0]^t$ which correspond to the functions p_0 and $p_1(x) = 2x$. p_0 is the derivative of $f(x) = x$ and p_1 is the derivative of $f(x) = x^2$.

(15) $M = \begin{bmatrix} 0 & 1 & 1 \\ 0 & 0 & 2 \\ 0 & 0 & 0 \end{bmatrix}$

SECTION 6.5

(3) In the first case it would have been a hyperbola and in the second the empty set. *Hint:* In attempting to cover all cases don't forget to consider the possibility that one or both of the eigenvalues might be 0.

(7) *Hint:* Use the quadratic equation to find the roots of the characteristic polynomial for the matrix of the quadratic form. Do not simplify your result too much!

(9) *Hint:* Look at the form of the quadratic form in the new coordinates.

(11) *Hint:* See exercise 10.

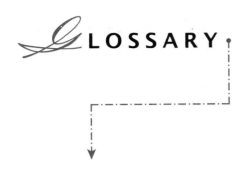

GLOSSARY

In this section we have collected what we feel are the most important definitions from the text. For more details on these concepts, as well as the meaning of other terms, the student should consult the index.

A

augmented matrix The augmented matrix for a system of linear equations is the matrix formed from the coefficient matrix by adding an additional column whose ith entry is the constant on the right side of the ith equation.

B

basis A basis for a vector space is a linearly independent set of elements which spans the vector space.

back substitution Back substitution is the process of solving a linear system using an echelon form of the augmented matrix by first determining certain of the variables and then substituting these into the system to determine more of the variables. One continues this process until all of the variables are determined.

C

characteristic polynomial The characteristic polynomial for an $n \times n$ matrix A is the polynomial $p(l) = \det(A - lI)$.

coefficient matrix The coefficient matrix for a system of linear equations is the matrix whose (i, j) entry is the coefficient of the jth variable in the ith equation.

cofactor expansion The cofactor expansion is the method of computing determinants by expanding along a row or column. (See formulas (3) and (4) in Section 5.1.)

column space The column space of an $m \times n$ matrix is the span of its columns in $M(m \times 1)$.

column vector See "vector."

complex eigenspace Let l be a complex eigenvalue for an $n \times n$ matrix A. Then the set of all $n \times 1$ complex matrices X such that $AX = lX$ is the complex eigenspace corresponding to the complex eigenvalue l.

complex eigenvector A complex eigenvector for an $n \times n$ matrix is a nonzero $n \times 1$ complex matrix X such that $AX = lX$ for some complex scalar l, which is referred to as "the corresponding (complex) eigenvalue."

complex eigenvalue See "complex eigenvector."

composite transformation Suppose that T is a transformation of a set A into a set B and S is a transformation of B into a set C. Then the composite of S with T is the transformation $S \circ T$ of A into C defined by by $S \circ T(x) = S(T(x))$ for all x in A.

consistent system A system of linear equations is consistent if it has at least one solution.

coordinate matrix The coordinate matrix for a basis of $\mathbb{R}^n$ is the $n \times n$ matrix C such that a point X in $\mathbb{R}^n$ has coordinate vector X' if and only if

$X' = CX$. The coordinate matrix may also be defined as the inverse of the point matrix.

coordinate vector See "coordinates with respect to a basis."

coordinates with respect to a basis Given a basis $\{X_1, X_2, \ldots, X_n\}$ of a vector space $\mathcal{V}$ and an element X in $\mathcal{V}$, the coordinates of X with respect to this basis are the scalars c_i such that $X = c_1 X_1 + c_2 X_2 + \cdots + c_n X_n$. The vector $X' = [c_1, c_2, \ldots, c_n]^t$ is the coordinate vector for X.

Cramer's rule Cramer's rule is the method of solving systems of linear equations using determinants, which is described on page 289 of Section 5.3.

D

deficient An $n \times n$ matrix A is said to be deficient over $\mathbb{R}$ if it does not have n independent real eigenvectors. Similarly, A is deficient over $\mathbb{C}$ if it does not have n independent complex eigenvectors.

dependent See "linearly dependent."

dependent system A system of linear equations is dependent if any one of the equations in the system is a linear combination of other equations in the system.

determinant of a matrix The determinant of an $n \times n$ matrix A is the number which results from applying the cofactor expansion to the first row of A. This definition is inductive in that it defines the determinant of an $n \times n$ matrix in terms of $(n-1) \times (n-1)$ determinants.

diagonal matrix An $n \times n$ matrix D is diagonal if its only nonzero entries lie on the principal diagonal.

diagonalizable matrix An $n \times n$ matrix is diagonalizable over $\mathbb{R}$ if it has n independent real

eigenvectors. Similarly, A is diagonalizable over $\mathbb{C}$ if it has n independent complex eigenvectors.

diagonalization A diagonalization for an $n \times n$ matrix A is an $n \times n$ invertible matrix Q and a diagonal matrix D such that $A = QDQ^{-1}$.

diagonalizing basis A diagonalizing basis for an $n \times n$ matrix A is a basis for $\mathbb{R}^n$ consisting of eigenvectors for A.

dimension The dimension of a vector space is the smallest number of elements it takes to span the vector space. If the vector space cannot be spanned by a finite set of elements, then it is said to be infinite dimensional.

domain See "transformation."

dot product The dot product of two elements X and Y in $\mathbb{R}^n$ is $X \cdot Y = X^t Y$.

E

elementary row operations Elementary row operations are transformations that one can apply to a matrix A of the following types: (I) interchange two rows (II) add a multiple of one row onto another and (III) multiply one row by a nonzero constant.

echelon form A matrix A is in echelon form if the first nonzero entry in any nonzero row occurs to the right of the first such entry in the row directly above it and all zero rows are grouped together at the bottom.

eigenspace Let l be a real eigenvalue for an $n \times n$ matrix A. Then the set of all X in $\mathbb{R}^n$ such that $AX = lX$ is the eigenspace corresponding to the eigenvalue l.

eigenvector A real eigenvector for an $n \times n$ matrix is a nonzero vector X in $\mathbb{R}^n$ such that

$AX = lX$ for some real scalar l, which is referred to as "the corresponding eigenvalue."

eigenvalue See "eigenvector."

element In set theory, an object which belongs to a set is said to be an element of that set.

equivalent systems Two systems of linear equations are equivalent if they have the same solution set.

F

free variable In a system of linear equations, the variables which are not pivot variables are free variables.

G

Gaussian elimination Gaussian elimination is the process of solving a system of linear equations by applying elementary row operations to the augmented matrix for the system.

Gram–Schmidt process The Gram-Schmidt process is the process which produces an orthogonal basis from a given basis of $\mathbb{R}^n$.

H

Hermitian symmetric matrix A matrix A with complex entries is Hermitian symmetric if $A = \overline{A}^t$ where $\overline{A}$ is the matrix obtained by forming the complex conjugate of each of the entries of A.

homogeneous system of linear equations A system of linear equations is homogeneous if it can be written in the form $AX = \mathbf{0}$ for some matrix A.

I

independent See "linearly independent."

image The image of a subset S of the domain of a transformation T is the set of vectors of the form $T(X)$ where X belongs to S. It is denoted $T(S)$. The image of T is the image of the whole domain of T.

identity matrix The $n \times n$ identity matrix is the $n \times n$ matrix I with all of the entries on the principal diagonal equal to 1 and all of the other entries equal to 0.

identity transformation On any set, the transformation I such that $I(X) = X$ for all X in the set is called the identity transformation.

inconsistent system A system of linear equations is inconsistent if it has no solutions.

invertible matrix An $n \times n$ matrix A is invertible if the equation $AX = Y$ has one and only one solution X for every Y in $\mathbb{R}^n$. This is the same as saying that the transformation defined by A is one-to-one and onto.

inverse matrix The inverse of an $n \times n$ matrix A (when an inverse exists) is the matrix A^{-1} with the property that $AX = Y$ if and only if $X = A^{-1}Y$. [*Note:* Many texts define A^{-1} as the unique matrix such that $AA^{-1} = A^{-1}A = I$ where I is the $n \times n$ identity matrix.]

L

Laplace expansion See "cofactor expansion."

least squares solution A least squares solution to a linear system $AX = B$ is a vector X with the property that $|AX - B|$ is as small as possible.

line A line is the set of all elements in a given vector space of the form $T_o + sX$ where X and T_o are fixed elements of the space and s ranges over all real numbers.

linear combination Let $\mathcal{V}$ be a vector space and let $A_1, A_2, \ldots, A_n$ be elements of $\mathcal{V}$. An element B of $\mathcal{V}$ is a linear combination of the A_i if there are scalars c_i such that $B = c_1 A_1 + c_2 A_2 + \cdots + c_n A_n$.

linear equation A linear equation is an equation in variables $x_1, x_2, \ldots, x_n$ which can be expressed in the form $a_1 x_1 + a_2 x_2 + \cdots + a_n x_n = b$ where a_i and b are scalars.

linearly dependent Let $\mathcal{V}$ be a vector space and let $A_1, A_2, \ldots, A_n$ be elements of $\mathcal{V}$. We say that the A_i are (as a set) linearly dependent if any one of the A_i is a linear combination of the other A_j. We also define the set consisting of just the zero vector as linearly dependent. If B is any element of $\mathcal{V}$, we say that B is linearly dependent on the A_i if B is a linear combination of the A_i.

linearly independent A set of elements of a vector space is said to be linearly independent if it is not linearly dependent.

linear transformation A transformation T from a vector space $\mathcal{V}$ into a vector space $\mathcal{W}$ is a linear transformation if for all X and Y in $\mathcal{V}$ and all scalars c, $T(X + Y) = T(X) + T(Y)$ and $T(cX) = cT(X)$.

lower-triangular matrix A matrix A is lower-triangular if its only nonzero entries lie on or below the principal diagonal.

M

matrix transformation A matrix transformation is a transformation T from $\mathbb{R}^n$ into $\mathbb{R}^m$ for which there is an $m \times n$ matrix A such that $T(X) = AX$ for all X in $\mathbb{R}^n$.

matrix A matrix is a rectangular array of numbers.

$m \times n$ matrix An $m \times n$ matrix is a matrix with m rows and n columns.

multiplicity of an eigenvalue An eigenvalue l_o of an $n \times n$ matrix A is said to be of multiplicity k if $(l - l_o)$ occurs exactly k times as a factor of the characteristic polynomial of A.

N

normalization Normalization is the process of dividing a vector by its length so as to produce a unit vector.

nullspace The nullspace of an $m \times n$ matrix A is the set of vectors X in $\mathbb{R}^n$ such that $AX = \mathbf{0}$. The nullspace of a linear transformation T is the

set of elements X in the domain of T such that $T(X) = \mathbf{0}$.

O

one-to-one A transformation T is one-to-one if for each Y in the image of T there is only one X such that $T(X) = Y$.

onto A transformation T is onto if its image is the whole target space.

origin The origin in $\mathbb{R}^n$ is the zero vector.

orthogonal complement The orthogonal complement of a subspace $\mathcal{W}$ of $\mathbb{R}^n$ is the set $\mathcal{W}^\perp$ of vectors X such that $X \cdot W = 0$ for all W in $\mathcal{W}$.

orthogonal matrix An $n \times n$ matrix A is orthogonal if for all X in $\mathbb{R}^n$, $|AX| = |X|$. This is equivalent with the statement that $A^t A = I$.

orthogonal projection Let B be an element of $\mathbb{R}^n$ and let $\mathcal{W}$ be a subspace of $\mathbb{R}^n$. The orthogonal projection of an element B onto $\mathcal{W}$ is the unique element B_0 such that $B - B_0$ is perpendicular to every element of $\mathcal{W}$. The projection is denoted $B_0 = \text{proj}_{\mathcal{W}} B$. The same definition applies if B is an element of a general scalar product space.

orthogonal set of elements In $\mathbb{R}^n$, a set $\{Q_1, Q_2, \ldots, Q_k\}$ of vectors is orthogonal if none of the Q_i are zero and $Q_i \cdot Q_j = 0$ for all $i \neq j$. More generally, in a scalar product space $\mathcal{V}$, a set $\{Q_1, Q_2, \ldots, Q_k\}$ of elements is orthogonal if none of the Q_i are zero and $(Q_i, Q_j) = 0$ for all $i \neq j$.

orthonormal set of elements In either $\mathbb{R}^n$ or, more generally, in a general scalar product space, an orthogonal set is orthonormal if each element of the set has length 1.

P

$\mathbf{P}_n$ The space of all polynomial functions of degree less than or equal to n is denoted P_n.

parametric form of a line The parametric form of a line is the description of the line as the set of all points of the form $T_o + tX$ where T_o and X

are fixed elements of $\mathbb{R}^n$ and t ranges over all real numbers.

perpendicular Two elements X and Y of $\mathbb{R}^n$ are perpendicular if $X \cdot Y = 0$.

pivot entry If A is a matrix in echelon form, then the first nonzero entry in any row of A is a pivot entry of A.

pivot variable In a system of linear equations, the variables which correspond to the pivot entries of an echelon form of the coefficient matrix are the pivot variables.

plane A plane is the set of all elements in a given vector space of the form $T_o + sX + tY$ where X, Y and T_o are fixed elements of the space, s and t range over all real numbers and X and Y are linearly independent.

point matrix The point matrix for a basis of $\mathbb{R}^n$ is the $n \times n$ matrix Q such that a point X in $\mathbb{R}^n$ has coordinate vector X' if and only if $X = QX'$. The point matrix may also be defined as the matrix whose columns are the basis elements.

principal diagonal The principal diagonal of an $m \times n$ matrix $[a_{ij}]$ is the sequence of entries of the form a_{ii}.

Q

quadratic form A quadratic form is a transformation f of $\mathbb{R}^n$ into $\mathbb{R}$ defined by $f(X) = X^t A X$ where A is an $n \times n$ symmetric matrix.

quadratic variety A quadratic variety is the set of points X in $\mathbb{R}^n$ such that $X^t A X = d$ where A is a given $n \times n$ symmetric matrix and d is a given scalar.

R

rank of a system of linear equations The rank of a system of linear equations is the number of equations left after eliminating dependent equations, one at a time. The rank of a system equals the rank of the augmented matrix.

rank of a matrix The rank of a matrix is the dimension of its row space. It equals the number

of nonzero rows in any row reduced form of the matrix.

reduced echelon form A matrix A is in reduced echelon form if it is in echelon form and, additionally, the pivot entries are 1 and the entries above and below the pivots are 0.

row equivalent matrices Two matrices are row equivalent if one may be obtained by applying a finite sequence of elementary row operations to the other.

row reduced echelon form See "reduced echelon form."

row space The row space of an $m \times n$ matrix is the span of its rows in $M(1 \times n)$.

row vector See "vector."

rule of addition A rule of addition on a set $\mathcal{V}$ is a well-defined process for taking arbitrary pairs of elements A and B from $\mathcal{V}$ and producing an element $A + B$ in $\mathcal{V}$. (This can also be called a "binary operation.")

rule of scalar multiplication A rule of scalar multiplication on a set $\mathcal{V}$ is a well-defined process for taking arbitrary real numbers c and arbitrary elements A from $\mathcal{V}$ and producing an element cA of $\mathcal{V}$.

S

scalar A scalar is a number. It can be either a real or complex number, depending on the context.

scalar multiplication Multiplication of an element of a vector space by a scalar is called scalar multiplication.

scalar product A scalar product on a vector space $\mathcal{V}$ is a function $(\ ,\ )$ that takes pairs of vectors from $\mathcal{V}$ and produces real numbers and satisfies $(U, V) = (V, U)$, $(U + V, W) = (U, W) + (V, W)$, $(cU, V) = c(U, V) = (U, cV)$ for all U, V and W in $\mathcal{V}$ and all scalars c. Furthermore, it must satisfy $(U, U) > 0$ for $U \neq \mathbf{0}$

scalar product space A scalar product space is a vector space together with a scalar product.

solution A solution for a system of linear equations in variables $x_1, x_2, \ldots, x_n$ is a sequence of scalars $s_1, s_2, \ldots, s_n$ such that each equation in the system becomes valid when each x_i is replaced by s_i. In this case, the vector $X = [s_1, s_2, \ldots, s_n]^t$ is also referred to as "a solution to the system."

solution set The solution set for a system of linear equations is the set of all column vectors X which are a solution to the system.

span Let V be a vector space and let A_1, $A_2, \ldots, A_n$ be elements of V. The span of the A_i (denoted "span $\{A_1, A_2, \ldots, A_k\}$") is the set of all linear combinations of the A_i. (See also "spanned by.")

spanned by Let V be a vector space. We say that V is spanned by elements $A_1, A_2, \ldots, A_k$ of V if V equals the span of the A_i. We also say that the A_i span V.

spanning vectors If the solution to a system of linear equations is written in the form $T_o + x_{i_1} X_1 + \cdots + x_{i_k} X_k$ where the x_{i_j} are the free variables and T_o and X_i are column vectors, then the X_i are the spanning vectors and T_o is the translation vector.

standard basis The standard basis for $\mathbb{R}^n$ is the basis $\{I_1, I_2, \ldots, I_n\}$ where I_j is the ith column of the $n \times n$ identity matrix I. The standard basis for P_n is $\{1, x, x^2, \ldots, x^n\}$.

subspace A non-empty subset W of a vector space V is a subspace of V if for all X and Y in W and all scalars a and b, $aX + bY$ belongs to W.

symmetric matrix A matrix A is symmetric if $A = A^t$.

system of linear equations A system of linear equations is a finite set of linear equations in the same set of variables.

T

target space See "transformation."

test for independence The test for independence is the theorem that states that

a set $\{A_1, A_2, \ldots, A_n\}$ of elements of a vector space is independent if and only if there are no nonzero scalars c_i such that $c_1 X_1 + c_2 X_2 + \cdots + c_n X_n = \mathbf{0}$.

transformation A transformation is a well-defined process for taking elements from one set (the domain) and using them to produce elements of another set (the target space). "Well-defined" means that each element of the domain is transformed onto only one element of the target space.

transformation defined by a matrix Let A be an $m \times n$ matrix. The transformation defined by A is the transformation T from $\mathbb{R}^n$ into $\mathbb{R}^m$ such that $T(X) = AX$ for all X in $\mathbb{R}^n$.

translation vector See "spanning vectors."

transpose The transpose of a matrix A is the matrix $B = A^t$ defined by $b_{ij} = a_{ji}$

triangular matrix A matrix is triangular if it is either upper-triangular or lower-triangular.

triangularization A triangularization of an $n \times n$ matrix A is an $n \times n$ invertible matrix Q and a triangular matrix T such that $A = QTQ^{-1}$.

U

upper-triangular matrix A matrix A is upper-triangular if its only nonzero entries lie on or above the principal diagonal.

V

vector A vector is a matrix which has either only one row (a row vector) or only one column (a column vector). In physics, a vector is a direction and a magnitude.

vector space A set V is a vector space if it has a rule of addition and a rule of scalar multiplication defined on it so that all the vector space properties listed on page 20 hold.

vector space properties The properties listed on page 20 of the text.

PHOTO CREDITS

Chapter Openers
© Chris Thomaidis/Tony Stone Images/ New York, Inc.

Chapter 1
Page 5: © Aldo Torelli/Tony Stone Images/ New York, Inc. Page 23: © Chris Thomaidis/Tony Stone Images/ New York, Inc. Page 56: © Ken Biggs/Tony Stone Images/ New York, Inc. Page 75: © Jeanne Drake/Tony Stone Images, New York, Inc.

Chapter 2
Page 92: M.C. Escher/© 1997 Cordon Art-Baarn-Holland. All rights reserved. Page 121: AP/Wide World Photos.

Chapter 3
Page 137: © Andrew Sacks/Tony Stone Images/ New York, Inc. Page 152: © Jon Riley/Tony Stone Images/ New York, Inc. Page 163: © '96 Charly Franklin /FPG International.

Chapter 4
Page 218: © Bill Bachmann/Photo Researchers. Page 245: © Alan Levenson/Tony Stone Images/ New York, Inc. Page 267: © F. Reginato/The Image Bank.

Chapter 6
Page 298: © Michael Quackenbush/The Image Bank. Page 348: © Hitoshi Ikematsu/The Image Bank.

INDEX